Sound Inventions

Sound Inventions is a collection of 34 articles taken from *Experimental Musical Instruments*, the seminal journal published from 1984 through 1999. In addition to the selected articles, the editors have contributed introductory essays, placing the material in cultural and temporal context, providing an overview of the field both before and after the time of original publication.

The *Experimental Musical Instruments* journal contributed extensively to a number of sub-fields, including sound sculpture and sound art, sound design, tuning theory, musical instrument acoustics, timbre and timbral perception, musical instrument construction and materials, pedagogy, and contemporary performance and composition. This book provides a picture of this important early period, presenting a wealth of material that is as valuable and relevant today as it was when first published, making it essential reading for anyone researching, working with or studying sound.

Bart Hopkin is a maker of acoustic musical instruments and student of instruments worldwide. He earned a B.A. magna cum laude from Harvard University in folklore and mythology specializing in ethnomusicology in 1974, and later received a B.A. in music education and a teaching credential at San Francisco State University. From 1985 to 1999, Bart edited the quarterly journal *Experimental Musical Instruments*. Since 1994, he has written numerous books on instruments and their construction, including the leading resource, *Musical Instrument Design*. He has produced CDs featuring the work of innovative instrument makers, including the very successful *Gravikords, Whirlies & Pyrophones*.

Sudhu Tewari is an electro-acoustic composer, improvisor, and bricoleur in sound, kinetic, and interactive art. He holds a B.A. in Music from Sonoma State University, an M.F.A. in Electronic Music and Recording Technology from Mills College, and a Ph.D. in Cultural Musicology from UC Santa Cruz. Sudhu builds audio electronics, acoustic instruments, kinetic sculptures, interactive installations, and sound sculpture. Dr. Tewari is Workshop Technician in the instrument building program at Mills College and teaches art, technology, and design at California College of the Arts, the University of San Francisco, and Expression College in Emeryville.

Sound Design

The Sound Design series takes a comprehensive and multidisciplinary view of the field of sound design across linear, interactive, and embedded media and design contexts. Today's sound designers might work in film and video, installation and performance, auditory displays and interface design, electroacoustic composition and software applications, and beyond. These forms and practices continuously cross-pollinate and produce an ever-changing array of technologies and techniques for audiences and users, which the series aims to represent and foster.

Series Editor: Michael Filimowicz

Foundations in Sound Design for Linear Media
A Multidisciplinary Approach
Edited by Michael Filimowicz

Foundations in Sound Design for Interactive Media
A Multidisciplinary Approach
Edited by Michael Filimowicz

Foundations in Sound Design for Embedded Media
A Multidisciplinary Approach
Edited by Michael Filimowicz

Sound and Image
Aesthetics and Practices
Edited by Andrew Knight-Hill

Sound Inventions
Selected Articles from *Experimental Musical Instruments*
Edited by Bart Hopkin and Sudhu Tewari

For more information about this series, please visit: www.routledge.com/Sound-Design/book-series/SDS

Sound Inventions

Selected Articles from *Experimental Musical Instruments*

Edited by
Bart Hopkin and Sudhu Tewari

LONDON AND NEW YORK

First published 2021
by Routledge
2 Park Square, Milton Park, Abingdon, Oxon OX14 4RN

and by Routledge
52 Vanderbilt Avenue, New York, NY 10017

Routledge is an imprint of the Taylor & Francis Group, an informa business

British Library Cataloguing-in-Publication Data
A catalogue record for this book is available from the British Library
Library of Congress Cataloging-in-Publication Data
Names: Hopkin, Bart, 1952- editor. | Tewari, Sudhu, editor.
Title: Sound inventions : Selected Articles from Experimental Musical Instruments / edited by Bart Hopkin and Sudhu Tewari.
Other titles: Experimental Musical Instruments.
Description: New York : Routledge, 2021. | Series: Sound design series | Includes bibliographical references and index.
Identifiers: LCCN 2020042903 (print) | LCCN 2020042904 (ebook)
Subjects: LCSH: Musical instruments. | Musical instruments--Construction. | Musical inventions and patents.
Classification: LCC ML460 .S68 2021 (print) | LCC ML460 (ebook) | DDC 784.1192/3--dc23
LC record available at https://lccn.loc.gov/2020042903
LC ebook record available at https://lccn.loc.gov/2020042904

ISBN: 978-0-367-43474-8 (hbk)
ISBN: 978-0-367-43473-1 (pbk)
ISBN: 978-1-003-00352-6 (ebk)

Typeset in Times New Roman
by SPi Technologies India Pvt Ltd (Straive)

The full archive of articles from *Experimental Musical Instrument* can be accessed at https://archive.org/details/emi_archive/

Contents

Contents by Subject Matter

Acknowledgements

This book owes its existence to the many people who contributed to the *Experimental Musical Instruments* journal during its 14-year run. There were many more than can be named here, but we do want to single out a few of the most generous and generative. These are people who contributed articles, reviews, commentary, ideas, constructive criticism and logistical support in consistent and ongoing ways over the life of the journal: René van Peer, Robin Goodfellow, Sasha Bogdanowitch, Mike Hovancsek, Mitchell Clark, Karen Jaenke, Donald Hall, Ivor Darreg, François Baschet, Reed Ghazala, Hugh Davies, John Coltman, Kim Johnson, Dean Suzuki, Janet Hopkin, Chip Dunbar, Warren Burt, Stephen Golovnin, Michael Meadows, Jeannie Filson, Colin Hinz... and this list could easily go on. To all of these and many more, our grateful thanks.

Our thanks also go to our associates at Routledge, particularly Hannah Rowe, Shannon Neill, Peter Hall, Thivya Vasudevan and Michael Filimowicz who have been knowledgeable, helpful, and supportive throughout.

Bart Hopkin & Sudhu Tewari

Part 1

Introductory Essays

1

Perspectives on New Instruments

Ideas and Questions that Shape the Field of Experimental Musical Instrument Making

Bart Hopkin and Sudhu Tewari

Through history and across cultures, musical instrument making has generally been a conservative practice. Most instrument makers are in the business of reproducing of existing types, rather than innovating or creating new types. Yet it needn't necessarily be so. As an alternative, people could approach instrument making the way that a sculptor produces sculptures, or the way architects design buildings: intending to create something new each time.

The conservative bent is not surprising; there are many reasons to stay with the standardized types. Consider, as an example, the violin: If you make a violin, you can be confident that there will be plenty of people ready to play it, many having spent countless hours learning to play just this type of instrument. There will be a preexisting repertoire awaiting the instrument, created by composers and arrangers with an experienced and refined sense of how to bring out the best in it. There will be in the surrounding culture an appreciation and respect for all that this type of instrument can do, all that it stands for, and all the history that makes it what it is. Those are some big, ready-made advantages! And though the specifics will differ, any of the standard instrument types, western or non-, will enjoy similar benefits.

But as for the alternative – the creation of unique and innovative instrument types – well, there are plenty of pleasures there too. And though the design of unconventional instruments may at first seem an obscure or esoteric activity, enough people have engaged with the creation of new instruments to give us a sense of how such a practice can take shape: what the possible aesthetics of the field are, what the important questions and issues are, and what the ideas are that motivate different builders. In this essay, we'll try to elucidate the most important of those themes. In a more expansive vein, this book as a whole can be seen as an exploration of them.

Tradition vs. Innovation; Standardized Types vs. One-of-a-Kinds

Anytime someone invents something new it is likely to be, at the start at least, a one-of-a-kind. The creator may then decide to produce it in quantity as a standardized type; or she may be happy to let it remain as a unique and individual object. There are plenty of people who have developed new instrument types with the idea of being the next Adolph Sax – that is, seeing their instrument, like the saxophone, join the class of recognized, widely played instrument types within the culture. Yet, with no disrespect to that approach, we can say that much of the energy in the contemporary world of creative instrument making – and certainly the focus of most of the articles in this book – is on the one-of-a-kinds approach, and on the pleasure of making and sharing things that are unique unto themselves.

All musical instruments, old or new, can be seen as products of innovation, in that they must have been newly invented at one time. But historically the development of instrument types and families has mostly been slow and incremental, analogous in many ways to the process of biological evolution. Just as the sudden appearance of an entirely new species without antecedents would be an anomaly in biology, the sudden creation of a truly new type is a rarity in the evolution of musical instrument families. Reflecting this, scholars of musical instruments have been much given to the development of classification systems for musical instruments types, often similar in form to the Linnaean taxonomies of biological species.

As an example of that rare creation that would seem to deserve credit as a genuinely new type, consider the Waterphone, developed by the American maker Richard Waters in the 1970s, and described in the Waterphones article in this book. The leap from the kalimba-like instruments that initially inspired him to create the water-modulated bowed rods of his fully developed instrument would have to be described as inventive rather than evolutionary. Notice, too, that this was not a communal process but clearly the work of an individual.

By contrast, as examples of instruments that have followed a more evolutionary path, we can point to several instrument types that developed recently enough that we were able to observe and have some record of the evolutionary process, even if that process was more rapid than is typical. Examples are the development of the modern drummer's trap set in the early 20th century, the rapid evolution of the Trinidadian steel pan in mid-century, and the way the pedal steel guitar, with its varying tunings and numbers of strings, has gradually taken shape over a period of decades. It's worthwhile to note, though, that even in these cases where there was more communal input, the role of inventive individuals was crucial, as with for instance the foundational ideas of people like Bertie Marshall, Alan Gervais, and Anthony Williams for the nascent steel pan. As for the aforementioned Adolph Sax, while he might seem to fit the model of the individual inventor type, we should note that his innovations too were very much incremental: in the matters of optimizing tone hole configurations and fingerings, refining bore shapes, and putting these things together into excellently conceived wind instruments, he was unsurpassed. But his ideas were closely derived from preexisting types and in that sense were not radically innovative.

For people making new instruments today, one of the pleasures is to be able to say "no one ever made anything like this before!" Such a claim frequently comes up short: most often when someone thinks they've come up with something completely new, it's only because they're unaware of the others who have already explored the same territory. Yet wonderful new ideas do come along now and then, and even short of entirely new species, there are both value and pleasure in exploring incrementally new varieties. In the world of musical instruments, even subtle differences in form, in sound, in the configuration of the notes, or in the ways the player interacts with the instrument can make a big difference in the character of the instrument and the sort of music it seems to want to make.

Two Approaches: Analytical and Prescriptive vs. Hands-on and Exploratory

Among reasons for designing and building new instruments, the first to come to mind might be a search for new sounds with musical potential. Surely we can freshen up the available palette a bit? When we speak of new sounds, as much as we are talking about the physical qualities of vibrations in the air, we are talking about the way the ear perceives such sounds, and the aesthetics of the human response. The question then is, can the maker come up with sounds that strike the ear in ways that are fresh and new? Many of the instruments touched upon in this book have proven fertile in this regard.

One approach to this challenge might be prescriptive and analytical: What kinds of sound characteristics is the maker after? What kinds of vibration patterns, on a physical level, will give rise to those characteristics? What kind of instrument design will bring those sorts of patterns about? This kind of thinking can be immensely rewarding and enjoyable, feeding directly into the maker's practice in designing instruments with particular musical and aesthetic effects in mind. It calls for some knowledge of the physics of musical instruments and their construction – although for many builders such knowledge might be more practical, or even just intuitive, than theoretical. It also suggests an approach to instrument design, including research, experimentation and prototyping processes, which is deliberate and intentional from the ground up, from conception through completion. One such maker is the French instrument designer Jacques Dudon, whose wide-ranging work is described in the article "The Photosonic Disk" in this book. His design process, as you'll see when you read his article, is very much a template for prescribing the character of the sound that is to result.

And yet, though this sort of prescriptive thinking might seem central to the experimental instrument maker's art, many makers do not follow such an analytical praxis.

To represent an opposite perspective, we can point to the found-sound approach. This is central to some builders' methods, and at least incidental to that of many. In its most obvious form, the found-sound approach amounts to exploring one's environs looking for preexisting objects that happen to make cool sounds, and finding ways to incorporate them into sound-making assemblages, be they sets of flower pot bells or rotating bicycle wheel spokes. The approach becomes richer and subtler for those who've developed both an ear for the potential of everyday sounds and the eye to know which improbable objects may harbor the most interesting sounds. Many are the experimental instrument makers who routinely carry a superball mallet with them as they explore the world; the junkyard is their playground.

But it's important to note that a found-sound approach need not be so literal. Even among makers who do a lot of thinking and planning as part of the design process, the periods of open-ended exploring and experimenting are crucial; the art is in listening always for what one may discover. In those moments of exploratory probing it is not unusual – indeed, it is often essential to the inventive process – to stumble upon some unforeseen sound that becomes the basis of the next instrument. Those hours of knocking about in the shop trying this or that, or just moving through the world with an open ear and a sense of discovery, are essential to the experimental instrument maker's work. To get a further sense of this, consider this observation from the allied field of electronic instruments: Several decades ago, when the art of sound synthesis was new, electronic sound designers were very much taken with the idea that you could electronically build from scratch any sort of sound you could imagine. While that remains as true now as it was then, something that early synthesists might not have expected followed. The practice of sound sampling came along. At some point sampling – albeit frequently with extensive electronic processing – came to the fore as the preferred approach for many people working in electronic and digital sound. The point here is that the sounds that we chance upon in the real world may often provide more subtlety, richness, complexity, and, shall we say, quirkiness, than whatever we may come up with in the antiseptic environment of our own abstract thought.

Intended Function, Expected Context, and the Possibility of Something New

For any promising new instrument idea, the maker could ask herself these basic questions: what is the purpose of this proposed new instrument, and how is it intended to function? A typical response might be: obviously, the intended function of any new musical instruments is music making, and it should fill that function just as traditional instruments do, by providing the machinery for players to produce melodies and harmonies and rhythms. But for someone with a real love of sound and sound-making, this response is too narrow. Let's open things up a bit. The composer John Cage famously suggested that all sounds, including sounds not normally thought of as music – urban traffic noise is the standard example – can be heard as musically meaningful if one brings the right sort of aesthetic awareness and appreciation to it. Whether or not you agree to such a radical expansion of the definition of the word "music," the point is that there's plenty of room for an aesthetic of sound unrelated to the musical parameters of melody, harmony, and rhythm. To this broader aesthetic, we might give the name "sound art," and many new musical instruments – or perhaps we should say "sound instruments" – are oriented to this broader aesthetic. This suggests an interest in sound for sound's sake, a pleasure in the distinctive and peculiar characters of different sorts of sound, a pure love of rustles, crackles, hoots, babbles, roars, rumbles, and whispers. A maker of sound instruments might try to capture the diversity and beauty of nature sounds, or the complexity of industrial noise. And then there's the possibility for sounds, be they harsh or gentle, comical or mysterious, that are not necessarily modeled after any real-world examples. The challenge is to find the most intriguingly ear-catching sounds one can with the materials and ideas and acoustic systems available – or, of equal value, to find ways to invite people to listen to seemingly commonplace sounds with new ears. While many more people choose conventional music than pure sound art when selecting what to listen to by choice, the aesthetic of pure sound can be immensely rich. Plus, there are contexts in which non-musical sound is precisely what's needed, including but by no means limited to the worlds of Foley and sound effects. A particularly rewarding area is the dusky territory between what the ear interprets as musical tone and pure noise.

This points to the related question: what is the prevailing aesthetic into which this new instrument's sounds are to fit? Is the instrument being designed to operate within a preexisting musical system? Or is the intent not limited by the conventions of a preexisting system, and instead more exploratory in nature? To exemplify the idea of an instrument made for an existing system, we can point to any instrument intended for playing "Für Elise," or "Melancholy Baby," or "Louie Louie" – in short, an instrument designed to play within a conventional musical scale, controllable by the player so as to produce the intended tones in the intended timings. This is a description of the great majority of musical instruments in the world. The alternative may seem less obvious but may also be quite liberating: what about instruments which create an idiosyncratic musical world of their own? Here the most obvious case would be an instrument that produces a non-standard musical scale – a topic we'll touch more on later. But there's also the more expansive set of possibilities: instruments made with the idea that they can create their own sound-world without reference to any of the usual musical parameters.

In this mode of thinking, a musical instrument may be seen as a bundle of open-ended possibilities in sound. It is then the role of the player to explore those possibilities and facilitate them, with an open ear to seeing what the instrument can do and "wants" to do. If the player does not impose too much in the way of preconceived notions, the results may be quite unlike either "Für Elise" or "Louie Louie;" they may involve sounds and sound-relationships unlike whatever other sorts of music may be currently abroad in the surrounding culture. Beyond the likelihood of non-standard scales, there may for instance be very unusual and uniquely characteristic sorts of note patterns or rhythmic patterns arising from the configuration of the instrument and the nature of the player-interface. Or there may be intriguingly irregular timbral relationships arising in different parts of the instrument. Or, on another level, the interaction of player and instrument may unfold in ways that create a different sense of what it means to be making music, as we'll see in the following paragraphs. The results might not always turn out to be interesting for most listeners – heaven knows, at concerts of new instruments one sometimes hears a few too many seemingly directionless scratchy sounds – but on the other hand, they may turn out to be interesting in ways one would never otherwise have stumbled upon. Much of the art, for the instrument maker and player, will be in finding one's way, through intelligent and open engagement with the peculiarities of the instrument at hand, to the sounds and sound patterns which are new and different, yet still engaging, meaningful, and appealing to the ear and mind.

This sort of thinking has allowed several experimental instrument makers to say something like "The instrument is the composition." By this they mean that each new instrument will present its own unique gamut of sound possibilities, and the work of designing the instrument can be understood as a sort of creative and curatorial process which opens this world for the player's exploration. The resulting music, be it pre-planned or improvised, can be seen as a composition arising naturally from the nature of the instrument.

Interface and Gesture

In exploring these ideas, makers of new instruments may find themselves rethinking on several levels the relationship between the player and the instrument. Most fundamentally, one can ask, "who's in charge here?" The ideal of the player's mastery of the instrument permeates our music culture – who wouldn't want to be seen as a virtuoso? But if the player thinks he already knows all there is to be known, that reduces the chances of finding something new. Newly invented instruments and one-of-a-kinds provide an excellent opportunity for a more exploratory approach. Some instruments are particularly promising in this regard; indeed, some acoustic systems are difficult enough to control that it's clearly more rewarding to let the instrument take the player for a ride than for the player to try to exert control. Such instruments may come in many forms, but as an example consider instruments such as Tom Nunn's Crustacean (described in his article in this book) and other like instruments with freely mounted sheet steel resonators. Those freely oscillating sheets, coaxed into vibration by bowing or other means, seem to have a mind of their own once they get going, and following them where they may take you will probably bring you to more interesting places musically than wherever you'd arrive if your intent was to maintain control.

Harry Partch (of whom we will speak more later) used the word "corporeal" to refer to – well, actually, he used it to refer to many inter-related things, but one of them was the physical nature of the instrument: the instrument as a real, present, physical body with which the player must interact. This sense of the physicality of the thing – something to touch; something subject to the laws of physics – is important for many instrument makers.

And this brings us to the question of the player interface: how the player interacts with the instrument. Interface design has been increasingly recognized as an essential consideration for all types of technology in recent decades, most obviously in computer applications. It is interesting to note that musical instruments provide some of the very earliest instances in human history of sophisticated interface design. We see this in the development of the organ keyboard in Europe, and to take one of many possible examples from elsewhere it can be seen as well in the early development of keying layouts in instruments of the kalimba and mbira families in southern Africa. From these two examples, we can see how choice of note layout – positioning and sequencing of pitches in the instrument's keying configuration – both influences and is influenced by the sorts of musical patterns players want to bring out in the instrument, very much coloring the nature of the music and thus shaping the character of the instrument. More subtly, the way the player interacts physically with the instrument is central to what the playing feels like for the player, thus helping to define the experience for both player and listener. From this perspective, we can move beyond the question of note layout for keyboard-like instruments, and into more broadly applicable issues in other instruments like orality, kineticism, bodily positioning, and ergonomics.

A key component of this is *gesture*: what sorts of physical movements go into the playing of the instrument, and how those physical gestures translate into the musical gestures of the resulting sounds. Makers of new instruments do well to take such considerations seriously in the design process, if only to ensure comfortable ergonomics. But the case can be stated more strongly than that: for many thoughtful makers, the nature of the interaction and the sorts of gestures involved – basically, what it feels like to play the instrument – turns out to be as essential to the character of the instrument as the sounds it makes.

Quite a lot of creative work has gone into the development of new ways of thinking about interface and gesture. Makers have explored, for instance, things like dance-controllers, wearable instruments, and instruments of sufficiently large scale that playing becomes a matter of moving from place to place within the room. This kind of creativity has been most prominent in the world of digital and electronic instruments where, with the use of various sorts of touch sensors or spatial sensors, interface design can be seen as a matter of how one maps inputs into outputs. This seemingly abstract question is, for musical instruments, very much anchored in the physical and human worlds. The degree of freedom afforded by digital technology is extraordinary, as almost any sort of sensor-read input can be mapped to almost any sort of sonic output. The possibilities multiply further when one considers the possibility of programmable mappings that can be rerouted to fit the user's inclinations or to fit a particular composition and may even operate variably and interactively in real time in response to different inputs. These questions typically play out less dramatically in acoustic instruments, but on the most fundamental level – what it feels like to play the instrument – the input-output question is equally central.

A Broader Range of Possibilities for Sound Art

Thus far still we have been implicitly working with a fairly narrow notion of what a musical instrument is or can be. It's time to broaden this a bit. A traditional conception for what an instrument is might invoke an image of someone sitting at or holding the device in question, manipulating it to produce intentionally organized sound as a means to musical pleasure for self or for audience. Most instrument makers produce things designed just for this kind of use. But one can also imagine other sorts of functionalities for instruments. Some of these fall under the rubric of sound sculpture. Whatever sound sculpture is – and the term has been applied to many different sorts of things – it does seem to open up possibilities beyond the image just described.

As a first example, there are site-specific sound installations. These are sound-oriented constructions that are designed not to be portable, but to reside in a particular location, often with form and function inspired by the features of the chosen site. Such a piece might be designed to work somehow with natural elements such as wind to produce sound without requiring human players. Or it may be designed for human interaction, with some way for members of the public, encountering the work, to engage with it in making sound. Still other installations might be provided with electronic sensors and activators, responding to the movement of people nearby by producing sounds. Whatever the activation mechanism, the sounds of such instruments might be conventionally musical, or they might be sonically interesting in other ways.

Another twist on the "who plays it?" question is automata – that is, mechanical or electrical self-playing instruments. This notion has a wonderful history extending back hundreds of years to early mechanical musical automatons and continuing through the heyday of commercial mechanical instruments in the early years of the last century. Today the idea has been given new stimulus in the possibility of digitally encoded, mechanically operated acoustic instruments. Recent work has taken forms ranging from industrial-looking machines in a steam-punk style, through anthropomorphic instrument-playing robots, to delicate assemblages of record turntables and tinker toys.

With many sound sculptures, one of the most engaging aspects of the work has to do with the observable mechanisms of sound production. Whether we're seeing a Rube-Goldbergian contraption or an elegantly observable exploration of acoustical phenomena, witnessing the connection between the mechanics and the resulting sound is always engaging for the mind, the ear, and the eye. One particularly fruitful area for exploration has been cymatics. This is the art of making acoustic phenomena observable. The most famous example is Chladni plates – metal plates with sand scattered on them, subjected to audio-frequency vibration. According to the natural vibratory modes of the plates, the sand seemingly self-organizes into fascinating visual patterns. In recent years, artists and scientists have been able to create equally enchanting observables using water, reflected or refracted light, and other resources.

Socially Interactive Instruments

Here's another angle on the question of how we expect musical instruments to function: Whatever sorts of sounds they may make, musical instruments typically operate in interaction with humans, and in that sense, they operate in some kind of social context. As a first distinction, we can note that an instrumentalist may play for others in a public sort of way or may play for herself in a very private way. Instrument makers may have this in mind in creating the design: an instrument intended for solo meditation, for instance, can be much quieter than a concert instrument, and it may have sounds that are particularly delicate, suited for a meditative sort of playing and listening.

Alternatively, the maker may see the instrument working in a group context. There are many forms this can take beyond the performer-and-audience paradigm. For instance, the maker may envision a highly interactive sort of music-making, with multiple players interfacing through the instrument to create an improvised sound composition that emerges from their interaction. Think of, for example, a reed instrument having two mouthpieces joined by a series of interconnected pipes and a single chamber. With a player on each mouthpiece, a third player operates valves affecting the pipe connections and lengths. (This is a description of Duo Capi, made by Oliver Di Cicco.) Or consider an installation of long interconnected sounding strings harnessed to the bodies of several players. As the players lean their weight this way or that, the string tensions and resulting sounds are in a state of constant interactive variation. (Instruments using such interactive string configurations have been created by Mario van Horrick and others.)

Another highly interactive approach – one which has a long history still much valued among makers and players of new instruments – is hocketing. This is the practice of having the notes of a scale not on a single instrument managed by a single player, but distributed, one or two notes apiece, among many players. Producing coherent music in this way requires a high degree of group coordination and awareness, and it is particularly gratifying when it works well. We see this in bell choirs of old, and in many distributable instrument sets conceived and made today, as with a set of long-sustaining, justly tuned tuning forks distributed to a participatory audience for a spacious group exploration of the chosen tuning (this describes a piece created by the American-Australian composer Warren Burt). Of course the individual instruments need not necessarily add up to a musical scale; it may be some other sound parameter providing the differentiating principle.

Here's a practice that would not be called hocketing but is related in its exploration of a distributed yet integrated social/musical interaction: one can pass out to a group of players a large number of like-sounding instruments to create among them an immersive, spatial sound atmosphere. The effect is quite different than hocketing's more differentiated functional approach. There is nothing else quite like the strangely sweet, insect-like cloud of sound made by classroom full of children all simultaneously playing simple, effortlessly assembled two-note plastic film-canister flutes (an idea explored by Darrell DeVore).

Sound Healing, Ritual, and Cultural Symbolism

A lot of interest and a number of new instrument designs have come from the world of sound-healing. People who believe that the right sorts of sounds can have therapeutic effects have a strong interest in instruments made to produce such sounds. Here again we see a human context – the relationship between healer and patient, or between individuals in a group with a shared interest in the salubrious effects of the music they make together. The social dynamic and the musical dynamic are integral to one another.

In cultures around the world, musical instruments have had a role to play in ritual, and non-traditional instrument makers may take an interest in how their instruments function in ritualized contexts. The key consideration here is: is the sound of the instrument to be enjoyed solely for its musical or acoustic qualities, or do those qualities feed into a larger symbolism? Sound can have profound subliminal emotional qualities, allowing the instrument as ritual object to take on special significance. Even for those with less of a spiritual bent, extra-musical associations of the instrument-as-object can enrich the response to the sound. For instance, an instrument design which has associations with ancient humanity may take on added meaning for many people, as might one which has particular ethnic roots. More prosaically, so does one which has contemporary associations like sci-fi or some recent pop-culture trend.

Democratizing the Making of Music

A moment ago we referred to sound installations made to be played by the public, and this reflects another interest expressed by quite a few makers of new instruments. The world of music today, these makers would say, is too much divided into active makers of music and passive consumers of music, with most people falling on the passive side. This is reflected in an artificial divide between musicians and non-musicians. One of the biggest factors here is the professionalization of music: the idea that to be a musician requires training and anyone without such training is not a musician. To counter this, and to make musicians of non-musicians, these makers focus on creating instruments that anyone can enjoy, no credential required. They strive to create a non-judgmental sense of freedom and play in their instruments. In addition to interactive public installations, these ideas may take shape as playground instruments – durable, all-weather constructs with simple and intuitive playing techniques, typically tuned to scales which require no harmonic sophistication and tend to sound consonant even in random playing. Some excellent makers have specialized in this, and there are a fair number of musical playground structures installed across the USA and abroad.

From the playground to the classroom: Many makers, as well as teachers, have also noted the pedagogical value of instrument design and construction. Indeed, instrument making in the classroom ticks quite a few boxes on the contemporary educator's agenda, and in an admirably interdisciplinary way. In addition to music and arts, a unit on instrument making can integrate math, physics, and shop skills; often computer and design skills as well, and it fits nicely with a project-based learning approach. As classroom projects go, instrument making tends to do quite well in keeping kids engaged. Such programs have been put in place for different age groups, preschool through high school and beyond. Countless books have been written featuring ideas for kid-buildable instruments. In practice, there's a challenge here, especially for younger kids: is it possible to create worthwhile, non-trivial instruments out of available, affordable, safe, workable, non-toxic materials, using minimal tools that are accessible and safe for children? These limitations are very real, but the answer turns out to be yes, there are some children's instrument making ideas that are quite workable and that kids can enjoy and take pride in. A common example is monochords. These are simple one- or two-string instruments easily playable with a slide which, in keeping with a centuries-old tradition of making such instruments, has the slide locations for the musical intervals marked beneath the strings. The positions for these locations are determined by simple length ratios, providing an excellent lesson in both applied math and musical scale theory. Many other, more imaginative kid-buildable instruments are possible, and if you put your mind to it perhaps you'll come up with more such ideas that no one has thought of yet.

Intonation Theory and Acoustical Exploration

And speaking of scale theory: Another crucial motivation for contemporary instrument makers has been a wish to realize non-standard tuning systems. This reflects the fact that the great majority of conventional instruments are designed to play in the 12-tone equal temperament scale that has been the near-universal tuning standard in the western world and beyond for the last couple of centuries. The same sort of person who might ask "why not build unconventional instruments?" might also ask "why not play in unconventional scales?" As a motivation for making new instruments, this kind of thinking was particularly strong through the second half of the last century, following Harry Partch's famous observation, "I am a philosophic music-man seduced into carpentry." Partch – now seen as a seminal figure for both instrument makers and intonation theorists – was saying that his motivation for instrument making was a need to have instruments that could play in the alternative tuning systems he was researching and developing. In more recent years, we have had synthesizers and computers programmable to any tuning, and this has seemingly made the need for instruments that can play new scales less imperative; yet the demand is still there for acoustic instruments capable of non-standard tunings, and this motivation remains strong for many makers.

Beyond intonation theory, for many makers there's an interest in pure acoustics: "Hmm … here's an interesting way of setting the up a vibration in the audible frequency range; let's explore it further, see what kinds of sounds it can generate, and see if we can find ways to control and manipulate it for musicality or sonic interest." To come up with entirely new acoustic systems that have the potential to be musically useful is quite a challenge; most of the brilliant new ideas one might come up with turn out to be either not so new (others have explored them already) or not so brilliant (they don't end up working especially well). But even short of something entirely new, there's quite a lot of fruitful ground to be explored and re-explored around the edges. As an example, consider the phenomenon of longitudinal vibration in solid materials. This is a type of vibratory movement that normally happens in open air or in the enclosed air columns of wind instruments. It also happens in solid materials, but rarely is this exploited in musical instruments. One reason is, to produce longitudinal vibrations of audible frequency in solid materials typically requires impracticably long components. This has led to the maker Ellen Fullman's groundbreaking work with her Long String Instrument, described in her article in this collection. It features strings over a hundred feet long, with both a powerful sound and a unique choreography in the playing of it, all arising integrally from the underlying physics. Ellen didn't "invent" longitudinal vibration, but in working with it, a combination of scientific and aesthetic interests led her to territory no one else had explored.

Pleasure in the Craft and the Beauty of the Work

And finally, here's one more answer to the question "Why make new instruments?" For those who have been "seduced into carpentry," there is the pleasure of pure craftspersonship. For many, the visual and sculptural aspects may be as important in the gestalt of the piece as are the musical qualities. But as with everything else in this essay, because we are talking about a diverse field populated by a lot of different characters, there are many different approaches. There are those who strive for refinement and beauty of the highest order in the final product and have the skills to bring that about. At the opposite extreme are those who enjoy a scrappy aesthetic and gravitate toward repurposed and junkyard materials. Despite divergences in style, we usually find a kind of aesthetic coherence in the look of the instruments that different makers come up with, as physical and musical requirements in combination with the character of the builder give rise to results that cohere visually.

A Very Brief History

We'll now attempt to provide some sense of how these motifs have played out over time, highlighting a few of the important names and signposts along the way.

People have been making musical instruments since the earliest times, but for the purposes of this essay, a useful historical starting point will be around the turn of the 20th century. This was a time when ideas about music were rapidly evolving toward more modern conceptions, and several threads were coming together which contributed to the themes we've been discussing.

One of these threads was a dawning recognition of the potential of electronics in producing and controlling musical sound. The advent of electronic technology of course made certain new instruments possible. More broadly, we can see in retrospect that electronic sound had a role in opening up people's ways of thinking about what musical instruments in general (not only electronic) can be and how they might function. As a first example, we can point to Thaddeus Cahill's *Telharmonium* (1896). The Telharmonium promised – and indeed for a number of years actually delivered – ethereal music piped into people's homes via telephone lines. This suggested a notable re-thinking of the relationships between player, instrument, and audience. The instrument was innovative also in the way it worked, introducing for the first time an electro-mechanical sound-producing system similar to what was later employed in the immensely popular Hammond Organ. Crucially, this system lent itself readily to the tone-building technique of adding harmonics to a fundamental frequency – an idea that had long been used in acoustic organ building, but which now, in this electronic context, foreshadowed the subsequent development of additive synthesis.

Soon after this, Ferruccio Busoni argued for freedom from the limitations of conventional instruments in his *Sketch of a New Esthetic of Music* (1907) and enthusiastically proclaimed the value of electronic instruments like the Telharmonium. Edgard Varése shared Busoni's view, envisioning a compositional sound-palette made possible by instruments yet to be invented. There seems to have been in the air a shared dream of liberation from traditional notions of what constitutes musical sound.

While Cahill's Telharmonium was equipped with a conventional keyboard, Maurice Martenot's *ondes Martenot* (1928), Leon Theremin's *Theremin* (1928), and Friedrich Trautwein's *Trautonium* (1929) used oscillators to create electronic sounds along a continuum of pitch, rather than discrete steps. These reflected a burgeoning interest in another kind of liberation, this time from traditional scales. In these instances, it was not so much an interest in the particular scales of just intonation – that was to come later – but rather an interest in freedom to explore the pitch spectrum free from fixed prescribed pitches, allowing for both stable and sliding pitches. Equally significantly, these designs reflected an openness to the new gestural qualities in the instrument interface – a theme we'll see again and again.

New possibilities in the approach to musical time were in the air too. In 1930, Henry Cowell collaborated with Leon Theremin to build the *Rhythmicon* (or *Polyrhythmophone*), an electronic device that could play complex rhythmic patterns based on the harmonic overtone series.

During this period, several people also explored sound-production techniques making use of electronic equipment in unexpected and inventive ways. One of these was tape music, the manipulation of magnetic recording tape and playback equipment in such a way that the equipment comes to function as an instrument. Many of these techniques are closely related to the development of *musique concrète* – the use of recorded sounds from the real world in compositions created by tape manipulations. Among the leaders working in the area were Halim El-Dabh, Pierre Schaeffer, Pauline Oliveros, and Delia Derbyshire. A related use of electrical equipment is drawn sound – the practice of using visible patterns on film to activate sound optically. A number of composers in Russia, Europe, and the USA explored this idea between the 1920s and 1960s. A notable instrument here is Daphne Oram's *Oramics Machine* (1957). See also Jacques Dudon's article in the current collection for another approach to light-activated sound. A similarly inventive extension of the potentials of an existing technology can be seen in the art of turntabling. Having seen some use in the early days of *musique concrète*, turntables found a new style and immense popularity in the 1970s and '80s in the hands of artists like Cool Herc, Grandmaster Flash, and Christian Marclay. In all of these instances, the physical nature of the equipment-as-instrument is central to the art: the forms and methods of the technology give shape to the compositional process, to the nature of the resulting music, and to the feeling of the interaction between player and instrument.

The later decades of this period saw the invention of the modular synthesizer, with Don Buchla (1963) and Robert Moog (1964) in the lead. Buchla in particular was interested not only in creating new sounds, but also in new and more expansive forms of interaction between player and instrument, and among players. Finally, in the closing decades of the century came the rise of digital sound, sampling, and increasingly sophisticated computer techniques now familiar to all.

Even as developments in electronic music were opening minds, parallel paths were taking shape in the worlds of acoustic instruments. We can start here with Luigi Russolo. In the early years of the 20th century, members of the Italian Futurist movement had championed the use of industrial sounds, notably in Russolo's 1913 manifesto *Art of Noise.* Russolo went on to make good on his proclamations, creating with Ugo Piatti the *Intonarumori*, an ensemble of instruments designed to make mechanical noises for the first Futurist concert of noise music in 1914. In some ways Russolo remained an outlier, yet by the late 1920s, both Edgard Varése and George Antheil were making waves in the music establishment by composing for percussion orchestras using such sound sources as sirens, whips, anvils, and airplane propellers (along with more familiar percussion instruments). A few years later, Lucia Dlugoszewski was composing for an orchestra entirely of her own invented percussion instruments.

1947 saw the publication of Harry Partch's seminal work, *Genesis of a Music.* This book has been, for many contemporary instrument builders as well as intonation theorists, an essential starting point. In some ways, Partch's work stood as a counter-balance to one of the major threads in the academic music world of the time: Since the early part of the century, the dominant trend in academia had been atonal musical styles, with serialism prominent among them. These forms were generally locked into the twelve-tone equal-tempered scale and often employed rule-based compositional techniques. With his emphasis on corporeality, Partch called not only for an alternative to the dominance of twelve-equal, but also for an escape from purely intellectual and abstract approaches to music. He favored, instead, an approach grounded in the physicality of the real world, the human ear's natural ways of interpreting sound and pitch, and a compositional practice engaged with meaning and feeling. These ideas foreshadowed a broader aesthetic sense that was to come increasingly to the fore in the '60s, '70s, and beyond: a reaction against abstraction and intellectualism in favor of more experiential engagement in the arts. Although it's not always articulated, this experiential orientation lies near the core of much of contemporary creative instrument making today. By directly engaging the physical thing that is the instrument, we get at the nature of sound and music more convincingly than one would with a purely abstract musical approach.

Also in the late forties, Conlon Nancarrow began composing for modified player pianos, experimenting with complex tempi and rhythmic relationships beyond anything a human could play. The real hey-day for mechanical instruments (of which player pianos are just one example) had been the early part of the 20th century, in the decades just before the recording industry took off. The diversity and ingenuity of popular, commercially made mechanical instruments at that time was marvelous. Nancarrow's work in treating such instruments as superhuman compositional tools represents one direction a composer or builder can take with such instruments (especially in later years with computer-controlled instruments). Another direction, which a number of more recent builders have followed, has more to do with the pleasure of the device itself, whether the resulting instruments are impressively elaborate and refined, as with the Youtube-famous *Marble Machine* created by the group Wintergatan, or quirkily simple as in the works of Ernie Althoff, Joe Jones, or Pierre Bastien.

Meanwhile, just after mid-century in France, the Baschet brothers (featured in an article in this collection) began their decades-long exploration of acoustic systems and timbre. In the process they developed an analytical schema for instrument design that was at the same time both practical and revelatory. Their work proved highly influential as it found its way into the art world and was widely shown and publicized internationally; for many it was a first introduction to the idea that one can take a creative and sculptural approach to musical instrument design. Much of the appeal for aspiring instrument makers lay in the way the instruments engaged and explored questions of acoustic design: to look at a Baschet instrument is to think about questions of sound radiation and transmission, impedance relationships, and modes of vibration – seemingly sophisticated stuff, yet invitingly plain to see, for anyone with a certain analytical bent, in the Baschet instruments.

Through the 1970s, '80s, and beyond, an increasingly important source of inspiration for many was instruments of non-Western cultures. Lots of people with no other connection to indigenous Australian culture, for example, took to playing the didjeridu, and some of them found they enjoyed making didjeridu-like instruments themselves. Kalimba-like instruments inspired many non-African makers. The ceramic wind instruments of pre-Columbian Central and South American cultures, with their strange and beautiful acoustic designs, were an inspiration for makers like Susan Rawcliffe and Brian Ransom (both of whom have articles in this collection). But the most widespread of these trends was what came to be called American Gamelan. This was an informal affiliation of instrument makers and composers profoundly inspired by Indonesian gamelan music, with its sonically enchanting ensembles of metallophones, winds, strings, and drums. Non-Indonesians strove to study, understand, and honor Indonesian traditions while at the same time allowing themselves creative freedom in "inspired-by" instrument making and composition.

Important figures in this movement included Barbara Benary, Jody Diamond, Daniel Schmidt, and the composer and instrument builder team of Lou Harrison and Bill Colvig. (In this collection you can find an article about the gamelan-based work of Daniel Schmidt, and a section on Indonesian scales by Larry Polansky in the article "Comparative Scales Chart".)

It was also in the 1970s and '80s that the interest in found objects and repurposed materials as sound sources began to flourish, reflecting the concern with the democratization of music-making. Prominent among the many who have been active in this area has been the group Music for Homemade Instruments, founded in 1975 by Skip LaPlante and Carole Weber. Bash the Trash, an organization founded in1988 by John Bertles and Carina Piaggio and still going strong, has a similar orientation and brings a classroom educational mission to the endeavor, as described in John Bertles' article in this collection.

The years just before the internet took over the world were a flourishing time for newsletters and small magazines. One of these in particular focused on many of the themes we've been discussing: *Interval Magazine* was founded in 1978 by Jonathan Glasier, who had worked as Partch's assistant for several years, as a forum for tuning theory, instrument making, and sound sculpture. Also important in the arena of the written word was the publication in 1984 of the *New Grove Dictionary of Musical Instruments*, a massive three-volume work. It devoted a fair amount of space to 20th-century instruments, mostly under the able and knowledgeable authorship of the late Hugh Davies (himself as well a maker of experimental instruments).

In the fine art world, the concept of sound sculpture, in its diverse manifestations, has gradually found purchase. Although the ideas inherent in sound sculpture have always been around, it was in the 1960s, often as an offshoot of kinetic art, that primarily visual artists such as Robert Rauschenberg, Jean Tinguely, and Harry Bertoia began to incorporate sound into their works. What was revolutionary then has gradually found acceptance and respectability: we now find that most art schools offer classes in sound art, and some even have sound art departments. Some of the prominent sound artists in years since have included Annea Lockwood, Paul Panhuysen, Bill Fontana, and Trimpin … and, as the field has grown, a great many more.

This very brief history has been by no means comprehensive or exhaustive, and inevitably there is an arbitrariness in what and whom we have and have not mentioned. New sound instrument design is a wonderful world – wonderful to be part of and wonderful in what it can offer even to those musicians and listeners who are not hands-on involved. We hope that this overview has given you some sense, in broad brush strokes, of what it's about and what motivates those who are part of it. The articles in this book will take you deeper.

2

The *Experimental Musical Instruments* Journal

A History, as Recalled by the Founder and Editor

Bart Hopkin

The notion of starting a journal devoted to the making of new and unusual musical instruments came to me some time in 1983 or '84. At that time I had enjoyed experimenting with offbeat musical instrument designs for – well, much of my life, though I hadn't always done it very seriously. But my knowledge of instrument making was pretty shaky, and my awareness of what others were doing in the field was essentially nil. I happened to hear on the radio an interview with someone named Becky Blackley who had started a newsletter for aficionados of the autoharp. It was called *The Autoharpaholic,* and under her leadership it went on to have a good and successful run for many years. The thought was born in my head: maybe somebody could do something similar for fans of inventive, one-of-a-kind musical instrument explorations. Without knowing at first whether I was really serious about this enterprise (and having no illusions about attracting a large readership), I began to research the question of what, if anything, other people might be doing in the field of creative instrument making, and whether other magazines, journals, or newsletters devoted to the subject existed. Because I was getting to like the idea, I managed to convince myself through my investigations that there was room for what I had in mind. So it came to pass in June of 1985 that I sent out the premier issue of *Experimental Musical Instruments* to a list of 40 charter subscribers that I had somehow drummed up. It was 16 pages long with a few line drawings but

Figure 2.1 Drawing by Mark Kelly.

no photographs. The choice of typefaces had been dictated by the availability of daisy wheels for the printer I was using, and the page layout was done by the hand paste-up method that was typical in those days. All of the articles in that first issue were written by me, but I didn't put bylines on them because it looked silly to have all the articles signed by the same person.

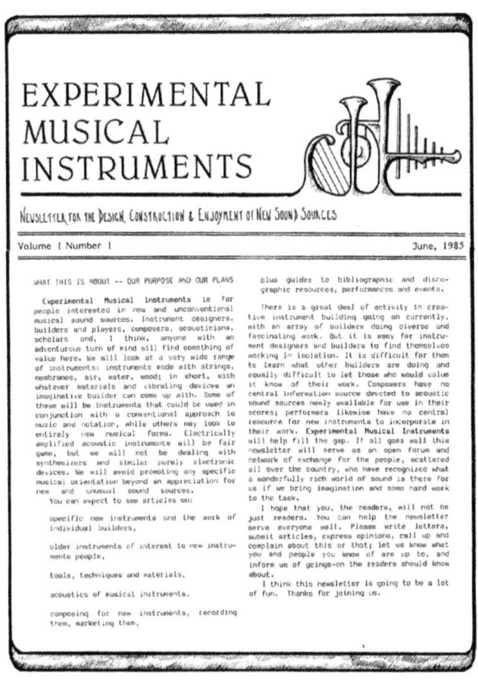

EXPERIMENTAL MUSICAL INSTRUMENTS

Newsletter for the Design, Construction & Enjoyment of New Sound Sources

Volume I Number I — June, 1985

WHAT THIS IS ABOUT -- OUR PURPOSE AND OUR PLANS

Experimental Musical Instruments is for people interested in new and unconventional musical sound sources. Instrument designers, builders and players, composers, acousticians, scholars and, I think, anyone with an adventurous turn of mind will find something of value here. We will look at a very wide range of instruments: instruments made with strings, membranes, air, water, wood; in short, with whatever materials and vibrating devices an imaginative builder can come up with. Some of these will be instruments that could be used in conjunction with a conventional approach to music and notation, while others may look to entirely new musical forms. Electrically amplified acoustic instruments will be fair game, but we will not be dealing with synthesizers and similar purely electronic devices. We will avoid promoting any specific musical orientation beyond an appreciation for new and unusual sound sources.

You can expect to see articles on:

specific new instruments and the work of individual builders,

older instruments of interest to new instruments people,

tools, techniques and materials,

acoustics of musical instruments,

composing for new instruments, recording them, marketing them,

plus guides to bibliographic and discographic resources, performances and events.

There is a great deal of activity in creative instrument building going on currently, with an array of builders doing diverse and fascinating work. But it is easy for instrument designers and builders to find themselves working in isolation. It is difficult for them to learn what other builders are doing and equally difficult to let those who would value it know of their work. Composers have no central information source devoted to acoustic sound sources newly available for use in their scores; performers likewise have no central resource for new instruments to incorporate in their work. **Experimental Musical Instruments** will help fill the gap. If all goes well this newsletter will serve as an open forum and network of exchange for the people, scattered all over the country, who have recognized what a wonderfully rich world of sound is there for us if we bring imagination and some hard work to the task.

I hope that you, the readers, will not be just readers. You can help the newsletter serve everyone well. Please write letters, submit articles, express opinions, call up and complain about this or that; let us know what you and people you know of are up to, and inform us of goings-on the readers should know about.

I think this newsletter is going to be a lot of fun. Thanks for joining us.

Figure 2.2 Front cover of *Experimental Musical Instruments* Volume I #1, June 1985.

Now comes the story of how, following this debut, *Experimental Musical Instruments* got to be a big success. But, of course, that's not what happened. The journal grew only very slowly in subscriber numbers and never achieved anything better than a wobbly, borderline sort of financial stability. But it did turn out that there were people in the world who liked having a journal devoted to inventive instrument making. The degree of support that EMI got from those people over the years was huge. Over time, the issues of the magazine itself grew quite a bit longer and a bit less homemade-looking. We started having lots of photos of course; more importantly we started receiving article submissions from people working in diverse facets of the field of instrument design and construction. We developed a network of knowledgeable people for expertise as needed on various aspects of musical instruments and instrument making, including our referee for all matters pertaining to higher acoustics, Donald Hall, Professor (now emeritus) at Sacramento State University. The editorial focus evolved a bit, from an emphasis almost entirely on new designs for acoustic instruments to an outlook that had room for historical and non-western instruments, sound sculpture and installations, scale theory and acoustic theory; occasionally some electronics. Gradually – I hope I'm not just flattering myself when I say this – something did seem to coalesce around *Experimental Musical Instruments*. The journal became a center for people around the world who were interested in creative instrument explorations and over time helped to give definition to a field which had previously been scattered and diffuse.

The journal kept going for fourteen years, putting out a total of 70 issues. By the time the last issue appeared in 1999, there were a variety of reasons for stopping. The appearance of the internet as an alternative method for disseminating information was one; my own need to reorganize my professional life was another. It most certainly was not the case that we were stopping because the field had become less interesting or was played out.

"If the pole-sitting craze of the 1920s had involved white elephants, then Neil Feather's music would have been precisely the strange love affair which was going on behind the oblivious elephant's back."

– John Berndt, in his article "The Funny Music of Neil Feather" appearing in the June 1998 issue of *Experimental Musical Instruments*.

Figure 2.3 Most of the writing in *Experimental Musical Instruments* articles was straightforwardly informational and factual. Yet even within that context there was room for the occasional flight of imaginative language.

In the course of those fourteen years and in years following, *Experimental Musical Instruments* as an organization branched out into several other activities beyond the print journal. With the completion of the first year of publication, we put out a cassette tape to let people hear the sounds of the instruments we had been writing about. This became an annual cassette series under the title *From the Pages of Experimental Musical Instruments*, and it continued through to the final issues. We put out a number of books and CDs, some published in-house and some produced in conjunction with outside publishers. Many of these were how-to books on the construction of diverse types of instruments, typically written by myself along with one or more coauthors chosen for their expertise in the particular type. In 1996, I worked with Ellipsis Arts publishing company to put out a book-and-CD collection called *Gravikords, Whirlies & Pyrophones*, featuring the work of some favorite inventive instrument builders, and the unexpected popular success of this collection brought a lot of welcome attention to the field. At some point, *Experimental Musical Instruments* got into the business of selling hardware for instrument makers – things like tuning machines, pickups, and fret wire. This work may have been less creatively gratifying than publishing, but it did a lot for the bottom line. Through all this, the journal itself, with its laborious, often exasperating, and always ultimately rewarding quarterly productions, remained our signature endeavor.

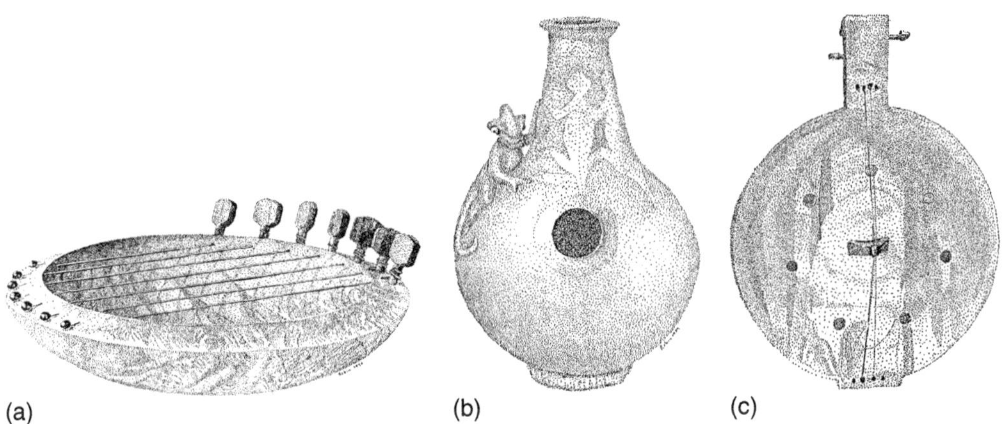

Figure 2.4 Robin Goodfellow employed several different graphic techniques in her many illustrations for the journal. Shown here are several of her stipple drawings.

This book presents some of the best articles that appeared in the *Experimental Musical Instruments* journal over the years. It does not do as good a job of conveying what went on in the journal outside of the major articles – the letters to the editor, the reviews of books and recorded music, the odd bits of filler here and there. And more: Among the regular features in each issue of *Experimental Musical Instruments* were not only letters and reviews, but also a list of relevant articles appearing recently in other publications, information on other, often obscure journals and organizations working in fields related to our own, public notices and "unclassified" ads, small commentaries and bits of information from our interactions with diverse characters on the scene, and a wide variety of other gems and throwaways. Much of the character of the journal was to be found in these interstitials.

The flavor of the journal was very much reflected in the letters to the editor. There were often, for instance, protracted back-and-forths on diverse topics that caught readers' imaginations, playing out over the course of multiple issues. Some of these were what you might expect: discussions of vibrational modes in musical glasses, the airing of ideas about alternative keyboard design, or an ongoing exchange concerning the acoustics of pyrophones. Others were more fanciful. One frequently recurring topic, for instance, was cat pianos and variations thereof. Readers turned up a quite number of historical drawings of keyboard instruments featuring a row of caged kittens which would yowl (on pitch?) when the pressing of a key caused some kind of stimulus to one or another of them. Comical contemporary variations were offered as well, such as a row of shower stalls, people inside scrubbing away, with a diabolical keyboardist controlling the flow of hot and cold to individual stalls. Another recurring topic was speed-bump or rumble-strip music. This arose from the observation by several readers that the track of grooved pavement along the side of some highways, serving to warn drivers who wander out of their lane by the sound of the wheels on the grooves, actually does produce a recognizable tone at the frequency at which the tires bump over the grooves. This led to speculation about how one could create melodies by the spacing of the grooves, and even some reports of actual experiments along these lines (Figure 2.5).

There were several people whose writings appeared in the letters column on a regular basis. Most notable among these was the late, great Ivor Darreg, whose extensive missives appeared in almost every issue up until his death in 1994. (One article of his is included in this collection.) Ivor was, in addition to many other things, a tuning theorist, most known for his interest in and championing of higher-order equal temperaments. He was also a wonderfully crotchety and colorful correspondent, often railing against a classical music establishment which seemed bent on oppressing outsiders. “What have the uptight perfectionists done for you lately?” he asked in one letter. “Don’t let them spoil your fun.” (Figure 2.6)

Following the letters in each issue was a section of the journal called “Notes from Here and There.” This was a catch-all for things that seemed worth making note of but weren’t sufficiently substantial to make a proper article. Lots of instrument

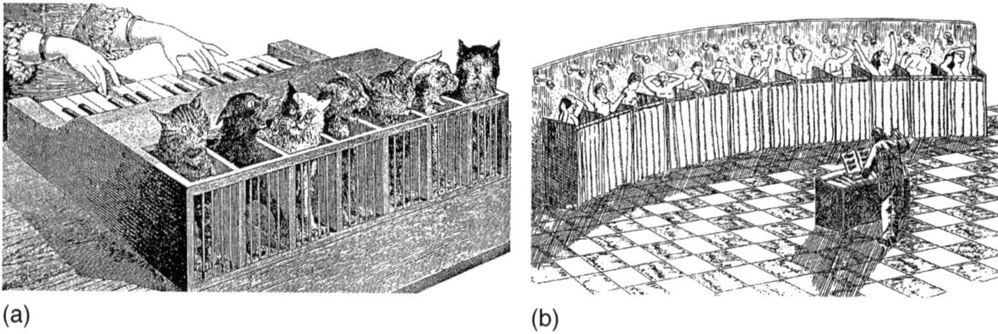

(a) (b)

Figure 2.5 On the left, from the February 1990 issue, the first of several historical Cat Piano drawings to appear in the journal, from a collection of public domain graphics; original source unknown. On the right, Peter Hurney’s depiction of the Cold Water Shower Reaction Device, or *The Scream*, from the June 1999 issue.

National Society for the Decriminalization of Microtones

Wenzdi 18 Novembur 1987

Figure 2.6 The letterhead from one of Ivor Darreg's letters to the editor.

ideas showed up here in brief description, often with accompanying photos, having been sent in by readers who saw the instrument somewhere, or representing some idea that a reader was developing and wanted to share. Despite the brevity of the presentation, many of these instruments and ideas were real gems (Figure 2.7).

A great many book and record reviews appeared in *Experimental Musical Instruments.* These are not included in this current collection, but some of them were

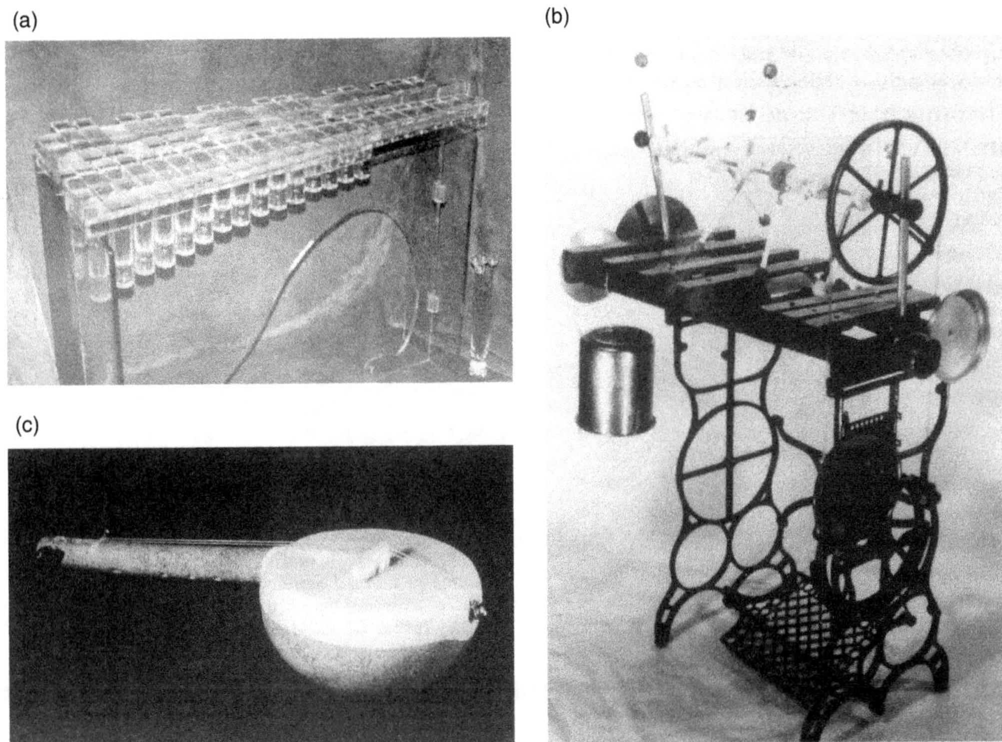

Figure 2.7 A few of the many instruments that were not covered in full articles but which appeared in the journal's "Letters" and "Notes form Here and There" sections. Clockwise from upper left: (a) Michael Meadows' Glass Marimba, (b) Frank Pahl's Automatic Marimba, (c) Barry Hall's Stone Fiddle.

quite substantial, having all the heft of real articles. Among them, for example, was Tony Pizzo's review of the then newly released English translation by Barclay Brown of Luigi Russolo's *Art of Noises.* "As I began reading *The Art of Noises,*" Tony began, "I fully expected to find in Luigi Russolo a mustachio'd Spike Jones with a manifesto." He progressed from there into an engaging and informative tour of Russolo's mind (which was not exclusively Spike-like, we discover), his noise instruments, and the cultural currents of his time and place. Another review looked at a book-length special issue of *Music Trades* Magazine called *History of the U.S Music Industry.* In commemoration of its 100th year of publication, the review reported that the people at *Music Trade*s had dug into their archives for the most telling stories and items they could find, such as this thought-provoking tidbit: In 1932 the American Association of Musical Merchandise Manufacturers proposed to Henry Ford that a plastic ukulele should be included with every new automobile sold, as an alternative to the new-fangled fad of listening to the radio in the car. Another book-length release from a periodical was the Percussive Arts Society's reprint of several of the catalogs from the Deagan company, released between 1910 and 1920. The pictures in these catalogs remain fascinating; Deagan (now remembered primarily for orchestral marimbas and vibraphones) was doing some very unusual stuff in those days. His ideas might have been called inventive, except that much of the most ostensibly innovative work involved taking out patents on instrument ideas stolen without credit from faraway lands. (Figure 2.8). Still other book reviews covered diverse writings such as Margaret Kartomi's *On Concepts and*

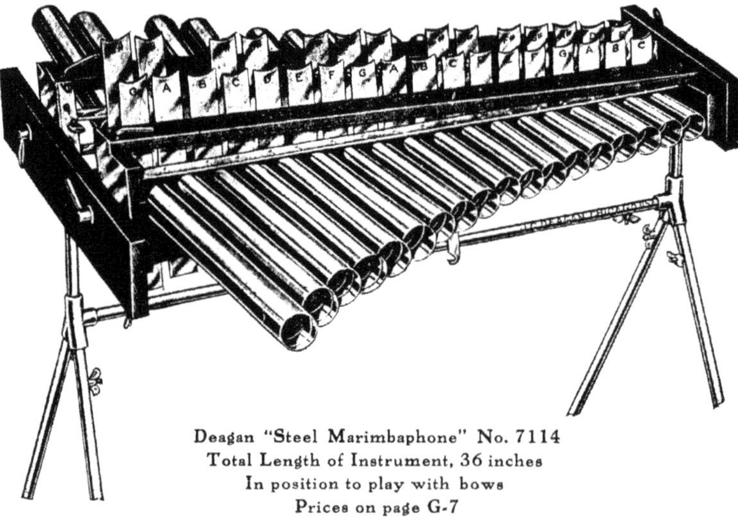

Figure 2.8 The Deagan Steel Marimbaphone, a steel bar instrument made for bowing (note the upright keys with curved upper edge), from one of the very early Deagan catalogs.

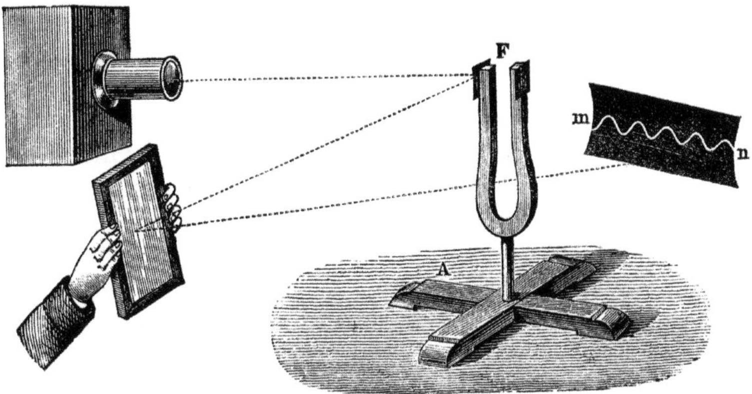

Figure 2.9 A device for making visible the wave form of a tuning fork, from Pietro Blaserna's *The Theory of Sound in its Relation to Music.*

Classifications of Musical Instruments, Reynold Weidenaar's *Magic Music from the Telharmonium*, and a look at some of the oldest surviving works on musical instruments from Michael Praetorius, Marin Mersenne, and Fillipo Bonani. Also old, but not quite as old: the 19th-century acoustics text *The Theory of Sound in its Relation to Music* by Pietro Blaserna, which we simply had to review because doing so gave us an excuse to reprint some of its wonderful woodcuts of early acoustic apparati (Figure 2.9).

As for reviews of recordings on vinyl, cassette, or CD: Thanks to the steady stream of submissions from a team of astute and culturally aware reviewers, there was no shortage of audio reviews, most of them brief but some more ambitious. They covered non-conformist musicians and groups with names like Nihilist Spasm Band and Dog of Lard, not to mention the Glass Orchestra, Spike Jones (that name again!), a release of the only known extant recordings of a castrato, Wendy Chambers' infamous performance of "New York New York" on the Car Horn Organ, and many, many more. Some recordings reviews were organized by theme, as in René van Peer's extensive series on nature sound recordings. Looking back now, it's interesting to note how a seemingly specific interest in unusual musical instruments provided the excuse for a remarkably capacious and inclusive take on worldwide audio culture of the time.

Experimental Musical Instruments also had reviews of other organizations and periodicals in related fields. In those pre-Internet days, there were quite a few such institutions, ranging from the Just Intonation Network, the Galpin Society, the American Musical Instrument Society, and the American Gamelan Institute through to *Sawing News of the World*, the *Newsletter of the National Association of Band Instrument Repair Technicians*, and my personal favorite,

Vierundzwanzigsteljahrsschrift der Internationalen Maultrommelvirtuosengenossenschaft (a newsletter for Jew's harp enthusiasts).

Experimental Musical Instruments occasionally reprinted articles that had appeared long ago in various now-forgotten journals – sometimes more for the fun or humor of it than for the intrinsic merit of the writing. One was "The Smell Organ" by Joseph H. Kraus, originally printed in the June 1922 issue of *Science and Invention*. It describes the work of the French chemist Dr. Septimus Piesse toward the development of his olfactory keyboard. An accompanying drawing showed the Liszt-like figure of a piano virtuoso playing the proposed instrument as scents, represented as wavy lines, waft toward an enrapt audience in evening attire. There is no indication that the instrument was ever built (Figure 2.10).

The number of deserving articles that appeared in *Experimental Musical Instruments* over the years was far greater than what we've been able to include in this collection. There were articles on musical instrument categorization systems, music of imaginary lands, overtone recipes for different instrument types, patenting for musical instruments, instruments played by whirling, blown chordophones, Apache violin, trumpet marine, mouthbows, wall harps, and slide whistles – along with, of

Figure 2.10 Septimus Piesse's Smell Organ, as depicted in a 1922 edition of *Science and Invention*.

Figure 2.11 Stick Figures illustrating whirled instruments of different sorts, from David Toop and Max Eastley's article "Whirled Music" in the August 1989 issue.

course, countless articles by or about individual makers and their respective instrument-making ideas – all sadly not included here. By the time the journal's fourteen year run was over, a huge number of people had contributed, either by writing or in other ways. You can see the names of a few of them recorded with immense gratitude in this book's Acknowledgments.

Much of this essay has been about all that appeared in *Experimental Musical Instruments* that we were not able to include in the current collection. You can explore this material more fully online at https://archive.org/details/emi_archive/, where we've posted scans of the entire series from start to finish (though, admittedly, not in the highest resolution). In the meantime, you hold in your hand this book, with its generous selection of some of the best that the journal had to offer. Please read and enjoy.

Part 2

Articles

3

Steel Cello and Bow Chimes

Designed and Built by Robert Rutman

Bart Hopkin

This article originally appeared in *Experimental Musical Instruments* Volume 1 #1, June 1985.

Have you ever fooled around with a thunder sheet?

Freely suspended flexible metal sheets are wonderfully efficient resonators. Strike them, shake them, flex them; they readily produce a remarkable array of sounds. For musical purposes, though, they are hard to discipline. The sheet by itself lacks mechanisms for controlling the exuberant sound it produces.

Over the past fifteen years or so, Robert Rutman has been evolving a series of instruments using flexibly mounted sheet metal resonators with separate initial vibrators – a bowed string in one case, and several bowed metal rods in another. The fact that the initial impulse comes from a source separate from the metal sheet allows for fuller control of the vibration but retains many of the metal's peculiar acoustic properties. The results are extraordinary.

The Single-String Steel Cello

Rutman's best known instrument is the steel cello. It employs a metal stand to support an eight foot by two-and-a-half foot sheet of stainless steel. The sheet is suspended from the top of the stand by wires at the upper corners, and anchored at the lower corners by two more wires – nowhere is it bolted or fixed in any rigid manner. The single string is an 18 gauge steel wire running from the top right corner to the bottom right, short enough to pull the ends together a bit, creating a curve in the sheet. The anchor wire below the lower corner where the string is attached has a turnbuckle, allowing for tuning by adjusting the tension on the string.

The player bows the string with a bass bow or other large bow. Rutman uses a picturesque, deeply curved bow that he has fashioned from bamboo. There is no fingerboard; the player varies the vibrating length of the string by pinching it between the thumb and middle finger of the left hand, or for faster passages by firm pressure of any of the left-hand fingers.

The resonating system is efficient enough so that the instrument projects very low pitches with generous volume and a feel-it-in-the-belly fullness. The same efficiency combined with the fact that the string is easily stopped by the left hand anywhere along its length means that very high pitches speak easily and are just as practical. Harmonic tones and fundamentals both sound readily.

While the steel cello's resonator dutifully reproduces the vibrations of the string, it definitely has a life of its own as well. Under some circumstances, the metal sheet develops standing waves separate from but coexisting with those of the string, producing two or more distinct simultaneous notes. The player can also bring out a larger movement in the sheet similar to the waves-crashing-on-the-shore mode of the thunder sheet, with large peaks traveling visibly over its length. With sustained bowing the string and resonator can be made to interact in such a way that wave patterns evolve and increase independent of changes in the bowing style or the way the string is stopped. All this creates a feeling that the instrument has a voice of its own; that the player is there to let it speak (Figure 3.1).

The Bow Chime

Rutman's bow chime uses five metal rods as initial vibrators. The rods are mounted upright on a horizontal bar set on a stand, attached to the bar by a mechanism which allows their length above the bar to be adjusted for tuning. The sheet metal resonator is bolted to the bar at the ends, forming a broad horizontal curve.

The player bows the rods at different points along their length with a broad, sweeping motion. The sound is controlled entirely through the bow, and, while it is not difficult to make it sing, the ability to bring out the variety of sounds it is capable of requires a lot of empathy with the instrument. Because metal rods are rich in partials, the five rods produce many more than five pitches. The overtones behave in a subtle but predictable manner, and an experienced player can learn what pitches are available and produce them at will by adjusting the pressure, speed, and position of the bow.

With its innate vitality, the resonator imparts something extra to the ethereal tone of the rods. The peculiar qualities of the free resonator described for the steel cello arise here as well, but in a more constrained form, since the sheet metal is more rigidly mounted (Figure 3.2).

The Builder and the Music

Robert Rutman was born in Berlin and has lived in England, Mexico, and the US. He currently lives in Cambridge, MA, working and performing in and around the Boston area. He considers himself an artist – a painter and sculptor – more than a musician, and has displayed his work in galleries around the country. Rutman has

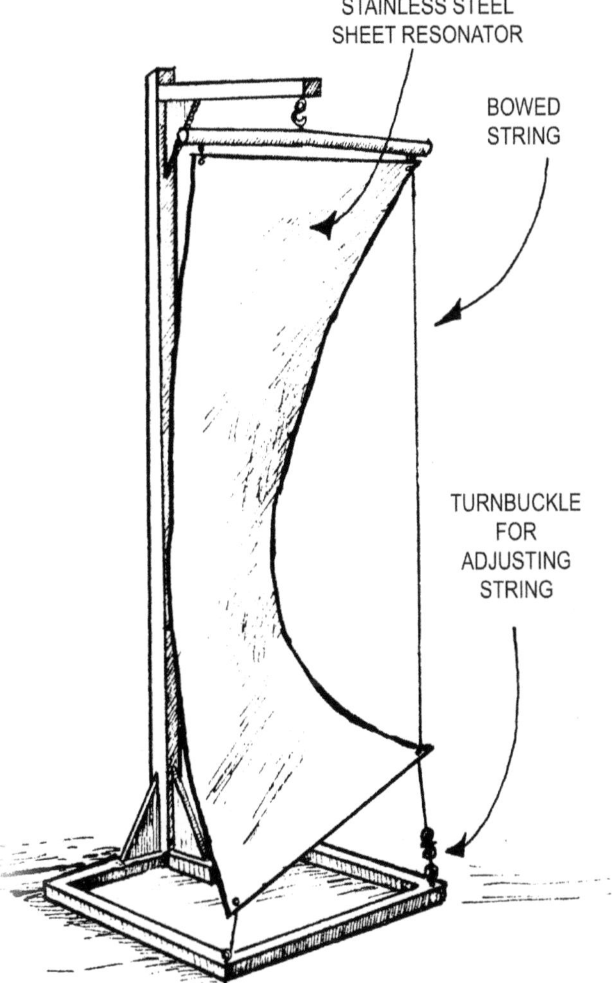

Figure 3.1 Steel Cello.

built a number of other instruments in addition to those described here, and he has cultivated the Tibetan technique of chordal chanting, in which a single voice produces a second and sometimes a third distinct pitch over a very low fundamental. His performing group is the U.S. Steel Cello Ensemble.

Shall I try to describe the music of his instruments? Taking a cue from Rutman's background as an artist, one could say that it has a sculptural quality to it. Dynamics, density, and tone color are essential shaping elements. The ensemble's pieces tend to be long, evolving, episodic; they have a quality of bigness. The players make no attempt to use particular scales or pitches, but pitch relationships, seemingly

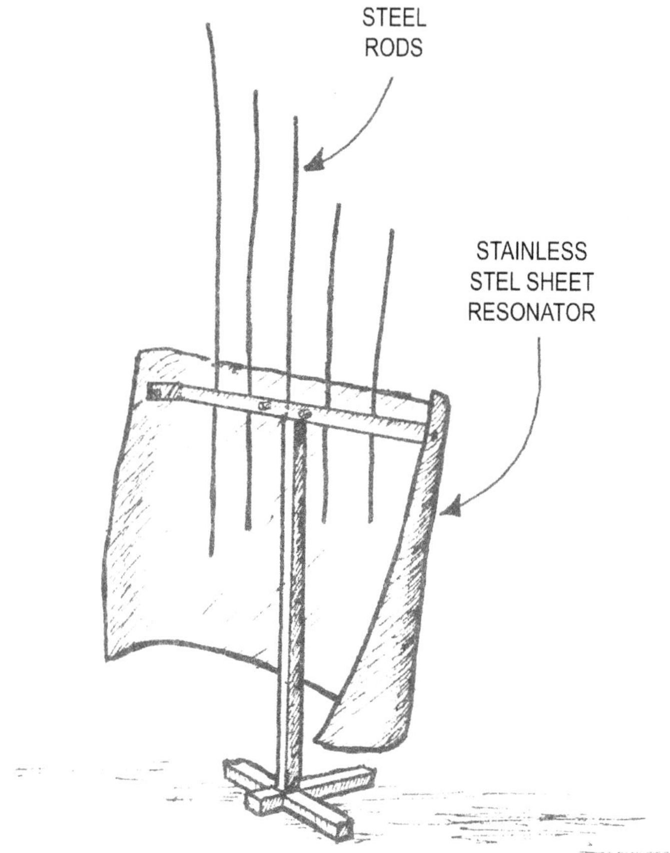

Figure 3.2 Bow Chimes.

random in one sense, are essential to the shape of the pieces as whole sections evolve around relationships between a few pitches and their overtones before moving on.

When Rutman talks about his music and instruments he passes quickly over the mechanics to focus on the relationship between the player and the instrument. New as they are and free of history, his instruments have no standard repertoire or playing technique. These things might evolve over time if people continue to play them, but for the time being their builder focuses on bringing out the personalities and possibilities of the instruments in the most flexible and sympathetic manner. He creates music for the steel cello and bow chimes not by conceiving a musical idea and applying it to them, but by doing his best to allow the instruments to suggest the content of the composition. One can hear this in the resulting pieces when what seems to speak through the wash of sound more than anything else is the emotional voice of the instruments themselves. This is not to say that the steel cello or bow chimes could not be reined in by a composer coming to them with established ideas about scale and rhythm. I suspect that the steel cello in particular could and that the results might be very beautiful. But Rutman leaves that job to someone else.

Related articles

Other articles featuring instruments with sheet metal sound radiators that appeared in *Experimental Musical Instruments* are listed below. These articles can be accessed through the Internet Archive at https://archive.org/details/emi_archive/.

"Holy Crustacean, Batman, That Beast Sings!" by Tom Nunn, in EMI Vol. 1 #4

"Structures Sonores: Instruments of Bernard and François Baschet" by Bart Hopkin, in EMI Vol. 3 #3

"More Baschet Sounds: A Mostly Pictorial Presentation of Architectural works, Museum Installations and Educational Instruments Built by the Baschet Brothers" by Bart Hopkin and François Baschet, in EMI Vol. 4 #1

"Fred 'Spaceman' Long: Troubadour from the 26th Century" by Walter Funk, in EMI Vol. 11 #4

"The 'Funny' Music of Neil Feather" by John Berndt, in EMI Vol. 13 #4

4

The Long String Instrument

Designed and Built by Ellen Fullman

Introduction by Bart Hopkin, additional commentary by Ellen Fullman

This article originally appeared in *Experimental Musical Instruments* Volume 1 #2, August 1985. It consists of an introduction by Bart Hopkin followed by Ellen Fullman's commentary.

The standard string instruments derive their sound from the transverse vibrations of the strings. In these strings, the predominant vibrational movement, the one which produces most of the sound, is that of the string moving back and forth in a direction perpendicular to its length. But transverse vibration is not the only dance that strings can do: there are other modes of vibration, most notably the longitudinal.

In longitudinal vibration, the direction of movement is along the length of the string. The vibration is caused by waves of compression and decompression of the material of which the string is made, comparable to waves in the air or in a liquid medium. In a string which is fastened at both ends, the waves will travel from end to end, reflecting back and forth, and causing any given point on the string to oscillate as the waves pass through. Longitudinal vibrations can be excited by an impulse given to the string in the direction that the waves travel, such as rubbing the string lengthwise with sufficient friction. The frequency of vibration – the rate at which a given point on the string moves back and forth lengthwise in response to the waves of compression and decompression – is determined by the length of the string and the velocity at which waves travel through the string's material, according to the formula f=V/2L. This means – and it does work out this way in practice – that the tension on the string has no effect on the rate of vibration, as long as the string is held taut. Given a string of a certain material, the sole variable determining the rate of longitudinal vibration is the length of the vibrating portion of the string.

By itself a string vibrating longitudinally cannot be heard, since it will move only negligible amounts of air. But, as with conventional stringed instruments, the string can be attached to a resonator such as a soundboard, which will move enough air to give ample volume. A string vibrating longitudinally readily makes an efficient coupling with a soundboard if it is attached perpendicular to the plane of the soundboard. The connection then is as direct as can be and, unlike strings vibrating

Figure 4.1 Ellen Fullman plays the Long String Instrument. (Photo copyright Pieter Boersma, Amsterdam).

transversely, the primary direction of vibration of the string is the same as that of the soundboard.

So why have instrument designers in the past not made more use of longitudinally vibrating strings? There is a very practical reason.

Longitudinal waves travel through different substances at varying, but generally very high speeds. The velocities at which they travel through the materials of which strings can be made are in the range of tens of thousands of feet per second. If you look back at the formula given above for determining the frequency of a string vibrating longitudinally, you will see the implication of these velocities: in order to produce fundamental tones which are within the range in which we can identify musical pitches, the strings must be outrageously long. Ellen Fullman provided the chart below, showing the length of string necessary to produce certain pitches for certain materials. (She has found the figures given in this chart to be somewhat inaccurate, but they do provide a broad picture.)

Table 4.1 Velocity of longitudinal waves through wires of various materials and wire-lengths required to produce a sample pitch

Material and diameter of the wire	*Velocity of the wave in feet/second*	*Length at which the wire sounds a sample pitch*	
0.012" Iron	22,421 ft/sec	A-440	25' 6"
0.0135" Iron	22,110 ft/sec	A-440	25' 3½"
0.0135" Bronze	11,513 ft/sec	A-220	26' 2"
0.014" Brass	10,908 ft/sec	A-220	24' 9½"
0.013" Brass	10,972 ft/sec	A-110	49' 10½"

It becomes clear that, in order to exploit the musical possibilities of longitudinally vibrating strings, one must have a very large parlor.

Fullman's Instrument – Design and Construction

With her Long String Instrument, Ellen Fullman has created a working longitudinal vibration string instrument. A description of the instrument follows, and Fullman's own account of the project's progress from a chance observation to a fully functional (though still evolving) musical construction appears a few pages hence.

The specific form of the Long String Instrument varies from one installation to the next. The most recent manifestation of the instrument has 14 strings in two groups of seven. Their full length (as opposed to their sounding length) is 90 feet. At one end the strings are attached to some stable surface, such as a wall. On earlier versions of the instrument, the strings were held taut at this end – but not tuned, since tension does not affect tuning – by means of several sets of machine pegs made for string basses. Fullman has since discovered that the job can be done more effectively by a ratcheted infinitely-turning winder, a device made in Belgium and used by European farmers for stringing fence wire.

At the opposite end, the strings are secured to a rectangular soundboard, eight inches by fifty-nine inches, made of spruce. Each string passes through a hole in the soundboard and then through a damper of felt and is tied to a guitar peg on the far side. The tension on the string pulls the peg fast against the soundboard, and the damper between prevents buzzing (Figure 4.2).

Fullman, with input from others, is currently redesigning the soundbox that supports the soundboard. The soundboard and sound box arrangement has taken several forms as the instrument has evolved. Earlier versions were far heavier and larger. But the smaller, lighter board, though seemingly out of proportion to the great length of the strings, is appropriately proportioned for the frequency and intensity of the vibrations it is designed to project. It produces a sound that is louder and

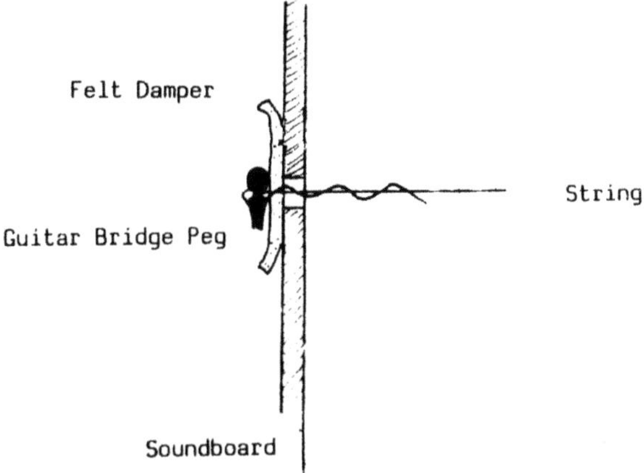

Figure 4.2 Attachment of the string to the soundboard.

warmer than larger boards and reproduces the detail of the strings' overtone content more clearly.

The strings are tuned by adjusting their sounding length. Fullman does this with C-clamps. They are clamped on the string and left suspended there at the point where the waves should reflect in order to produce the desired pitch. One can calculate this point using the frequency formula given above. It seems surprising that the clamp, resting on the string and affixed to nothing rigid, would be sufficient to stop the vibration in such long strings. But if the mass of the clamp is equal to or greater than that of the string, it can function as a nearly-fixed termination, and it does not take a very large clamp to equal the mass of the relatively thin (.0125"diameter) string (Figure 4.3).

Playing Technique

To sound the instrument, the player uses violin bow rosin on both the strings and the hands and strokes the strings lengthwise. The strings sound best with a moderately

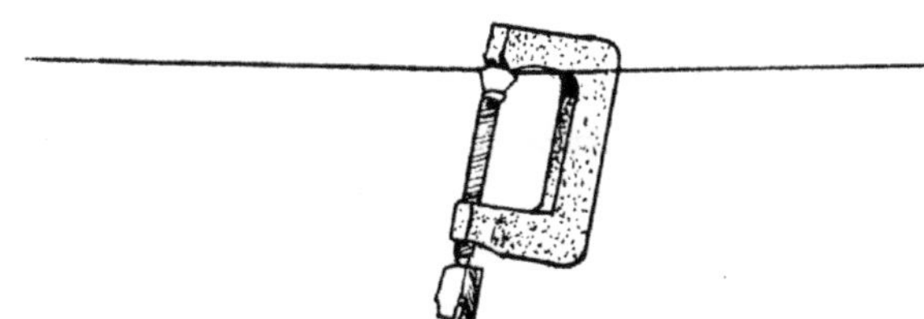

Figure 4.3 The tuning mechanism for the Long String Instrument.

light touch. Fullman compares it to the way one coaxes the tone from a wine glass by running a finger around the rim. The optimal speed for the stroke is, to use another of Fullman's metaphors, slow like the movements of T'ai Chi. The player walks back and forth along the strings as she plays, giving performances a choreography shaped by the music.

Tuning

Fullman has evolved a tuning system designed to fit the form of the instrument and its playing technique, and which feeds directly into her processes of composition and improvisation. The strings are grouped in two or more sets of several strings – the precise numbers have varied from one installation of the instrument to the next. Within the sets, the strings are spaced closely enough that two or more strings can easily be played simultaneously. Fullman tunes the strings within each set to intervals which she determines will work well together for her compositional purposes, considering the relationships both of fundamentals and overtones. Adjacent strings are not tuned to adjacent scale steps, but to selected harmonic intervals. The other sets of strings are tuned to the same relative intervals, but at different pitch levels, allowing for a form of modulation and for some complex harmonies.

Timbre

The Long String Instrument derives much of its flavor from a peculiar attack and decay and the equally peculiar behavior of the overtones. The fundamental – the object of all those calculations concerning the rate at which the wave travels through the medium – is always clearly present. But a mix of very prominent overtones shifts dramatically according to where the player strokes the string. The nearer the strokes to one or the other end, the higher are the overtones that predominate. Toward the middle of the string, one gets all the lower overtones very clearly and recognizably. If the stroke begins very near one end and moves toward the center, a waterfall of high overtones tumbles out over the fundamental in the first moments of the tone, descending rapidly. The descent continues, but ever more slowly, with the movement toward the center, as the fundamental remains more or less constant.

Like a bowed instrument, the long strings sing as long as the stroke lasts. But unlike with transversely vibrating strings, whose vibration will die away slowly if they are not muffed, the longitudinal vibration ceases abruptly after the stroke ends. In Fullman's instrument, a quieter echo of fairly fast decay can then be heard; it is the vibration in the soundboard giving up the ghost a trifle more reluctantly.

The Music

The Long String Instrument lends itself to long tones and sustained harmonies, and this is reflected in Fullman's composition. Her pieces are built around predetermined sets of harmonic relationships within which performers can improvise using certain strings and their overtones for a period of time and then moving on to the next set of relationships. Two players usually perform together, engaging in their T'ai Chi walk along the length of the installation as the music dictates. Fullman has performed at galleries in New York and has brought the instrument – dismantled for the journey, needless to say – to performances in Europe as well (Figure 4.4).

The Evolution of the Long String Instrument

By Ellen Fullman

For the past several years, I have been developing an installation, "The Long String Instrument." In 1981, in St. Paul, Minnesota, I was stretching long lengths of string using various materials and tying them to metal containers. The containers acted as resonators and were amplified with contact microphones. I bowed the

Figure 4.4 Arnold Dreyblatt and Ellen Fullman at the Long String Instrument. (Photo by Peter Cox, Eindhoven, Netherlands).

string and put water in the containers, moving them and listening to the resonance change. One day I brushed against one of these strings and it made a loud clear sound. I began stroking it lengthwise with my hands. I sensed that what I was discovering had a lot of potential but I needed to learn what was happening scientifically to be able to control the sound produced. I was unable to find the kind of information that I needed in Minnesota, although I'm sure it exists there. I saw evidence of there being more integration of art with technology in New York City and decided to move there.

For about a year in New York, I took false steps in relation to the project. I wanted a warm, low sound, and to be able to tune the strings. I tried using a better contact microphone and tried to modify the sound by electronic filtering. I was now using very large containers of water and setting up the strings on the roof of my building, the only place large enough. One afternoon, Steve Cellum, an engineer, explained to me how the string was vibrating. We had on hand a large board, originally used to reflect the sound. Steve suggested that we attach the string directly to it. We drilled a hole, put the string through, tied it to a washer, then tightened it against the board. The string produced a loud, rich sound without amplification.

Soon after, I set up the project in a better location and began building test resonator boxes. At this time, my friend Arnold Dreyblatt, a composer, brought over his friend Bob Bielki, an engineer, to look at what I was doing. Steve was also there, and we tried several experiments which greatly clarified the procedure I was to take. In a physics handbook, they found a formula which was to become my method of tuning. We also discovered that by using brass wire, I could lower the frequency produced. It seemed that the next step was to build a large box resonator that would sustain the sound longer than only a board.

My next studio space was the Terminal New York Show which was to begin a month later. I spent the intervening time reading about musical acoustics, planning the box, and gathering materials. In the show I had a very large area to work in and built a large plywood box. I suspended the strings from this in clusters of four, tuning the groupings to equally tempered chords. I spent this period listening to different combinations of tones and thinking about the musical possibilities.

When the show was over some friends, let me use their basement to work in. The strings ran through a doorway and into another room with the bass section extending down a long hallway. At this time I met David Weinstein, a composer, and we began a series of sessions in which David taught me about just intonation. Since overtones are so clearly present in this instrument, a tuning system using just intervals seemed more appropriate and more interesting.

I tuned the instrument in various ways, listening to ancient systems and generating my own. I settled on a just 12-tone chromatic scale based on F. Rather than in chromatic sequence, the strings were laid out in a pattern in which each string has a simple harmonic relationship to its adjacent strings. This was done so that while playing one string, others beside it could be touched also, adding a harmonic density. I added a second section of 12 strings in the same pattern of adjacencies but tuned in a perfect fifth relationship to the first section.

David and I began playing long, sustained, slowly shifting tones. Our playing was really random, as I didn't know much about musical intervals. In time I began learning more, laying out charts in which one could see the mathematical relationships in chord structures. I realized that, since the overtone series is swept through in each string as it is played, complex, shifting chordal relationships would occur when two or more strings are played at once. I began building chord sequences in which I stroked the same strings walking away from the sound box and returning, listening to the shifting of the chords in the overtones. Now I'm interested in dealing with time in a more precise way than that delineated by the player's footsteps.

The project has become for me my personal music school. It leads me to read and study, as the information I seek gets put to use in very practical ways. The piece is like a microcosm of the history of music. The lessons I learn materialize in a very graphic form. There is a quality of its being a science project that displays principles of musical acoustics. I am an outsider to music, and it's as if now I am seeing the inner workings, the gears, pulleys and bricks that build music, and it's my intention to affect the listener the same way.

Related articles

Other articles pertaining to Ellen Fullman and her instruments which appeared in *Experimental Musical Instruments* are listed below. These articles can be accessed through the Internet Archive at https://archive.org/details/emi_archive/.

"Ellen Fullman's Long String Instrument" by Mike Hovancsek in EMI Vol. 13 #3

Briefer references to Ellen's work and related instruments appear in "Conjoined String Systems: Reports From Builders" by Mario Van Horrik and Paul Panhuysen in EMI Vol. 7 #1 and in the review "Books and Recordings: Long String Installations" in Vol. 3 #3

5

Holy Crustacean, Batman, That Beast Sings!

Tom Nunn

This article originally appeared in *Experimental Musical Instruments* Volume 1 #4, December 1985.

What is 34" in diameter, flat, shiny with bronze antennae and balloon legs with bucket feet, and sings? You guessed it; the Crustacean. How did you know? Well, you may have run into one at some point (there are about twelve of them out there), but most likely you are wondering what the hell I'm talking about.

The Crustacean is, generically, what I call a "space plate." (I originally called them "balloon-mounted rodded metal sound radiators" or BMRMSRs for short, but it wasn't short enough!) (Figure 5.1).

Space plates consist of (1) a steel (preferably stainless) plate, (2) bronze rods brazed to one surface of the plate, and (3) inflated toy balloons in small buckets used to support the plate in a table-like fashion. The balloons are highly elastic, allowing the plate to vibrate freely as the rods are bowed or struck.

The Crustacean, in particular, is a stainless steel disk with bronze rods supported by three inflated balloons. The sound of the Crustacean is extremely resonant. (One tone from a single bow stroke can sound for nearly a minute.) Two bows are used, one for each hand, and the placement of the rods on the plate is symmetrical, allowing two bowing trajectories for each hand. The rows of rods are curved for the same reason a bowed string instrument bridge is curved, so that one string (or rod) may be bowed at a time, or all bowed in a sweeping motion.

There is much variety in the timbre of the bowed rods. It is determined by their length, diameter, and shape, along with the point of contact of bow on rod, bowing pressure and bowing speed. The instrument can be made to sound like a choir, a trumpet, or an electronic instrument; an explosion, or a traffic jam, among numerous other possibilities. I can add that the Crustacean is effective for resonating the voice to sound as if in a large cathedral. This is especially effective when amplifying the instrument (with a voice mike about 1/2" from the underside of the plate).

This instrument is hypnotic. Once one gets involved in bowing, it's difficult to stop. With its harmonic character, the Crustacean tends to take its own course, and the player goes along for the ride. I believe this happens because the harmonic

Figure 5.1 Tom Nunn's Crustacean.

combinations and changes are unpredictable, yet always harmonically related. Each rod is capable of from two to six or more different tones based on the fundamental and harmonics of the rod. The rods vibrate in sympathy with one another to such an extent that non-harmonic combinations of tones will not last but will be superseded by harmonic ones. Let's say I bow a rod and let it ring, then bow a second rod whose tone is nonharmonic to the pitch(es) of the first rod. The second rod will either shift the sound of the first to one of its harmonics which is related to those of the second rod, or will dampen the first rod entirely.

The harmonics of the rods are changed when the rod is bent or curved. The curved "antenna" rods have more harmonics than the short straight rods.

In 1981, Chris Brown and I were commissioned to build an instrument to be exhibited at the San Francisco International Airport during the New Music America '81 Festival. We built a space plate called the "Sun Sing Plate," which was an eight-foot by three-foot sheet of stainless steel with rods arranged in a sunburst pattern. This instrument was suspended along a wall and, because of its size, did not require balloons to maintain its resonance. However, when supported by balloons, its resonance is greater. Chris's *Wing* is a single-balloon, wing-shaped space plate which uses tuned straight rods played with a double bass bow.

Most recently, I made another space plate using a five-foot diameter steel picnic table – the kind with the hole in the center for the umbrella pole. I've named this instrument the Fleur d'Esprit. It utilizes only straight rods, which are arranged on the plate to accommodate two players facing one another. It has somewhat greater resonance because of the larger, thicker metal plate.

There are many possibilities yet to be explored in the genre of space plates. Perhaps others will become interested in this idea and help further the cause of new music and experimental musical instruments.

Related articles

Other articles by Tom Nunn which appeared in *Experimental Musical Instruments* are listed below. These articles can be accessed through the Internet Archive at https://archive.org/details/emi_archive/.

"Meet Mothra" by Tom Nunn, in EMI Vol. 1 #3

"Jacque Dudon's Music of Water and Light" by Tom Nunn, in EMI Vol. 3 #5

"'Bugbelly'-A T-rodimba EPB" by Tom Nunn, in Vol. 6 #1

"Conjoined String Systems: More Reports From Builders" by Jeff Kassel, Tom Nunn, and Bart Hopkin, in EMI Vol. 7 #2

"Improvisation with Experimental Musical Instruments" by Tom Nunn, in EMI Vol. 8 #1

6

The Megalyra Family of Instruments

Designed and Built by Ivor Darreg (1917–1994)

Ivor Darreg and Bart Hopkin

This article originally appeared in *Experimental Musical Instruments* Volume 2 #2, August 1986.

Introduction

by Bart Hopkin

Ivor Darreg is a writer, composer, theoretician, and instrument builder from Southern California. In his long and active life, he has done everything in his power to advance the cause of microtonality, broaden our musical horizons all around, and generally serve as a thorn in the side of the classical music establishment. Perhaps in a future issue EMI will take a look at the broader spectrum of his work in all its color and variety; for the present, we focus on a particular set of his instruments.

In the article that follows, Darreg presents the vital information on his Megalyra family of instruments and gives the thinking behind the designs. Three instrument types make up the family: the Megalyra, the Hobnailed Newel Post, and the Drone. All are steel-string instruments with electric guitar pickups. The strings are stopped with a hand-held metal bar, or "steel," in the manner of a Hawaiian or pedal-steel guitar. The strings and pickups are mounted on flat boards or posts, without separate necks or resonators. Two special characteristics make these instruments noteworthy: The first is that sets of strings are mounted on both or, in one case, all four sides of the instruments. Aside from serving to balance the string tension on the body, this has the effect of making multiple string-tuning systems available on a single instrument. The second noteworthy characteristic is the emphasis on the use of visual guides in the form of fret-lines painted beneath the strings to indicate where to place, the steel for particular pitches. Where the pedal steel has a single set of lines indicating the location of the tones of the 12-tone equal temperament octave, instruments of the Megalyra family use a graphic marking system designed for flexibility in indicating a variety of just and equal tuning systems. It makes for easy playing in any number of intonational systems and encourages musical and intellectual exploration of the wider range of tonal possibilities (Figure 6.1).

Figure 6.1 Ivor Darreg with the Megalyra and Drone.

A Brief Description of Each of the Three Instruments

The Megalyra is a two-sided instrument of about seven feet long, generally played in a horizontal position with one or the other playing side up. There are sixteen strings on one side, fifteen on the other, arranged in multiple-string courses – that is, groups of several closely-spaced strings tuned to the same note, intended to be played together to produce a fuller sound than that of a single string. Each side of the instrument has several of these multiple-string, single-note courses. Not only are the individual strings of the Megalyra not played one at a time, but neither, normally, are the separate courses. Rather, all the strings of all the courses on one side are usually played together. Their pitches comprise what is to be heard as a single composite tone in the style of an organ stop which brings several ranks of pipes into play. The steel bar used to stop the strings at the desired length is, accordingly, made to cross all the strings at once, and the tuning of the strings is selected with traditional organ stops in mind so as to be appropriate for the composite tone. As one might imagine, between the great size of the instrument and the chorusing of so many strings, the tone of the Megalyra is most impressive.

The Drone could be considered a smaller version of the Megalyra – a two-sided instrument of about five to six feet long, designed for a slightly higher range and perhaps less imposing tone. On one side are twelve strings tuned to produce a single composite tone as on the Megalyra; on the other are a smaller number of individual strings tuned to a scale.

The Hobnailed Newel Post is smaller and may be played either upright or on its side. The body of the instrument is a four-by-four post, with twelve to sixteen strings mounted on each of the four sides. Less emphasis is placed on obtaining the great big sound of multiple strings here: the primary purpose of the Newel Post is versatility and flexibility in available harmonic systems. According to the tuning of the strings and the placement of the visual fret-lines, a different harmonic system of the player's choice can be available and accessible on each side. There is no standard tuning; Newel Posts may be tuned and retuned, and their fret-line patterns altered, to fit the purposes of an individual player or a particular performance.

Megalyra, Drone, and Newel Post

Article by Ivor Darreg

The Hawaiian or Steel Guitar has been around for about a century now. According to a dictionary of Hawaiian music, one Joseph Kekuku discovered how to slide the back of a knife-blade along the strings to get the gliding effect. However, there are other claimants to the invention, among whom James Hoa and Gabriel Davion were mentioned in that dictionary. Some writers have supposed that the idea was suggested by voyagers who had heard a similar effect in India. There is some plausibility to that.

During the thirties, at its half-century mark, there was a surge of Hawaiian-guitar popularity, brought on largely by the development of magnetic pickups and amplification. Now this method of producing musical tones could compete in volume with the other instruments in a band, and in turn, the resulting popularity of steel guitars promoted the idea of amplifiers. Since then, the steel guitar and amplification has been closely associated in the public mind – except for the Dobro and a few special instruments of that genre, steel guitars are always amplified.

The late Leopold Stokowski mentioned the advantages the flexible tone of the Hawaiian steel guitar would bring to the regular symphony orchestra if it were admitted there, but nobody ever took him up on that. Well, what would you expect from the hidebound Musical Establishment? Open arms!? Now really.

I bring this up because I want to explain my motivations in developing the Megalyra Group of Instruments. I was started on conventional music training over fifty years ago – piano, organ, and cello – and took part in choruses, ensembles, and orchestras. I saw and heard the limitations of conventional instruments. Then and there I was determined to do something about it, but that took decades, not just years.

"Find a need and fill it" – frequent advice to inventors. What does the conventional orchestra lack? Power in the bass. What's the matter with the pipe organ, and its successor, the electric organ? They have power enough in the bass, but they are too rigid. All they can do is turn notes on and off. Nearly all keyboard and fretted instruments are rigidly tied to the 12-tone equal temperament tuning. The new synthesizers have it built into their vitals. The steel guitar, on the contrary, can play any gradation of pitch whatever, just like the human voice. Its frets are mere painted or inlaid lines on the board underneath the strings and function only as guides. But we must be fair: What's wrong with the regular steel guitar? It hasn't progressed. It has been held back for decades – for a while, it almost faded and went out of the picture. Decade after decade went by, during which it was used for syrupy mawkish trivia. The older models were tuned to an A-major-chord – no way to get an augmented triad or diminished seventh tetrad out of that! You had to retune it if you wanted minor scales. Then how would you play a major chord? A few models appeared with two necks to get around that. Then a while ago somebody thought up the Pedal Steel. Call in the bicycle-maker and a sewing-machine mechanic and put in various gizmos to tighten and loosen the strings so that the tuning can be rapidly changed. Also put more strings on for variety. Ingenious, and it does work for country music and similar styles. But it's complicated and expensive. And a considerable portion of the standard guitar strings' length is "eaten up" by the Pedal Steel's tightening and loosening mechanisms that the pedals and knee-levers operate on to change the tuning.

In particular, the bass strings do not have enough actual useful sounding-length, as compared with a conventional Spanish guitar. You can make up for this with amplification, but there is really no substitute for long bass strings, as anyone who hears a concert grand immediately realizes. Getting back to the pipe organ mentioned above, what does it have that the ordinary steel guitar doesn't enjoy? Why, depth and dignity. Great depth but still ringing brilliance. So what did I do? Think Big. What if a steel guitar were made four times the size?

I had a board 205cm (6ft 8ins) long. I fitted it with pin-blocks, tuning-pins, bridges, pickups, and strings. Room for fifteen strings on each side. Why both sides? To balance the tension so the long board wouldn't warp. Otherwise one would need a heavy metal beam and careful bracing: we are talking about tons of tension here. For the tones to sustain, the strings must be massive enough to store energy. But if they are thick enough, then they must be tightened until they behave as strings, rather than rods. The right size and tension were arrived at by examining various instruments, all the way from ukuleles to pianos. My forty-six years of piano tuning, applying such tension, was a help too. Experiments determined the average breaking point of the music wire used, and so the factor of safety to allow. That gave the maximum sounding string length on one of the instrument's sides.

For example, the designs for both the Megalyra Contrabass and the full-size Drone instrument call for several strings in unison sounding Tenor C 132Hz. So the natural question was, what is the longest string that will sound Tenor C without breaking frequently? Experiment showed that for thin steel wire of the kind used for first or second strings, on guitars, that the length was 60 cm or 5 feet, 3 inches. Some of the strings tested would withstand Tenor E-flat 156 Hz before breaking. Memorize this, it's useful to store in the back of your mind: Doubling the frequency, or taking the pitch of a string up an octave, requires QUADRUPLING the tension! Twice the tension raises the pitch half an octave. The minor-third factor of safety just considered, then, comes to about 1.4 times the tension to go from Tenor C to E-flat above, about 1/4 octave. You might feel more comfortable playing safe and settling for a shorter Tenor C of 152cm (5 ft) long. But if the strings on a Megalyra or Drone are too loose, several disadvantages appear. They do not stay in tune as long, and they give under the steel or wooden bar too much. Moreover they lack the snap, fire, and brilliance that set these new instruments above the dull plane of trivia.

There are no stock strings suitable for the Megalyra – you will have to make them out of steel music wire. Sizes will be given at the end of this article. Similarly for the Drone if it is to have a wide compass and the timbre of long strings. The Newel Post can be designed with shorter strings, so that the standard guitar strings will be long enough for it. On the Megalyra and Drone I used loop ends rather like those used on harpsichords and banjos. These are easy to make and books on harpsichords usually tell how to.

Since we are using high string tension and many strings (a full size Megalyra needs about 30 strings, and a Drone 20 to 24), the gear heads used on guitars will not stand the tension and would cost a fortune. So I recommend regular piano tuning-pins which can be placed fairly close together and tuned with standard tools. If a pin becomes loose, there are larger sizes for replacement. The much smaller tuning-pins used on zithers, harpsichords, and clavichords are not suitable for the heavy wire that will be used on the Megalyra, although you might use them on a Newel Post. Generally a Drone Instrument should have fairly heavy wire for some strings. The tuning-pin-blocks are traditionally made of five-ply maple or other hardwood, but this can be very expensive. We won 't be using the very high tensions found in a piano, such as 170 lb (77kg) but more like half that tension, so can get along with less expensive plywood. For the common fir plywood, the holes should be somewhat tighter than the standard size for maple. About one-half the length of the threaded portion of the tuning-pins will traverse the pin-block – the rest goes into the main board. Drill the holes clear through.

To avoid weakening the board or beam, the pin-blocks for the two sides should be at opposite ends of the board, even though this means having to tune one side at one end of the instrument and the other side at the other end. (In case of the four-sided Newel Post, of course, two sets of tuning pins should be set at right angles, not opposite to each other, and the other two sets at the other end.) The hitch-pins or other fastenings must be staggered with respect to the tuning-pin block opposite it again not to weaken the board. If the strings on opposite sides are the same length, they must be offset considerably for similar reasons.

Many popular designs will require different string lengths, and this turns out to be more of a solution than a problem. For example, the usual Megalyra design will make the strings on the bass or accompaniment side much longer than those on the solo or melody side. This automatically solves the problem of fittings on one side being opposite those on the other.

For the Megalyra Contrabass, there should be from three to five strings for the principal "courses" as some luthiers call groups of strings. The full array is: five strings on Contra C 33 Hz; five strings on Great C 66 Hz; two strings on Great G 99 Hz; four strings on Tenor C 132 Hz. This is for the solo or melody side. This is derived from the traditional appointments of the Pedal Organ in a large organ, so it has been tested out for some centuries. The steel is ordinarily laid across all these strings and they are strummed at once to produce a compound tone having the 1st, 2nd, 3rd, and 4th harmonics in suitable proportions. This is how you can get both brilliance and depth. Also, you get a chorus effect, as you do with three strings per note on the piano, or several ranks of pipes in the organ. On the bass or accompaniment side of the instrument, there should be five each of strings tuned to Contra C 33 Hz, Contra G 49.5 Hz, and Great C 66 Hz. There is a space left between the groups of course, so that one can play one group without sounding the others if desired.

The tuning of the Drone will vary with the user. My scheme is to have eight very thin strings in unison to Tenor C 132 Hz and a slightly separated group of four thicker strings for Great C 66 Hz next to them – in all twelve strings on the melody side for a rich chorus effect and the ability to play in octaves when needed. A special bridge can create a sitar-like buzz for this portion of the instrument. These pitches are for a Drone five to six feet (say 150 to 180 cm) long. For a shorter drone how about F or G above those C's? The Drone strings themselves can be on the backside of the board or divided between the other two sides. For five drone strings, a pentatonic scale C D F G A would be suitable. For a sixth drone add a B-flat.

The tuning of the Newel Post's four sides becomes a personal matter, just as the tuning and pedal layouts on the Pedal Steel are highly individual, so not much use giving details here. I have just-intonation chords and use major, minor, harmonic seventh, and minor seventh, as well as a harmonic series or minor-ninth chord. Your choice will depend on whether you want a performance instrument or a Composer's and Arranger's Studio/Harmonic Laboratory. Thousands of variations are possible, even with the same stringing. Twelve to sixteen strings will be on each of the four sides, spaced as you desire.

The steel used on these instruments can be a rod or thick-walled tube preferably somewhat longer than the width of the board. Also you will need a pair of wooden bars about as big as hammer handles or broomsticks. When you do not want the gliding or sliding effect of the steel, take up a wooden bar in each hand and strike all the strings on the solo side simultaneously, using right and left hands in alternation for a clear-cut articulation of the notes without sliding between them. One reason for using both hands is to prevent undesirable ringing of the open strings between notes. The wooden bars must be held down firmly while each note sustains. With practice you can play quite rapidly.

All these instruments require amplification with magnetic pickups. Each side of the Megalyra requires two pickups, placed at about one-twelfth and one-sixth of the open-string length, respectively. The Drone Instrument can get along with one pickup for the melody strings and one for the drone strings. Each of the four sides of the Hobnailed Newel Post can have one pickup, which should be located near the bridge – say one fifteenth of the open-string length, to pick up the high harmonics. On these instruments played with a steel, we have to take precautions not to pick up too much of the sound from the portions of the strings on the far side of the steel. A small amount of such tones will not matter, but they must be kept down. Muting with the hand or having a felt muting-strip glued to the steel might help.

All these instruments in the group use a pattern of visual fret-lines as a guide to the player. Before the bridges are fitted, the board should be painted with the ground-color of your choice. If you prefer a natural finish, it can be coated with one of the new heavy transparent finishes. On top of this, the fret-line patterns are painted over a range of two octaves. For the Drone, only the melody side bears the pattern; for the Megalyra, both sides have it; for the Hobnailed Newel Post, all four sides. Tables indicating the fret-placement distances for a variety of string-lengths are available from the author. Briefly, the standard pattern is as follows: A row of green lines (unless the ground color of the instrument is green) is laid down for the conventional 12-tone chromatic scale. Alongside that first row, a second, multi-colored set of lines is laid down delineating an untempered just scale. It is made up of a row of black lines (white in some cases) for the naturals, red lines for the sharps, and blue lines for the flats. Beyond these patterns extends a set of yellow or orange lines clear across the width of the board, showing harmonic nodes. On the Megalyra, it is worth the trouble to carry the harmonic markings up to the sixteenth harmonic since the long strings will clearly sound them and with four or five strings tuned to Great C you can do a chord of harmonics in your demos. On the Drone take them to the tenth, and there should be room on the Newel Post for up to the eighth.

The pickups may hide some of these markings, of course. To alleviate this problem, you might have the markings at the pickup area extended to the edge of the board and leave out the center portion where you know the pickup will be installed.

The fret-lines may be painted on with a sign painter's brush. On the Newel Post where the lines have to be narrower, you can use the colored draftsmen's charting tapes available in various widths. Another possibility is to rule the lines on a plastic or special drafting sheet, which is later secured to the board. (It is recommended that you place the conventional 12-tone and the just markings in some permanent fashion; if you want to have charts of other tunings such as 19- or 31-tone, they can be on overlay charts). Don't wear yourself out trying to be over-precise; these lines have to be much wider than the real frets on a guitar. The steel can't be placed with great accuracy – correcting is done by ear.

Before the bridges are attached, you can rout channels for the wires to the pickups. Or if you like, do the wiring on the surface. I made my own pickups but if you prefer commercial pickups, low-impedance types are recommended. Bring the wiring from where the pickups will be installed to a terminal strip at the end of the board, using different colored wiring to be able to trace connections later.

The Megalyra strings should be one inch (2.5cm) above the board, allowing for striking them with wooden or metal bars, and for playing with mallets if the regular steel is used. On the Newel Post, the strings can be much closer on the board, as with ordinary steel strings. The Drone is an intermediate case.

All these instruments must have end-pieces so that when they are laid on the floor, a table, or stand, the strings on the bottom side of the moment will be clear of the floor or table. I use square end pieces from ten to twelve inches square (25 to 30 cm square). This permits standing the instrument on either end and provides a place for jacks or even an onboard preamp.

One reason for treating these three instruments together is that hybrids are possible, as well as further variations of the basic idea.

Sizes of Wire Used in the Megalyra Instruments

(These figures are for high-carbon spring-steel music wire of the kind used for pianos, guitars, harpsichords, and the like.)

The special Music Wire Gauge system for designating wire diameters is being phased out. We give the numbers here, but usually at a hardware store, they will want you to identify what you want by diameter in thousandths of an inch. This is just as well: Music Wire Gauge runs in the opposite direction to the American and British gauges for other kinds of wire, such as B&S for copper.

MEGALYRA. The figures given here are for the average size Megalyra seven feet (231 cm) long. Shorter instruments may use the next size larger (see table).

Figure 6.2 Ivor Darreg with the Megalyra. (Photo courtesy of Jonathan Glasier and *Interval* Magazine).

Contra C	No. 22	.049"	1.25mm
Contra G	No. 17	.039"	.99 mm
Great C	No. 12	.029"	.74 mm
Great G	No. 8	.020"	.51mm
Tenor C	No. 3	.012"	.30mm

DRONE. These figures are for an instrument of 5½ to 6 feet long (183 cm). Shorter drones will use larger sizes.

Great C	No. 12	.029"	.74mm
Tenor C	No. 3	.012"	.30mm

NEWEL POST. Nearly all Newel Posts will be about the length of a large guitar and will be capable of using standard guitar strings, although the actual sounding length on one or two of the four sides will be somewhat greater than the standard

650 mm (25-5/8"). Wound strings are required for the lower notes and plain for the higher, and all these sizes are widely available at music stores. Some strings are shorter than usual and may have to be extended by threading a steel-wire loop through the loop or the ball end. Figure on a somewhat higher tension than the guitar normally uses. N. B.: You may have to drill some of the tuning-pin holes out larger to pass some of the wound strings through!

Tuning-pin sizes

Size 1/0 appears in lengths of 2" and 2½". All others are 2¼", 2⅜", and 2½". Size 2/0 is the one usually installed in pianos at the factory, so they can often be obtained from junked pianos.

Size	*Diameter – Thousandths of an inch*	*Diameter – Millimeters*
1/0	.276	7.01
2/0	.282	7.16
3/0	.286	7.26
4/0	.291	7.39
5/0	.296	7.51
6/0	.301	7.65

Related articles and letters

Author Ivor Darreg was a regular correspondent in the Letters section of *Experimental Musical Instruments* in the years up to his death in February 1994, and you can find many of short commentaries from him by perusing the issues leading up to that date.

EMI Volume 9 #4 contained an "In Memoriam" section with several other correspondents' notes on the man and his life

Ivor also wrote or contributed to two more short articles in the journal, including "Comments From Ivor Darreg" (on refretting for guitars) in EMI Volume 3 #6, and "Notes from Ivor Darreg on Theremin Making" in Volume 8 #3

These writings can be accessed through the Internet Archive at https://archive.org/details/emi_archive/

7

Daniel Schmidt's American Gamelan Instruments

Bart Hopkin

This article originally appeared in *Experimental Musical Instruments* Volume 2 #2, August 1986.

In recent years, a number of American composers and instrument builders have developed an interest in Gamelan music. A Gamelan is an Indonesian orchestra composed primarily of metallophones (metal sounding bar instruments) and gongs. These are usually complemented by a smaller number of instruments of other types, including bowed and plucked string instruments, drums, a bamboo flute, and a vocalist. A gamelan is a complete set of such instruments, and the word refers to the instruments themselves, not the players or the music. The music itself has, to Western ears, an exotic, floating quality which many find utterly enchanting; Debussy is known to have been profoundly moved and lastingly influenced by his one encounter with it. Within its own culture, it is deeply respected as the culmination of a long and highly refined tradition, replete with painstakingly taught performance practice and theory, and a rich surrounding culture.

The relationship of the American musicians to Indonesian models varies considerably. Many have considered it important to remain as true as possible to the Indonesian traditions in composition, instrument building, and less tangible aspects of the music culture. But an increasing number of people on this side of the Pacific have taken the position that respect for the tradition does not demand literal emulation. With this attitude taking hold, new musical styles have begun to arise out of the US-Pacific confluence. The love of the Indonesian music, and its methods, structures, timbres, moods, and surrounding culture, has been leading to increasingly independent creative work here. Some of the new music resembles traditional gamelan music closely, and some is very different.

In the mid-seventies, there was only a handful of people exploring a non-traditional approach to gamelan music. They included Paul Dresher (now mostly working with interactive performance-oriented electronics); Jody Diamond (now director of the Bay Area New Gamelan and the Mills College Gamelan); composer Lou Harrison, working in conjunction with builder Bill Colvig; and Barbara Benary in New York, director of the Son of Lion gamelan (which was using instruments derived from found objects, designed by Dennis Murphy). Daniel Schmidt was among these people too, building and composing for gamelan-like instruments in non-Indonesian ways. The music was a bastard child then: the prevailing aesthetic in new music was still running in a post-serialist highbrow vein; ethnomusicologists meanwhile were primarily concerned with preserving existing traditions in unadulterated form; and non-traditional gamelan didn't get much respect. Things are a little different now, and that is a credit to the work and the imagination of the early new gamelan people mentioned above. The movement for new approaches to gamelan music has grown in size, gained some respectability, and established validity.

Daniel Schmidt has been building gamelan instruments since 1975. In that time, he has earned a substantial reputation, and his instruments have spread far and wide: the gamelan used at North Texas State University and Sonoma State University are his, as well as those played by musicians of the Berkeley Gamelan and the Bay Area New Gamelan, to mention a few. His gamelan are traditional in this sense: they comprise complete sets of the metal instruments (strings, winds, and drums are not included), with instruments corresponding to a Javanese gamelan in range, number, and, loosely speaking, type. Traditional gamelan music can be performed effectively on them. But in designing them, Schmidt makes decisions based on his own ear, rather than on the dictates of tradition. The resulting instruments differ from Javanese and Balinese instruments in several respects. Most prominently, Schmidt makes the sounding bars and gongs of aluminum and brass, machined to shape, while builders in Indonesia use forged bronze. The cross-section shape of Schmidt's bars and gongs are his own. And he makes more extensive use of tuned air resonators than is traditional. Schmidt's approach to instrument building is disciplined and hard-working, tending to the cerebral. The instruments themselves show a high degree of understated craftsmanship. They are never showy – visual effect is only a secondary concern – but matters which affect the acoustic result receive detailed attention. Tuning, both for the relationships between fundamental pitches of different bars and for timbre within a single bar, is perhaps Schmidt's primary interest. The resulting instruments are quietly handsome to look at, and in sound, virtually flawless. Descriptions of the basic instrument types follow.

Trough-Resonated Instruments

The highest pitched instruments of the traditional gamelan are the *Sarons* (Figures 7.1 and 7.2). They are metallophones set over a trough without individual resonators, usually made of bronze or iron. There are three sizes and pitch ranges, given the names *Peking*, *Saron,* and *Demung*. The instruments of Schmidt's corresponding trio are similar to the traditional models, and he retains the same names. The trough for each is made of wood and sits low on the floor. (With all of the gamelan

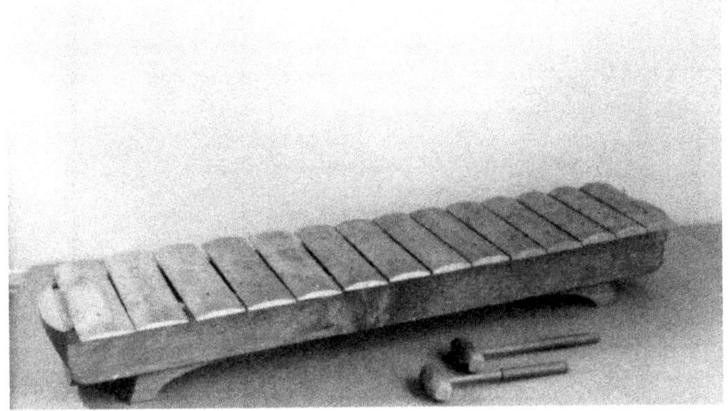

Figure 7.1 Brass *demung* (Photo by Daniel Schmidt).

instruments, traditional and in Schmidt's ensembles, the musicians play sitting on the floor.) The bars rest at their nodes on pads on the rim of the trough. The range of the instruments has been increased to two octaves from the traditional one. Schmidt makes the bars in three shapes, shown end-on in Figure 7.3 (all are rectangular when seen from above).

The bars are of brass or aluminum. In both cases, they are played with moderately hard mallets of balsa wood, or wood wound over with yarn. The aluminum bars produce a light tone with relatively few overtones. Brass bars are closer in sound and look to the bronze of the traditional instruments. Their emphasis on the high partials is a valuable trait, in Schmidt's thinking, in the upper voices. That brightness

Figure 7.2 Sarons (Photo by Daniel Schmidt).

Figure 7.3 Schmidt's bar shapes for *Peking*, *Saron* and *Demung*, seen end-on.

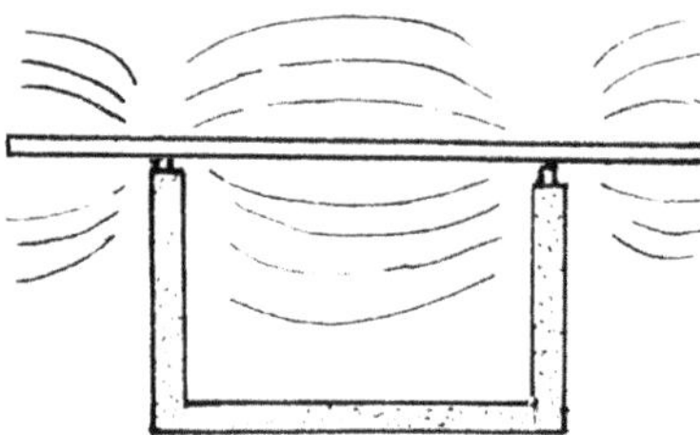

Figure 7.4 Isolation of out-of-phase waves (shown in side view of the bar and trough).

of tone is also enhanced by the fact that the *Sarons* lack individual resonators for each bar which would bring out the fundamental at the expense of upper partials.

Since any enclosed or partly enclosed chamber will have its resonating frequencies, the troughs tend to favor some overtones over others. In extreme cases, this has the effect of partially cancelling certain frequencies. Schmidt reduces the problem by venting the troughs – drilling holes in the bottom to break up the standing waves. One can also achieve a similar result by simply leaving off the ends. This has the effect, once again, of encouraging more overtones and brightening the sound. In spite of all this, the trough does have an important function. When a metallophone bar or marimba bar vibrates, the ends and the middle of the bar are 180 degrees out of phase with one another, creating the potential for some cancellation in the surrounding air. The vibrations in the air above and below the center section of the bar are likewise exactly out of phase (in terms of compression and decompression of the air). The walls of the trough serve to isolate the opposing waves, allowing them to operate without interference (Figure 7.4).

Instruments with Individual Resonators

In the Javanese gamelan, there are two types of metallophone having an individual tuned resonator for each bar: the *Gender*, and its larger cousin, the *Gender Panembung*, or *Slentem* (Figure 7.5). Below these in pitch in the traditional gamelan are three sets of large, vertically suspended gongs (smaller gamelan may not have full sets of all three). The *Kempul*, the smallest of the three (about 38 cm in diameter), appears usually in sets of five or more individual gongs with a pitch range of about 105-150 hz. The larger *Gong Suwukan* is a set of three or four, pitched around

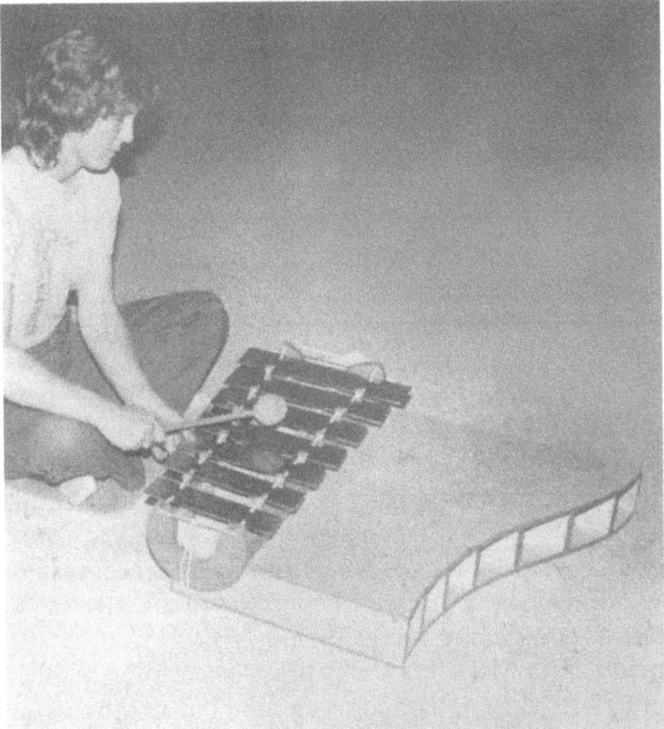

Figure 7.5 *Slentem*. (Photo by Susan Lindh photography).

70–85 hz. The *Gong Ageng*, usually appearing in pairs, is the largest and lowest, with a diameter of 85 to 100 centimeters and a harmonically complex tone with a fundamental of about 40 to 60 hz.

In Schmidt's re-creation of this lower end of the gamelan, there are no traditionally made gongs. For instruments corresponding to both the *Genders* (the resonated metallophones mentioned above) and the gongs, he uses resonated metallophones. His *genders* and the smaller *slentems* are played with lighter mallets and may have several bars and resonators joined together. The gongs (in the form of metallophones, remember) are played with heavy, soft mallets made of yarn wound over rubber balls. Like the traditional gongs, these instruments are made separately, rather than being mounted in groups in a single frame.

Schmidt uses quarter-wave resonators, meaning that the length of each resonator corresponds to one fourth the wavelength for the frequency of its bar. Quarter-wave resonance is characteristic of resonators that are open at one end and stopped at the other. (This creates a node at the stopped end while the point of maximum amplitude occurs at the open end, thus enclosing 1/4 of the entire cycle within the resonator.) Quarter-wave resonators yield more fundamental and tend to suppress overtones. This is because while the fundamentals of the bar and resonator are deliberately tuned to reinforce one another, the upper partials of the bar and those of the resonator do not correspond and the same reinforcement does not occur. This helps Schmidt's resonated instruments to yield the pure tone of a near-perfect sine wave.

Schmidt's resonators are made of wood, with an air column that is square in cross section. They are tunable by means of movable stoppers inserted in the end away from the bar. The stopper must have sufficient mass to effectively stop long waves, a smooth and rigid enough inner surface to reflect the wave efficiently, and a seal tight enough to prevent leakage but still allow for sliding to adjust tuning. Schmidt's design is shown in Figure 7.6. Tunability in the resonators is essential. It is the acoustic coupling between the bar and resonator that creates a satisfyingly rich tone; naturally it will only occur if the two are tuned correctly relative to one another. When temperatures vary, the bar and the resonator detune themselves in opposite directions – the bar flattening as the metal warms up and becomes larger and less rigid, the resonator sharpening as the waves travel faster in the warmer enclosed air. With the movable stopper the resonator can always be retuned to the bar, and the acoustic coupling restored (Figure 7.7).

The question of coupling can be a tricky one. As marimba builders know, perfect agreement of pitch between the resonator and the bar may actually have the effect of killing the vibration in the bar. The solution is to deliberately detune the resonator microtonally. This allows the resonator to effectively store energy from the bar and feed it back to the bar over a longer period, creating the full richness of the resonated tone and ample sustain. Different builders have different ideas on whether to set the resonator sharp or flat of the bar. Schmidt sets it flat for the practical reason that atmospheric conditions in performance tend to raise resonator pitch and flatten the bar; with the resonator set below to begin with, the two are likely at least to remain in the same ball park.

For the lowest-pitched bar-gongs the resonators are, as one might imagine, very long – long enough, in fact, that their size and placement present some problems. When Harry Partch built his giant Marimba Eroicas, he was forced to place the players on stools in order to be high enough to reach the bars above the long resonators.

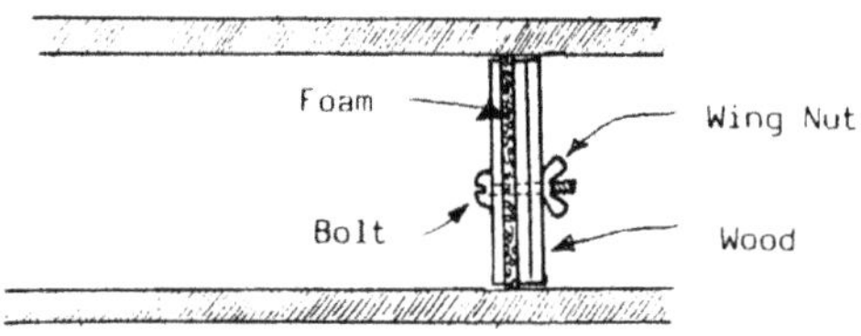

Figure 7.6 Schmidt's resonator stopper design.

Figure 7.7 Schmidt's lower-pitched gongs (actually metallophone bar instruments) are similar to these in design, but longer.

Schmidt was reluctant to take this approach, in part, he says, because in the tradition of Indonesian gamelan, "sensitivity is great about 'aboveness.'" He found that it was possible to orient the resonators horizontally and lay them on the floor, with the sounding bars placed above the resonator, as shown in Figure 7.8.

Orienting the long resonators horizontally on the floor presented another sort of space problem, occupying as much floor space as they did, so in Schmidt's more recent designs, the resonators have a 90-degree bend in them (Figure 7.9).

To continue our description of the gongs in Schmidt's gamelan: the bars are mounted individually in small wooden frames. The frames are independent, detachable units, and fit over the opening on the upper side of the end of the resonator. Within the wooden frames, the bars hang by strings at the nodes from four wooden posts. The tone of these instruments is extraordinary. It is almost entirely pure and devoid of overtones. Played well with soft mallets, the attack is free of non-harmonic noise. Without seeming loud, the sound is indescribably full, spacious, and resonant, and the sustain seems inexhaustible. In the low gong tones especially, for fullness, richness, and understated power, the sound is a composer's dream.

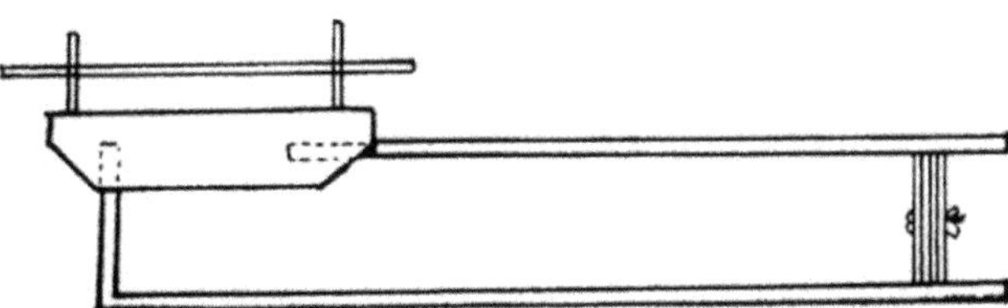

Figure 7.8 Gong with the long resonator extending along the floor.

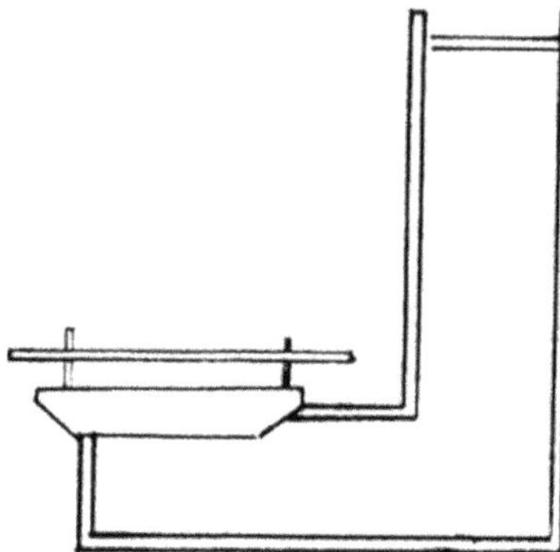

Figure 7.9 Gong with a 90-degree bend in the resonator.

Tuning Methods for the Metallophones

Schmidt's approaches to tuning reflect a primary concern with uniformity of timbre throughout the range of the instrument, and a secondary concern with uniformity of appearance. The primary factors governing the tuning of a metallophone bar are weight at the ends of the bar and rigidity at the middle. Shortening the bar or thinning the ends raises the pitch by removing weight; thinning at the middle lowers it by reducing rigidity. Thinning uniformly over the bar's length likewise has the predominant effect of reducing rigidity and lowering pitch. The width of the bar affects amplitude but not pitch as long as the width remains substantially less than the length.

Along with Indonesian builders, Schmidt believes that the greatest uniformity of timbre can be achieved by making the bars uniform in weight, with the weight of the highest-pitched bar actually equal to that of the lowest. This can be achieved by making the higher bars progressively thicker. The manipulability of this variable makes for considerable flexibility in what is probably the most important visual factor, namely bar length. Schmidt opts for bars of uniformly graduated length over the range of the instrument, with the potential decrease in weight of the shorter higher bars compensated for by greater thickness.

As for the width of the bars, the trade-off is between the greater amplitude of the wider bar and the greater playability and convenience in overall instrument size with narrower bars. Schmidt has found that bars whose width is less than about 1/5 of their length leave something to be desired in volume and tone. His procedure is to make the bars progressively narrower in agreement with the progressively shorter lengths of the ascending scale – that is, keeping the width-to-length ratio constant, generally at 1:4 or 1:5.

With pencil and paper, Schmidt is able to generate possible workable arrangements – that is, sets of dimensions which will abide by these rules and produce the desired pitches – before cutting and machining the bars. He then manufactures the bars and does fine tuning by thinning the bars at the middle or the extremes. This final thinning is kept to a minimum, since great variation or highly irregular shapes is sure to create irregularity of timbre, in addition to being unsightly. In the end result, in Schmidt's instruments, it is virtually imperceptible visually.

Sets of Horizontally Mounted Gongs

In addition to metallophones and vertically hanging gongs, the Indonesian gamelan includes the *Bonangs* – instruments made up of sets of smaller gongs mounted horizontally in a wooden frame, without individual resonators. Schmidt's corresponding instrument uses brass or aluminum rimless disks in place of the traditional gongs. (This sort of disk-gong was originated by Paul Dresher, who made them in aluminum.) A hemispherical boss is raised in the middle. This can be done simply by hammering the metal cold. The player strikes the gong directly on the boss. Raising the boss in the disk causes the pitch to rise, as it forces the disk as a whole into a slightly conical shape, making it more rigid (Figures 7.10 and 7.11).

This is how the gongs are tuned. The rise in pitch as the boss is raised is very dramatic, and the disk seems to sound best at a final sounding pitch about an octave above the original pitch of the boss-less disk. The overtone spectrum of the disks' sound, in contrast to that of the resonated instruments, is rich in partials.

Ideally these rimless gongs vibrate in a circularly symmetrical pattern. The center portion rises and falls as the peripheral portions fall and rise along a circular node at some distance from the center. In practice, this node is usually far from truly circular, following an irregular path as it makes its circumference.

Figure 7.10 Rimless disk gong with boss.

Figure 7.11 More hammering, creating a more conical shape, raises the pitch of the disk gong.

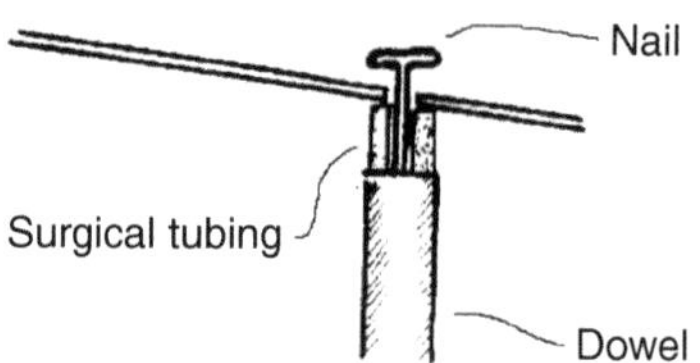

Figure 7.12 Mounting at the nodes for the disk gongs.

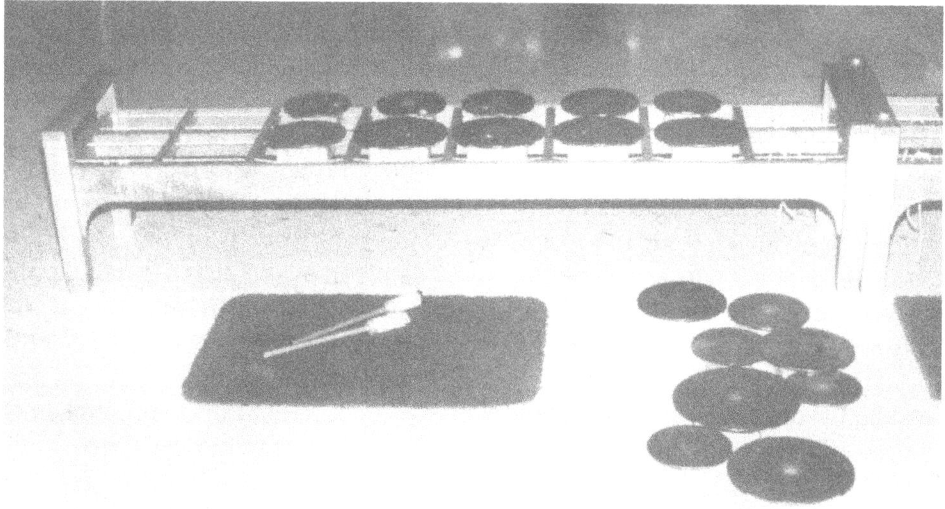

Figure 7.13 *Bonang* (Photo by Daniel Schmidt).

Each disk gong rests on three or four upright dowels rising from the main framework of the instrument. In the top of each dowel is a small nail or wire brad, with a segment of surgical tubing over it to serve as a pad and prevent rattling. The nails and tubing pass through holes drilled in the disks at points along the nodal circle (Figure 7.12).

Locating the nodal circle, with its unpredictable irregularities, is difficult. Schmidt uses the technique often used with marimba bars, placing sand on the vibrating surface and exciting it. Where the surface vibrates actively, the sand dances around and eventually dances off; at the nodes, the sand should remain stationary and tend to collect there. This test is a bit awkward with the disk gongs though, because the slightly conical shape of the gong makes all the sand want to slide off due to gravity (Figure 7.13).

Intonational Systems

It may come as a surprise to learn that among Indonesian gamelan there is no uniform intonational system. There are two universally recognized scale types – *pelog* and *slendro* – which are readily identifiable and have consistent characteristics. But, within certain limits, the actual relative pitches of the scales, as they are manifested in the tuning of the instruments, vary from one gamelan to the next. Needless to say, this is not due to a failure of Indonesian musicians to recognize the discrepancies. It arises from an attitude toward tunings and scale systems which differs from ours. Where in Western music we have felt a need to create universal pitch standards and intervallic relationships, Indonesians approve of and appreciate the individuality of each gamelan, regarding it as a valuable aesthetic trait. Indeed, before the construction of a new gamelan begins, the master musician in charge spends some time considering possible tunings, arriving eventually at a unique tuning for the gamelan-to-be, which he then communicates to the builder.

This means that an American builder of gamelan has a model to work from if he chooses, but it is a flexible model. Daniel Schmidt has responded to this situation by evolving, and continuing to evolve, a system which uses identifiable *slendro* and *pelog* scales, but with a mathematical basis in the extended harmonic series and just frequency ratios. It is a carefully devised, mathematically precise, and rather cerebral system. This may not be the place to discuss it in depth, but certain aspects of its acoustic implications are worth looking at here.

As discussed earlier, Schmidt has substituted low, but definitely pitched, resonated metallophones for the indefinitely pitched traditional gongs. The tone of these instruments is extremely pure – essentially devoid of overtone content – and with soft mallets they can be played with virtually no extraneous noise in the attack. With these gongs as basic material, Schmidt is able to practice what amounts to a natural acoustic synthesis of sound. Playing two or more simultaneously, with carefully controlled relative volume, he can build up timbres as he chooses, in a manner reminiscent of electronic synthesis. On a simple, imitative level, this means that he can approximate the effect of the more acoustically complex Indonesian gongs by playing several selected pitches together. Just as often he uses this approach to create timbres of his own devising. The specifics of the tuning system become very important in this process. Twelve-tone equal temperament, the tuning system commonly used in the West, uses tempered intervals which, unlike just intervals, have frequency ratios that do not reduce to a simple fraction. For reasons both physical and cultural/psychological, such tempered intervals do not work very well when presented as components of what the ear is to hear as a single timbral mix. With Schmidt's carefully chosen just intervals the ear will much more readily hear a mix of selected pitches as a single timbral blend. With the extremely accurate intonation of Schmidt's instruments, and their rich, lucid tone, that blend takes on the quality of a new sound altogether. The whole is more than the sum of the parts, and it is also very different from the sum of the parts, allowing Schmidt to create a palette of effortless, room-filling, nowhere-and-everywhere sonorities.

Understanding the frequency relationships of his available pitches and their acoustic implications as intimately as he does, Schmidt is also able to work with combination tones in a very deliberate manner. He is even able – I heard it with my own ears – to manipulate the phase relationship between two low-frequency bars played together in a slow roll.

Some Additional Notes on Materials and Construction Methods

Most backyard workshop people are comfortable with wood. Metal is somehow a little more intimidating. The raw materials are less readily available and their varieties less well understood; tools for working it less commonly accessible and more expensive; and the methods for working it less familiar and perhaps more dangerous. Daniel Schmidt comments that when he first began metal work he was as much a novice as anyone, and only learned what he needed to know through a lot of nosing about, asking questions, and trial and error.

Aluminum, brass, and other metals are available in various forms at supply houses in most areas which can be found by asking around, checking the yellow pages, and making some phone calls. Employees of large industrial suppliers, which will usually have the best prices and selection, are often surprisingly willing to take the time to talk to the little guy, fill small orders, and answer questions.

The materials are generally available in sheets of various gauges, and in extrusions. The word "extrusion" refers to the manner in which rods, tubes, and similar elongated shapes are formed. The molten metal is forced through something analogous to a cake-decorating nozzle to create a particular cross-section shape. Materials formed this way can be problematic as vibrating elements in musical instruments, because the extrusion process creates tensions in the metal due to differences in density between the surface and inner material of the piece. Large sheets of metal, on the other hand, are rolled out and are generally free of such irregularities and inner tensions.

Aluminum and brass are both available in several alloys. The different alloys vary in hardness, workability, and, of course, sound quality. Several alloys of aluminum produce acceptable musical results. Schmidt recommends avoiding the one designated as 7075, which is too difficult to cut. For his aluminum instruments, he generally uses 6061 T6, which is not too hard to work easily, but not so soft that it suffers in tone quality or clogs tools. For the brass, Schmidt recommends any of the alloys labeled "half-hard." In practice, there is some variation in the material sold under this designation. All will work well but Schmidt does suggest being sure that on a given instrument the same alloy is used throughout.

Aluminum can be dangerous to work. Bandsaws are the safest for cutting it and work well at woodcutting speeds, with blades of 6-8 teeth per inch.

A Few Words About the Maker of the Instruments

Daniel Schmidt's primary work is composing – instruments are not the thing; they are only tools. But he has always been confronted with the challenge of hearing the sound he wants in his head and having no means of producing it. Earlier in his career, he turned to electronic music to bring the sounds to life, and he has built acoustic instruments all along. In his acoustic explorations, he initially worked in a relatively free manner, incorporating a lot of found objects into his work and not emphasizing specific tunings. After 1970, he became increasingly concerned with control of pitch and timbral relationships, and the randomness of found objects took a back seat.

He had worked with metallic percussion for quite some time before he began to apply the label 'gamelan' to his instruments. By that time, he had studied the Indonesian model extensively, and just as painstakingly, examined his own ideas on composition, timbre, and tuning systems. He continues to study traditional Javanese music with K.R.T. Wasidodipuro and has had extensive contact with performing Javanese musicians.

In addition to his composing and instrument building work, Schmidt currently directs the gamelan at Sonoma State University and the Berkeley Gamelan. He is active on the advisory board of *Balungan*, the publication of the American Gamelan Institute.

Related articles

Other articles pertaining to American Gamelan which appeared in *Experimental Musical Instruments* are listed below. These articles can be accessed through the Internet Archive at https://archive.org/details/emi_archive/.

"The American Gamelan Institute"(a review of the organization and its newsletter) in *Experimental Musical Instruments* Vol. 2 #2

Review of Will Ditrich: *The Mills College Gamelan: Si Darius and Si Madeleine* in *Experimental Musical Instruments* Vol. 9 #2

"Bill Colvig" by Sasha Bogdanowitsch *Experimental Musical Instruments* Vol. 9 #4

8

The Waterphone

Designed and Built by Richard Waters

Bart Hopkin

This article originally appeared in *Experimental Musical Instruments* Volume 2 #3, October 1986.

By any of several measures, the Waterphone is one of the most successful of new acoustic instruments. It has been taken up and played by many musicians, composed for by many composers, and heard by vast audiences – though most in those audiences may not have been aware of what they heard. It has been used widely in recordings, performances, and movie and TV scores. It has sold commercially in respectable quantities. But the primary consideration for any new instrument must be its sound – and the sound of the Waterphone is incomparable.

Some Background and History

The inventor and builder of the Waterphone is Richard Waters of Sebastopol, California. One day in 1968, Richard chanced to see a kalimba in the hands of a street musician in a Haight Ashbury parade. He was attracted to the instrument for both its sound and its appearance. He also was drawn to it by the fact that a world of lovely sounds was available to anyone picking up the instrument for the first time: relatively little stood between the novice and the music.

Richard applied the skills and tools of a sculptor to what he had seen – for sculpting was his stock in trade, and where his training had been. He put together some tin-can sansas, spidery-looking things with tongues of wire welded to metal resonators. He did not have the kind of academic musical background that would lead him to produce specific scales; he was simply after objects of interesting appearance capable of producing the sounds he was drawn to.

Another chance encounter which had taken place earlier now enabled him to go a step further. At the house of a friend, he had had the opportunity to see and play a Tibetan water drum. There are different types of instruments commonly called "water drums"; the one that Richard played consisted of a bronze, flattened sphere with an aperture, and water held inside. Richard spent a long evening exploring the possibilities of the instrument. He was taken with the way in which the moving water within brought about a continual bending and shifting of resonances in the body of the instrument after the tone was first sounded. In the meeting of the memory of this experience with the welded metal sansas he had been making, the idea for the first Waterphone took hold.

Richard began welding metal rods to metal resonators which could be filled with moving water. The extraordinary acoustic effects of the Waterphone began to appear. He found that, while striking the rods and the body of the instrument produced fine results, the most ravishing sounds were to be obtained by bowing the rods. Time passed, and the Waterphone design and construction methods continued to evolve and improve.

Before too long, people outside of Richard's immediate sphere began to take notice of the instrument. The first well-known and influential person to pick up on it was jazz drummer Shelley Manne. Through him the Waterphone found its way into the collection of percussionist Emil Richards, and from there became known to an increasing number of studio musicians operating in and around Los Angeles. Its widest exposure came with Jerry Goldsmith's soundtrack for the movie *Startrek*, but that is just one of a very respectable list of recordings, compositions, performances, and sound tracks in which the Waterphone has figured. Richard himself has performed on the Waterphone with a number of ensembles, most notably the Gravity Adjusters' Expansion Band, and more recently, Totem.

The Waterphone is one of very few truly new and unconventional acoustic instruments that has been sold commercially with reasonable success. While they may not appear to be taking over the world, Waterphones, still made individually and by hand, have sold in sufficient quantity to cover costs and give the maker some modest recompense for his efforts. As great as the appeal of the instrument may be, this did not happen without considerable on-going effort and stick-to-it-iveness on the part of Richard and his wife Christine. One of the greatest difficulties has been preventing others from selling imitation Waterphones to the potential buyers the enterprise depends on. Richard has responded to this in several ways. He has promoted his work carefully with brochures, flyers, demonstration cassettes, and albums of music. He has kept a step ahead of other would-be builders by continuing to improve the instrument and developing manufacturing methods which are difficult to imitate. And he has patented the instrument, and made the effort necessary to keep the patent active (patents do go stale with time). Please notice this – readers of this article are asked to respect the patent and refrain from attempting to capitalize on the construction principles here described.

The Waterphone as it is Today – a Description

The Waterphone as Richard makes it today is a refinement of the idea that originally took shape through the sansas and water drums. The acoustic principles are the same; the materials, formal details, and construction methods are considerably advanced. With the process of evolution and refinement, a family of instruments has grown up. Waterphones now appear in several sizes, with some variation in shape. The following description refers to the most widely used of the group, the Standard Waterphone.

The body of the instrument is made up of two stainless steel bowls, 7½ inches in diameter, welded together to form a globular base. From the top of this form rises a stainless steel cylindrical column an inch and a half in diameter and eight and three quarters inches high. Around the periphery of the globe are welded forty-two bronze rods. They stand not quite upright, leaning slightly toward the stainless steel column at the center. The tallest is just under a foot long, the shortest about four inches, and they are graduated in length in an irregular pattern in keeping with Richard's tuning methods (more on that later).

Water is placed inside the body of the instrument. Large amounts have the effect of dampening the vibration, so the quantity of water is kept to about ¼ cup – just enough to cover the bottom shallowly when the instrument is upright, and run up the sides a bit when it is tilted (Figure 8.1).

Playing the Waterphone

These three elements – the stainless steel body, the bronze rods, and the water within – form the essential instrument. They can be played any of several ways. Because of the extraordinary acoustic properties of this combination of elements, almost any means of exciting a vibration anywhere in the instrument will produce an intriguing sound of some sort. The most common means of excitation is by bowing the rods. Richard uses a standard bass bow rosined in the usual way. The instrument also is often played by striking the rods or the body of the instrument with the hands, or with soft rubber mallets. (As accessories to the instruments, Richard sells a fiberglass bass bow and pairs of superball mallets.) On the larger, flat-bottomed instruments, the player can also use the superball mallets, with their high traction, to sound the instrument by friction, dragging the ball slowly along the bottom with moderate pressure.

As the instrument is sounding, the player tilts, rocks, or rotates it, so that the water inside moves and sways. This alters the natural resonating frequencies of the body and brings about the peculiar pitch bending and timbral shifts that are so characteristic of the sound.

The sound of the bowed Waterphone is strikingly reminiscent of whale songs. Waterphones have been used in whale watching and research expeditions in Mexico, Hawaii, the South Pacific, and elsewhere. Whales have indeed been found to respond to the sound, gathering and displaying apparent interest and curiosity. An alternative bowing technique has been developed for this application. The Waterphone is

Figure 8.1 The Waterphone Family. Above: R.S.G. (Revolving Sound Generator). Below: Bass, W.R.F.B. (Wide Range Flat Bottom), and Standard Waterphone. (Photo by Lenny Cohn).

taken into the water, inverted, and bowed with a non-oily, wet finger. This type of hand-bowing can also be used for performance in swimming pools or just a bucket full of water.

Tuning

By bowing the Waterphone's rods at different points along their length, one can isolate different overtones as the sounding pitch. On a single Waterphone rod, there are about four or six readily producible pitches, ranging from extremely high to subsonic or nearly so. The forty-plus rods on the instrument, each with its several modes of vibration, create a profuse and diverse intervallic palette.

Although the overtones of the Waterphone's rods may seem unruly at first, and pitch bending is an essential part of the instrument's character, it is possible to produce deliberate, controlled definite pitches. A few Waterphone players have worked in this mode, seeking out and labeling the rods for their pitches and in some cases tuning them to particular scales by shortening individual rods to raise their pitch. Most players, on the other hand, have not chosen to produce definite scales on the instrument, but to concentrate on exploring its remarkable timbral possibilities, accepting the available pitches as given.

Richard himself gives a great deal of thought to the question of tuning as he manufactures the instruments. But he has resisted deliberately tuning to standard diatonic or chromatic scales. In what could be described as a sculptor's approach to sound, he tunes each Waterphone individually to a mixture of smaller intervals (approximately one semitone or less) and larger intervals (up to about a minor third), chosen not for specific pitch relationships but for an overall intervallic sweep that is attractive, interesting, challenging, or flowing.

He tunes for timbral quality as well. Since there is a lot of communication between the rods through the body of the instrument, the natural resonances of one will affect the resonating behavior of others. As a result, the tuning of the whole set influences the timbral quality of the individual rods. Metal rods tend to have many very prominent partials, and conflicts between dissonant partials occur frequently. Much of Richard's tuning process aims to reduce such conflicts and bring out constructive relationships between the partials.

As part of the tuning process, Richard also considers the pattern created in the arrangement of the graduated lengths of the rods. Over the eight years, he has used a number of different arrangements of longer and shorter rods. One that is frequently used currently incorporates two overlapping patterns: the set made up of every other rod comprises a row of progressively longer rods, creating an upward spiral as it circles the perimeter of the globular body. The complementary set of every other rod becomes progressively shorter, creating a downward spiral within the upward one.

Figure 8.2 Waterphone played by Richard Waters.

The Acoustic Behavior of the Waterphone

As the Waterphone is usually played, the stainless steel body serves primarily as a resonator for vibrations generated in the rods. But the two are not really so very separate or independent in their functions. Their operation is highly interactive. This is one of the keys to the extraordinary acoustic effect of the instrument. All of the metal elements – the stainless steel body and the bronze rods – are welded firmly together to create a single, monolithic piece. As a result, vibrations are transmitted readily throughout the instrument: when one rod vibrates, the entire instrument sings. The flat base of the instrument, on which the water sits, seems to have an important role as a partially independent diaphragm. When the water shifts and as a result covers the diaphragm to a greater or lesser extent, it has profound effects on the vibrations as they travel through the metal.

The effects of the water on the overall vibration seem to be of two sorts, which can occur simultaneously. Both result from the fact that as it shifts position it changes the natural vibrating frequency preferences of the body of the instrument, independent of the rods. This can be demonstrated by holding the instrument upright, tapping the body and listening to the pitch; then tilting it (altering the position of the water within) and, after the water has become still, tapping again. A new, higher pitch is produced. The steeper the angle of tilt, the higher will be the pitch of the tapped resonator. In normal playing conditions, discrete pitches of this sort in the resonator rarely come into play, but as the player causes the water to sway and swell, the natural vibrating frequency of the body of the instrument is in a continuous state of flux.

The first effect of this resonance frequency shifting is to cause timbral shifts in the tones from the sustained vibration of the rods. For any given position of the water, and corresponding set of frequency biases in the resonator, certain overtones in the vibrating rods will be enhanced while others are de-emphasized. As the water moves, and with it the predilections of the resonator regarding which frequencies to respond to most readily, the overtone make-up of the tones originating in the vibrating rods changes as well. The second effect of the water movement and resulting resonance frequency shift concerns vibrations occurring primarily in the body of the instrument. In most cases, these are vibrations initiated in the rods and communicated into the metal of the body. The flux of the water does not cause much change in timbre in the body tones. Rather, it causes them to bend in pitch, and quite dramatically at that.

These two effects – timbral shifts in the rod tones and pitch bending in the body tones – can and do happen at the same time. The vibration initiated in the rods and communicated to the body continues to resonate in both places. Changes that the now partially independent vibrations go through may affect one another, but the two don't seem inclined to damp one another out. The acoustic result of the simultaneous occurrence of the two acoustic behaviors – overtone shifting and pitch bending – is somewhat analogous to flanging, an effect found in many signal processing units available for electrified musicians. But in the Waterphone, there is an unpredictability, a wildness about it, which give it an entirely different feeling. And when many rods are initiating many vibrations, often at microtonal intervals, all undergoing these multiple transformations, the effect is bewitching.

Related articles

Other articles by Richard Waters or pertaining to Waterphones which appeared in *Experimental Musical Instruments* are listed below. These articles can be accessed through the Internet Archive at https://archive.org/details/emi_archive/.

"The Superball Mallet" by Richard Waters, in Vol. 5 #5

"Process and Development of the Waterharp" by Richard Waters in EMI Vol. 7 #6

"Bamboo: The Giant Musical Grass", parts 1-3, by Richard Waters, in EMI Vol. 10 #3, Vol. 10 #4, and Vol. 11 #1

"Bamboo and Music", parts 1 &2, by Richard Waters, in EMI Vol. 14 #2 and Vol. 14 #3

Richard Waters contributed to the "Letters" or "Notes" sections in EMI Vol. 4 #3, Vol. 4 #6, Vol. 5 #4, and Vol. 10 #4

"Playing Music with Animals: Four Passages from Dolphin Dreamtime" by Jim Nollman in EMI Vol. 6 #4

"Events" (includes a review of a performance by Richard Waters' group Totem), in EMI Vol. 1 #3

9

The Musical-Acoustical Development of the Violin Octet

Carleen M. Hutchins

This article by Carleen Hutchins (1911–2009) appeared in *Experimental Musical Instruments* Volume 2 #6, April 1987.

The Violin Octet consists of eight experimental instruments of the violin family designed and constructed on acoustical principles to carry the timbre, tone, and playing qualities of the violin into seven other tone ranges from the tuning of the bass to an instrument an octave above the violin (Figure 9.1).

The concept of a balanced consort of violins, just as there is a consort of viols, is not a new one, for we find a description of eight "geigen" or violins, in nearly the same tone ranges as the Octet instruments in Syntagma *Musicum* (1619) by Michael Praetorius. In the museums of Europe, one can see many violin-type instruments of various sizes that conceivably could have been part of such a family. In fact, Lowell Creitz, on a two-year sabbatical in Europe, did find instruments approximating those of the Octet (Creitz, 1977, 1978).

The present effort to create a violin octet was brought into focus by Henry Brant, the composer, when he was at Bennington College. Brant came to me in 1956 hoping he could find some violin maker oddball enough to try to make him a set of violin-type instruments that would carry the tone and playing characteristics of the violin into seven other tone ranges – one at approximately each half-octave from the double bass to an octave above the violin. Brant did not want the beautiful blending qualities of the viola or the cello or the husky sound of the bass. He wanted the clarity, brilliance, power, and ease of playing of the violin on all four strings to be projected into the larger and smaller instruments.

Background

At the time I had been working with Harvard Professor Emeritus Frederick A. Saunders, who had pioneered violin research in the USA during the 1930s and 1940s at the Harvard Laboratory and continued in his retirement. I had made my first viola mostly following the directions in Heron Allen's classic treatise *Violin Making As It Was and Is* (Heron-Allen, 1895) and it was rated as the work of "a good carpenter"

Figure 9.1 The Violin Octet. From left to right: Baritone Violin, Small Bass Violin, Contrabass Violin, Tenor Violin, Alto Violin. On the floor: Mezzo Soprano and Treble violins. (Photo by John Castronovo).

which was about my speed in 1949. Some friends introduced me and my viola to Saunders. He looked the instrument all over, blew in the f-holes, tapped around on the top and back and said "Young lady, I'll be interested in seeing your next one." At the time I hadn't planned to make a "next one" for all I wanted was one to play myself. But by 1950 I had had the opportunity of working under the supervision of Karl A. Berger, a Swiss violin maker with a shop in the Steinway Building near

Carnegie Hall, and in the next few years had made several reasonable violas. Also I had made several experimental violas for Saunders to work with, cutting and changing the wood as he indicated. This involved making instruments with deep ribs, shallow ribs, flat tops and backs, high arches, tiny f-holes, large f-holes and the like to try to discover clues to some of the acoustical characteristics of the violin. On one viola with flat top and back, we did more than 125 experiments, keeping careful records of the effects of the changes. This was the instrument where a channel to simulate the purfling groove, cut around the edges of the top and back, changed it from a poor sounding box to a reasonably good sounding viola.

To make a very long story short, by the time Brant came to me in 1956, I had done enough experimenting with moving violin and viola resonances up and down the scale and changing constructional features, so that within a half an hour's conversation, I agreed to try to do what Brant wanted. Later Brant and his friend Sterling Hunkins, the cellist, brought some existing instruments they had adapted – a small size cello for the tenor range between viola and cello, and a child size violin strung an octave above the viola. These two instruments had the proper range but left much to be desired in tonal qualities.

The Ten-Year Project

The Search

The first problem to be faced in designing the new instruments was to try to find the distinguishing acoustical characteristics of the violin as different from those of the viola, cello, and bass. This led to a search through records of several hundreds of tests made not only by Saunders (1875–1963), but by Herman Backhaus (1885–1958) and Hermann Meinel (1904–1977) in Europe during the 1930s and 40s, by John Schelling (1892–1979), a retired Bell Laboratories research director and cellist who was working with us, and by me. Finally one distinguishing characteristic emerged. In many fine violins, the two main resonances, so called "air" and "wood", lie very close to the frequencies of the two open middle strings – the "air" near D 239Hz and the "wood" near A 440Hz.[1] In fact, we found that these two big resonances were exactly on the two open middle strings of the Guarnerius del Gesu belonging to Jascha Heifetz that Saunders tested some years back. In the viola and cello, these two resonances were found to be three to four semi tones higher in frequency in relation to the tuning of the strings. For example, in the viola, the "air" resonance is usually around the B to B♭ above the open G string and the "wood" resonance around the F to F♯ above the open D string. In the basses which we could test, these two resonances were found to be even higher in relation to the tuning of the strings. Obviously the viola, cello, and bass would have to be larger than the present ones. But how much?

A search was made for instruments that might have dimensions such that they could be adapted to have the desired parameters. One child's cello was found that had its "wood" resonance about right for our projected viola, but its "air" resonance was much too low because of the deep ribs. We were able to purchase this so that I could start slicing down the ribs a few inches at a time to try to bring the "air" resonance up to the open second string. This meant several rounds of taking the back off, sawing the ribs down, gluing the back on, and checking for the position of the "air" mode. Calculations of the air cavity gave some indication of how much to cut off, but because of the flexibility of the box (compliance), the calculations were never accurate enough, so I had to cut and try. The desired resonance placements were finally achieved with the little cello (21 inches body length) looking like a very large, shallow viola!

The Dautrich Instruments

About this time we were most fortunate in learning of the work of a violin maker named Fred Dautrich who had lived in Torrington, Connecticut during the 1920s and had spent many years working to develop instruments that would "fill the gaps in the violin family." He described his work in a little booklet of this title and had succeeded far enough to have a demonstration of his five instruments played for Arturo Toscanini at Carnegie Hall in New York. I was able to locate his son, Jean Dautrich, who was kind enough to let me borrow and test the in-between instruments, a small bass tuned a fifth below the cello, a tenor, tuned a fourth above the cello, and a 20-inch viola played on a peg. Without the help of these instruments, the project would have taken at least another five years. Finally, Jean Dautrich was willing to sell me the three instruments so that I could try to adapt them to our desired acoustical parameters. The small bass had the right placement of the "wood" and "air" resonances, but the response was not particularly good. The viola had the right "wood" resonance, but the "air" resonance was too low because of the deep ribs. The instrument between viola and cello had its resonances placed about right but did not have good tone quality. If only Fred Dautrich had had the benefit our technical information and our method of tuning the free top and back of the instruments before assembly, his life's dream could have been realized!

The Acoustical-Musical Development

I was able to adjust rib heights and tune the free plates of the Dautrich instruments so that they not only had their resonances where we wanted them, but had good tone and playing qualities so that they were able to take their places in our new series. A small violin that Karl Berger had made as a "violino piccolo" a fourth above the violin was used in our first demonstration of the middle five. This left the big bass and small bass at the bottom of the range, and the instrument an octave above the violin to be developed from scratch. Even a large Prescott bass loaned by Brant did not have its resonances low enough for the large bass. By 1962, the middle five were ready for a trial run. While I was working to get the actual instruments to have the parameters we wanted, Schelling was working on scaling theory for overall body length and other parameters for each new instrument. In this way, we were able to project a chart with a curve of tuning frequencies versus body lengths so that it was possible to extend the curve at each end and say that the large bass should have a body length 3.6 times the length of the violin and the treble a body length of 0.75 of the violin length (Figure 9.2).

The big bass with a body length of 52 inches and an overall height of seven feet took three years to develop and construct. Without the help of Donald Blatter, a bass maker in Erie, Pennsylvania, who actually set up the wood for this big fellow, made the ribs and roughed out the top and back plates, the job would have taken even longer. Blatter also provided a nearly finished small size bass so that I could tune the plates and adapt it for the instrument between cello and bass tuning – the small bass.

Henry Brant was delighted with the sound of the whole octet and composed a piece which he conducted in 1965 at the Young Men's Young Women's Hebrew Association on 96th Street in New York City as part of Max Pollikoff's program "Music in our Time," which was very well received.

Overall Design

With the eight instruments doing the musical job we had projected, the next step was the creation of an overall design characteristic so that the Octet would not only play properly, but look well as a balanced consort. We had been working with musicians who wanted a shorter string length on the big viola so they could use viola fingering. They wanted a longer string length on the tenor so fingering would be closer to cello mensure. (The term "mensure" refers to the proportioning of string, neck, and fingerboard lengths.) Bassists were willing to settle for a 46" string length, but no longer. A Stradivarius violin pattern was used as the basic design and projected into the other sizes. Many adjustments had to be made in basic outlines and overall mensures to accommodate player needs, while at the same time respecting basic violin parameters. For almost two years, our living room walls were covered with cardboard outlines of each instrument so we could look at them as changes were made. (If one is not careful, a poorly designed instrument can look very much like a fat lady with a string around her middle!)

Those Who Help

This whole development represents the work of many people. In the first full-scale article I wrote about the Octet (Hutchins, 1967), I listed 180 names of those who had contributed in this development on way or another. They represented chemists,

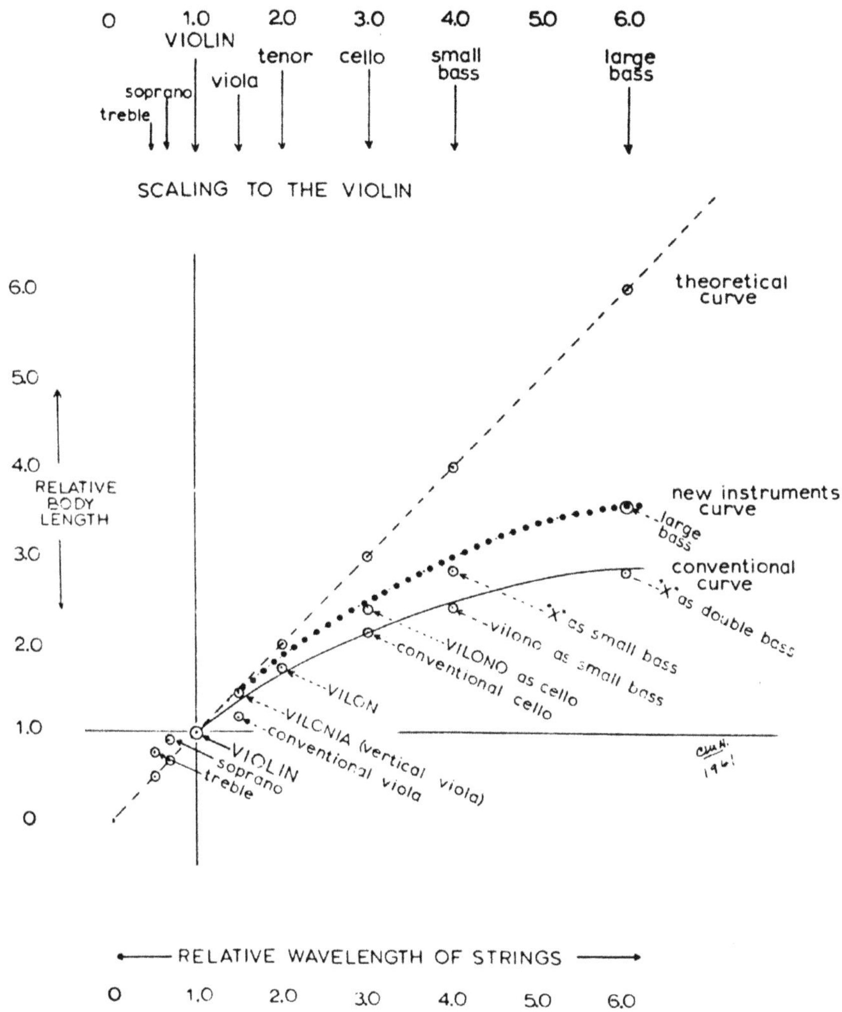

Figure 9.2 Projected scaling for the Octet Instruments.

architects, electronic engineers, translators, editors, photographers, artists, lawyers, general consultants, secretaries, violin experts and makers, violinists, cellists, bassists, composers, and conductors. In other words, people from a broad spectrum of the music world including some famous names of the 1950s and 60s were interested enough in the project to contribute their time and abilities. Funding for the project came largely from the Martha Baird Rockefeller Fund for Music, the Harriett M. Bartlett Fund of the Catgut Acoustical Society, Inc., two fellowships to me from the John Simon Guggenheim Memorial Foundation, and my own working time for ten years.

Today six sets of the Octet are completed except for one large 7' bass, which is well under way. Two sets have been sold abroad; one went to England where it spent three years at the Royal College of Music and one to the Swedish Academy of Music, which resides in the Stockholm Music Museum where the instruments can be on display as well as used for performances. A third set has just spent three years at the University of California, San Diego, under the direction of Bertram Turetsky, and is now with the Metropolitan Symphony Orchestra of Boston for a series of concerts. I have taken the Octet to over 150 lecture-demonstrations for college Physics and Music departments around the USA and Canada.

Musicians find that these new instruments have very exciting tonal characteristics: the clarity and sweetness of the high sounds of the soprano and treble violins, the power and projection of the mezzo violin and particularly the full dynamic range and clarity of the alto, tenor, and baritone violins on all four strings. After hearing the alto (viola) in our first concert in 1965, Leopold Stockowski said: "That is the sound I have always wanted from the violas in my orchestra." William Berman, formerly violist in the N.Y. Philharmonic and professor at Oberlin Conservatory, who now lives in Seattle, has played the alto violin in orchestras and chamber ensembles all over the world for twenty years. Berman reports that whenever the conductor wanted more sound from the violas he could provide it, but that also he could play the alto in string quartets without overpowering the other instruments. Daniel Kolbialka tells us that he is finding the high sounds of the treble and soprano violins very rewarding in his music.

We found that having one baritone in the cello section and one large bass in the New Jersey Symphony greatly enhanced the sound of the two sections. A large bass, made by Hammond Ashley to our specifications, is presently in the Seattle Symphony and Opera Orchestra. Ashley reports that when it was first introduced, the other bass players were very critical of having it in the section because the big bass could produce more sound than all the other basses together, but it is now well accepted. A bassist recently described the tone of the big bass as one "bassists only dream of."

The real problem for performers on these new instruments is that they are truly new instruments, not only in their sounds and overall dimensions, but also in the changed placement of their resonances giving different dynamics from those of the standard instruments to which the players are accustomed. Even expert professional performers find that it takes time in order to realize and be able to exploit the full potential in each of the new Octet instruments.

The Future

The whole project, which started as an interesting acoustical-musical experiment, seems to be getting a foothold in our musical world. Composers particularly are excited over the possibility of composing for eight octaves of balanced string tone, and instruments which have the clarity and power of the violin on all four strings so that they can be heard clearly through a thick musical texture or can speak pianissimo when desired. Several composer contests have been held and the Catgut Acoustical Society has over fifty pieces composed or arranged for the Octet. The Society has an illustrated brochure giving the musical and constructional features of each instrument (CAS, 1981) as well as a set of full-scale drawings for the eight instruments with suggested plate tuning frequencies for violin makers, that can be purchased through the office. About two dozen of the individual instruments have been sold separately, particularly the alto violin and the baritone violin. These can be used in standard orchestral and solo repertoire.

Whether the Octet as a whole will eventually become a real part of the musical scene or whether some of the individual instruments will find their place is a matter for conjecture. Hopefully there will be enough players who will learn to play the instruments well, composers to compose for them effectively, and audiences who will feel the thrill of the new sounds which the Octet instruments are capable of producing, so that the musical potential of the first family of orchestral instruments based on a consistent theory of acoustics can be fully realized.

Note

1 "Air" and "wood" are in quotes because we now know that each is a resonant combination of wood and cavity modes.

References

Catgut Acoustical Society, Inc., Violin Octet Brochure, 1981.

Creitz, Lowell, "How to Play the Tenor Violin Now. A brief Compendium of Resources," *Catgut Acous. Soc. Newsletter* 28, Nov. 1977, pages 17–18.

Creitz, Lowell, "The New and Old Violin Families – An Organological Comparison," *Catgut Acous. Soc. Newsletter*, #30, Nov. 1978, pages 18–22.

Hutchins, C.M., "Founding a Family of Fiddles," *Phys. Today* 20, 1967, page 23.

Related articles

American Luthier, a biography of Carleen Hutchins by Quincy Whitney, is available at www.quincywhitney.com.

Other articles on instruments closely related to violin which appeared in *Experimental Musical Instruments* are listed below. These articles can be accessed through the Internet Archive at https://archive.org/details/emi_archive/.

"Shape and Form, Contemporary Strings, Part I: Fred Carlson, Francis Kosheleff, Susan Norris, and Clif Wayland" by Bart Hopkin, in Vol. 4 #5

"John Maluda's Instruments for the Montessori Classroom" by Bart Hopkin & John Maluda, in Vol/ 5 #3

"Instruments from the Marx Colony" by Bart Hopkin, in Vol. 9 #1

Several articles by Cary Clements on Stroh Violin appeared in *Experimental Musical Instruments*; one of them, "Augustus Stroh and the Famous Stroh Violin," is included in the current collection. Others appeared in Vol. 10 #4, Vol. 11 #1, Vol 13 #2, Vol. 14 #3, and Vol 14 #4

10

Structures Sonores

Instruments of Bernard and François Baschet

Bart Hopkin

This article originally appeared in *Experimental Musical Instruments* Volume 3 #3, October 1987.

Background and Introduction

In Paris and its environs, two brothers have been designing and building new instruments since the 1950s. Bernard and François Baschet (1917–2015 and 1920–2014), pursuing their own brand of practical acoustic research, arrived at a set of unique acoustic systems and created from them an orchestra of completely original instruments. Their work has included concert instruments, sound sculpture, children's instruments and environments, and large-scale public musical sculpture. It has been a long road for the two, but in the long haul they have been rewarded with well-deserved recognition and some very exciting commissions. If their work is not as well known in the United States as it should be, well, perhaps this writing will make some small difference. For the *Sculptures and Structures Sonores*, as the Baschets call their works, are some of the most inventive, inspiring, and beautiful instruments of our time (Figure 10.1).

To understand the nature of the Baschets' contribution, it helps to understand an attitude that permeates their work. François and Bernard have spent many years engaged in a kind of practical, hands-on experimentation. They have manipulated acoustic vibrating, conducting and amplifying systems in countless ways, observing the results and modifying further, continually seeking beautiful sounds, beautiful forms, and elegant methods. Their investigations have not been in the service of any theory, any dogma, or any particular musical style. The real motivation behind *Structures Sonores* has been, simply, a love of sound for sound's sake, form for form's sake, and exploration for exploration's sake.

This is not to say that the Baschets do not like to philosophize about their work and their purposes. They have definite ideas about what their work should achieve.

They believe first and foremost – and this is not always a popular belief – that it is the function of artists like themselves to make our world more beautiful, and that beautiful things are worth striving for.

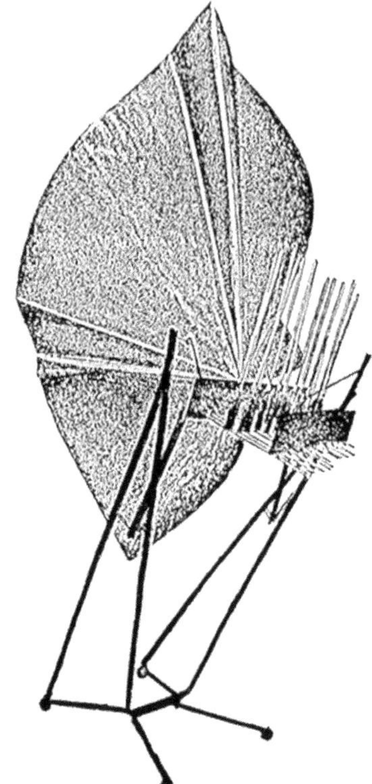

Figure 10.1 The Cristal.

They also devote a good deal of thought and effort to the ways in which the structures function socially. They design with human interaction in mind, and they watch to see the ways in which big people and small people, musically trained and musically untrained people, interact with the *Structures Sonores*, and with each other through them. They make the instruments durable, to withstand a thousand curious hands in public places, and they work to make most of them accessible and playable by anyone. In short, the products of the Baschet Brothers' explorations are made to allow for further explorations on the part of whoever wishes to touch and hear.

In their larger public works, the Baschets have leaned especially toward structures which have a role to play in people's day-to-day life, and which can bring music as well as visual beauty into daily routine. They have built, for example, a school bell tower, a number of public musical clocks, and some elaborate kinetic musical fountains.

Finally, the brothers have designed a number of instruments with children in mind, including classroom instruments oriented toward free sonic exploration, and musical playground environments.

It took a little time for the Baschets to arrive at the exploratory method that unlocked so many doors for them. François began his explorations with a study of conventional instruments:

> If it is true that genius is ten percent inspiration and ninety percent copying over the neighbor's shoulder, then alas, here there was no neighbor! All acoustic research in my time was directed towards electro-acoustics and electronics.
>
> For months and months I collected all sorts of instruments from the flea market. I took them apart and analyzed them. I repeated experiments I had read in library books. I attempted to improve clarinets, trumpets and cellos, but without great result.[1]

The "without great result" part of all this began to change when François developed a new conceptual scheme for the purpose of thinking about musical instruments. The scheme was designed to account for existing musical instruments and then point the way to possible musical instruments. It begins with a functional definition of the term "musical instrument":[2] A musical instrument is the joining of at least three of the four following elements:

1. vibrating elements: strings, vibrating rods, reeds, etc.
2. energizing agents: bow with rosin, wind, percussion, etc.
3. modulating devices: modification of length, modification of tension, multiple tuned vibrating elements, etc.
4. amplifying devices: sound board, sound box, etc.

As familiar examples of the application of this set of requirements, François Baschet looks at the violin and flute. For the violin: 1) vibrator – strings, 2) energizer – bow, 3) modulating device – plurality of strings and fingerboard to vary their length, and 4) amplifying device – sound board and box. For flute: 1) vibrator – the air flow over the aperture, 2) energizer – the wind of the player, 3) modulating devices – fingerholes and the levers and pads to operate them, and 4) there is no separate amplifying device (remember that according to the model, only three of the four elements need to be present). "Obviously the next step," Francois continues, "was to establish a table listing:

1) All forms of vibrators
2) All energy generators
3) All scale making devices, and
4) The amplifiers."

> By combining these elements one by one, one could establish a sort of table, as Mendeleiev did for the basic elements in chemistry. This table would describe

> all the instruments already made and those to be made in the future. This illumination gave me a deep happiness which boiled within me for a week and, though it subsides a little bit each day, still remains strong to this day.

As you shall see, this kind of thinking freed the Baschets' imaginations in a manner that produced some wonderful results. To understand better how it has worked for them, we should make note of one more aspect of the Baschets' mechanical approach.

With their instruments, as with instruments in general, the life story of the sound begins with an initial vibrating element and some means of exciting it. Sometimes the initial vibrator takes a form such that it readily transmits its vibrations to the surrounding air and from there to the ear, as with the air column of a flute or the membrane of a drum. In other cases, the initial vibration may be of a form that does not transmit effectively to the air, as with a vibrating string (strings, having so little surface area, are not capable by themselves of moving much air). Most of the Baschet instruments fall in this latter category. In these situations, if the instrument is to be heard, the vibration must somehow be transmitted to some form of sound radiator which can move more air. For most instruments, the link between the initial vibrator and the radiator is very direct – for example, strings are usually attached directly to a soundboard. Central to the Baschets' approach, however, is the fact that many materials, and hard metals in particular, are extremely efficient in conducting sound vibrations. They may do so inaudibly as long as no effective radiating surface is present, but they can do so none-the-less, even over long distances, often with amazingly little loss of energy.

The Baschets have become masters of simple vibration-conducting technology. They can initiate a vibration in one place, have it conducted to another locus for some sort of modification and then send it to a radiator to let it be heard. This approach has allowed the Baschets to mix and match the various particulars of the table of organological elements above and made most of their innovations possible.

OK, with all that said, let us look at some of the Baschet's specific innovations.

Initial Vibrators and Modulation Devices

The Baschets have built many instruments using familiar sources of vibration, particularly strings and various sorts of metal percussion. An initial vibrator they have used often is a little more unusual: steel rods set in vibration by glass rods stroked with wet fingers. This original system for exciting a vibration and modulating its frequency is one of the Baschet's finest contributions. It is used on several of their instruments, including the Crystal shown in Figure 10.1. It may take any of several forms. What follows is a description of its most common arrangement, called by the Baschets an "L fitting."

Glass Rods and Threaded Metal: The L Fitting

The initial vibrator for this system is a tuned set of pairs of rods. Each pair consists of a vertical rod mounted upright upon a more massive horizontal one. The vertical rod is glass; the horizontal is threaded and made of steel. At a point farther along the threaded rod is a weight, in the form of another upright piece, a small bar of steel. The position of the glass rod and the weight on the horizontal piece are adjustable. Sets of these assemblies – the threaded rods with their two vertical appendages – protrude in a row from a horizontal bar of steel or aluminum, called the "gum" (holding, as it does, a set of "teeth") (Figure 10.2).

OK, what's going on here?

The vibration is started in the glass rod. The player excites it by stroking with wet fingers. Now, it is important to understand that the glass rod in itself is not tuned, beyond the fact that smaller rods are used for higher pitches. It merely is the source for the vibrating energy. That vibrating impulse is communicated to the metal rod below, setting it into vibration at its own natural frequency. The Baschets compare the metal rod to a violin string, the glass one to the bow: the bow provides the vibrating impulse; the string determines the frequency.

Accordingly, if tuning is to be done, it must be done in the horizontal metal rod. Several factors come into play in obtaining the desired frequency and allowing it the greatest possible resonance. They are 1) starting with a rod of appropriate length, 2) injecting the vibration at the optimal point, and 3) forcing a node at an appropriate place, ideally about 2/3 of the length of the rod away from the gum. Adjustment for #1 can be done at the point at which the threaded rod is mounted in the gum. For #2, it can be done by shifting the location of the glass rod, and for #3 by shifting the location of the weight. It turns out that #3 – the placement of the node created by the presence of the weight – is most effective for fine tuning, while the overall length of the metal rod (#1) and the positioning of the glass rod (#2) can make a world of difference in resonance and resulting richness of tone.

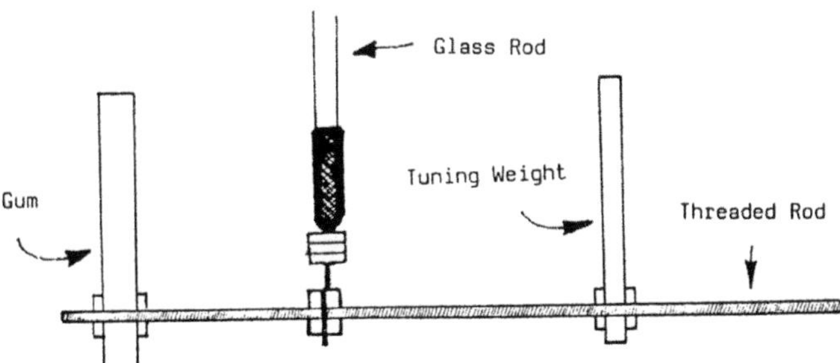

Figure 10.2 The L Fitting.

The net result of this arrangement is a fully tunable, easily played friction idiophone. It needs a radiator of some sort to communicate its vibration to the atmosphere, but the Baschets are at no loss in that department, as we shall see a bit later. The resulting tone is difficult to describe – resonant, singing and full, but retaining a light and floating quality.

Transmission and Isolation of Vibrations

Transmission

We spoke earlier of the importance in the Baschets' work of efficient transmission of vibrations from the initial source, through whatever intermediary steps apply, to a radiator or radiators which may be far away or connected somewhat indirectly. One of the keys to this process is that the vibration, to travel well, should be traveling through a medium whose impedance is high – that is, it should be a low-displacement-amplitude, high-pressure wave in a medium that is dense and rigid.

The heavy metal gum in many of the Baschet instruments serves well for this purpose, as a gathering point for vibrations from many possible sources. The vibrations of the initial vibrating system are fed from the metal rods into the gum directly. If there are to be other sources of initial vibration or additional devices for reverberation, they too are affixed to the gum so that their vibrations feed back into it. "In short," Francois says, "we simmer a complicated vibration inside the bar."

Typically in the *Structures Sonores*, steel bars or rods run from the gum to the radiator. All the connections in the route the vibration travels within the instrument must be made fast, eliminating unwanted rattles and minimizing energy loss in the system. The radiators, in whatever form they may take, can then take the signal and communicate it broadly to the surrounding air. In doing so they convert it to a vibration of larger amplitude and lower pressure – which is to say, they convert the high-impedance vibration in the metal to low-impedances vibrations in the air.

If all is as it should be, it is remarkable how effectively the system can conduct and then project the oscillation, with great efficiency and over long distances. In one very large structure, the Baschets were able to use resonators joined by a steel pole to a source of initial vibration thirty-six feet away and retain an impressive output.

Isolation

There is another possible source of unwanted noise and energy loss to be dealt with. The high-impedance oscillation, given its low amplitude and the minimal surface area of its carrier, will lose little energy to the surrounding air, but it will conduct effectively to anything rigid it meets. That means that instead of sending the full strength of the vibration to the intended resonator, it may dissipate it off into the floor or whatever else the body of the instrument is in contact with. Accordingly, the Baschets have found it important to isolate and insulate their vibrating systems. How? They have depended primarily on two methods.

One is string or chain. The vibrating parts of an instrument are suspended from the instrument stand. Arranged properly, the string conducts far less vibration to the stand than a rigid connection would. It also conducts very little to the surrounding air.

The other isolation device is balloons. Balloons will readily communicate low-impedance vibrations to the air (that's why they work, as we shall see, as radiators) but no balloon, aside perhaps from the proverbial lead one, will have the capacity to start something as massive as the floor vibrating. Thus, balloons can serve the dual purpose of low-impedance radiators and high-impedance insulators. Many of the Baschet instruments actually sit upon balloons – practical considerations giving rise to a peculiar visual effect. In other cases, balloons insulate between different parts of an instrument, most often between the vibrating elements and an inert instrument stand (Figure 10.3).

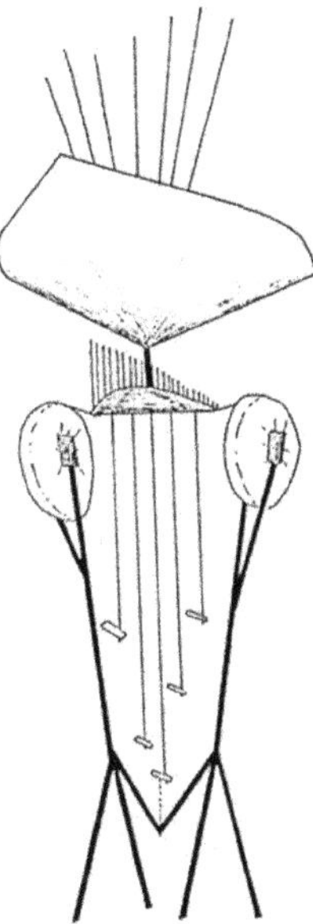

Figure 10.3 Manflower. The sounding elements are isolated from the framework and the floor by a combination of balloons and string or chain.

Sound Radiators

When the Baschets first looked over their table of organological elements, they recognized that acoustic resonators were the element showing the least variety in the standard instruments, and the one most ripe for innovation. Accordingly, it was in this area that their innovations began.

Balloons

The Baschets' first step away from standard design entailed the use of flexible inflated bladders – that is, balloons – in place of the standard soundboard resonator. This was first realized in the form of an inflatable guitar, originally conceived as a convenient travel instrument. It had standard neck, fretting, strings, tuning pegs, and so forth, but the soundbox was replaced by an appropriately shaped plastic balloon held in place by a minimal skeletal framework of wood and metal. The bridge pressed down against the surface of the balloon, communicating to it the vibrations. The arrangement turned out to work well, and for quite some time, François used it in cabaret performances. In those early days, as the Baschets proceeded to develop increasingly adventurous instruments, they used the balloon resonators frequently and in all manner of different circumstances, feeding them vibrations originally excited in rods, springs, strings, and various idiophonic metal percussion materials (Figure 10.4).

The greatest difficulty in working with balloons was finding ways to mount them securely and to effectively transmit vibrations from another source to them, all in such a way that they remained able to vibrate freely. Ultimately an arrangement which came to be called "lobster pincers" proved superior. It involved a U-shaped metal piece with "bridges" on the inside ends. The bridges contacted the balloon, as if holding it between the tines of large tuning fork. Whatever it was that produced the initial vibration was made to communicate it to the metal of the pincers, which in turn communicated vibrations to the balloons. Setting both sides of the balloon in vibration in this way increased the acoustic output noticeably. Many of the early sculptures used several of these resonator-balloons, producing once again, along with their acoustic output, a very peculiar visual effect (Figure 10.5).

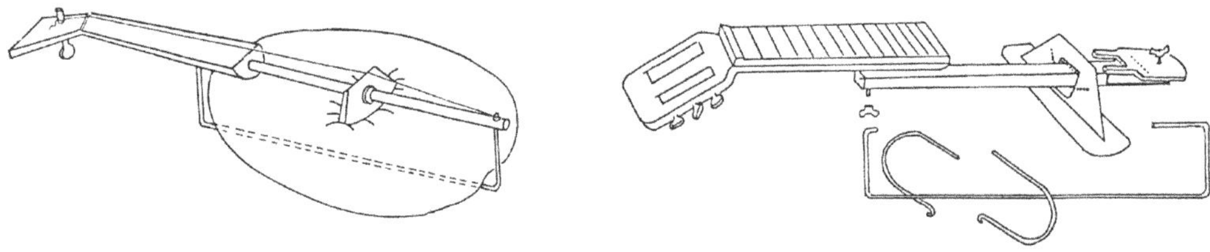

Figure 10.4 Left: Balloon Guitar. Right: the elements (sans balloon) of a more elaborate version.

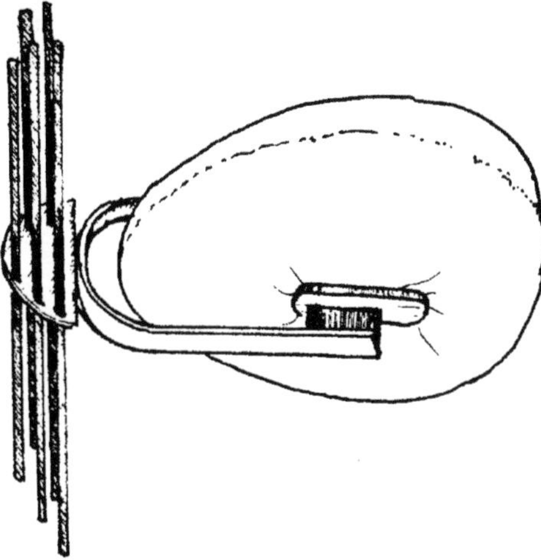

Figure 10.5 Lobster Pincers, employed in a hand-held percussion instrument used by dancers in a ballet.

Cones and Flowers

In time the Baschets evolved another form of resonator which eventually supplanted the balloons for the most part. In its most basic form, it is a sheet of metal or cardboard, cut and bent to form a cone shape to give it some rigidity and project the sound effectively. Since the Baschets have never been known to stand still, the cones have metamorphosed to a variety of shapes, cut, bent, and folded into a garden catalog of floral forms. The cones and flowers are usually mounted at the ends of metal rods reminiscent of flower stems. The visual effect of several of them rising in graceful curves from an instrument is lovely.

The cones may be mounted on the stems in either of two positions. In cases where the acoustic energy is not strong, the stem joins the cone at the seam along the side, allowing the cone to act like a soundboard. Where the vibration being fed in is stronger, the rod can be attached at the point of the cone, so that it acts more like a loudspeaker (Figures 10.6 and 10.7).

Piston Pipes

A very different approach to resonators is the piston pipe. The piston pipe is a system for efficiently transmitting a vibration from an outside source to an air column in a pipe of the same resonating frequency. A pipe of moderately large diameter (up to 42 cm) is used. One end remains open; the other is closed air-tight by a rubber membrane stretched and held on snugly by rubber bands. Two disks of plywood

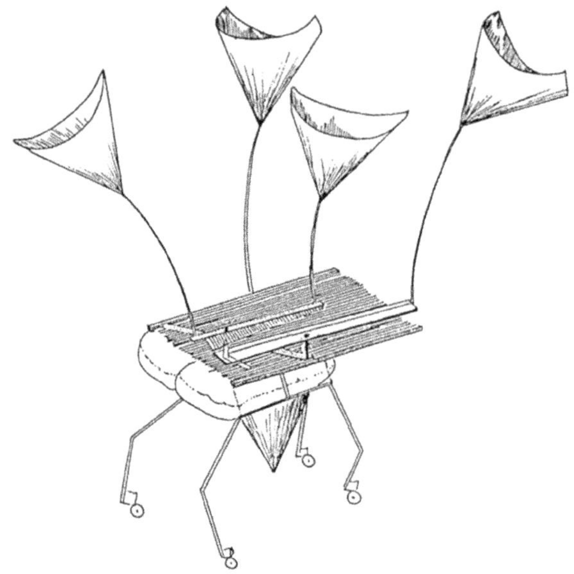

Figure 10.6 The Nanny Goat, insulated by balloons and radiated by cones.

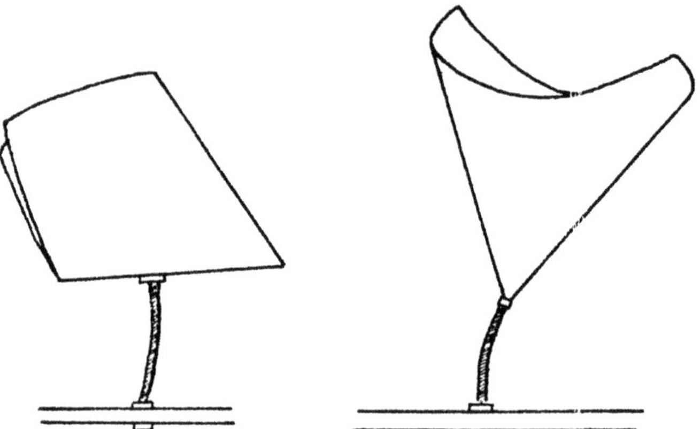

Figure 10.7 Two methods For mounting sound-radiating cones.

or aluminum reinforce the membrane on both sides, forming a sort of membrane sandwich. The diameter of the disks is slightly smaller than that of the pipe, so that they can move in and out of the opening on the flexible spring of the surrounding membrane.

The vibration is transmitted to the membrane sandwich by a wire, metal bar, or something similar which is affixed to the center of the disks. Normally this attachment is taut, so that it stretches the disk slightly outward from the pipe. Any longitudinal vibration in the attachment will be communicated to the disk. If the resonant frequency of the pipe agrees with that which is being input, coupling will occur and the sound will be effectively amplified.

The piston pipe is designed to handle vibrations from any number of sources. In practice, the source of the vibration applied to the pipe has usually been the glass rods and threaded rods assembly described earlier, with the vibration transmitted through metal bars, rods, or wires.

If a scale of many pitches is to be available in instruments using this system, either there must be many pipes or some system must be found to vary the lengths of a smaller number of pipes to bring their resonating frequencies in line with the desired pitches. Since many large diameter pipes quickly become cumbersome, at one point, the Baschets devised systems of telescoping pipes with a mechanism operated by the player, manipulating their lengths as vibrations of different frequencies are fed into the pipes. The arrangement made it possible for about three pipes to handle a fairly large range. The mechanism was complicated though and proved impractical, and the instrument is now a museum piece (Figure 10.8).

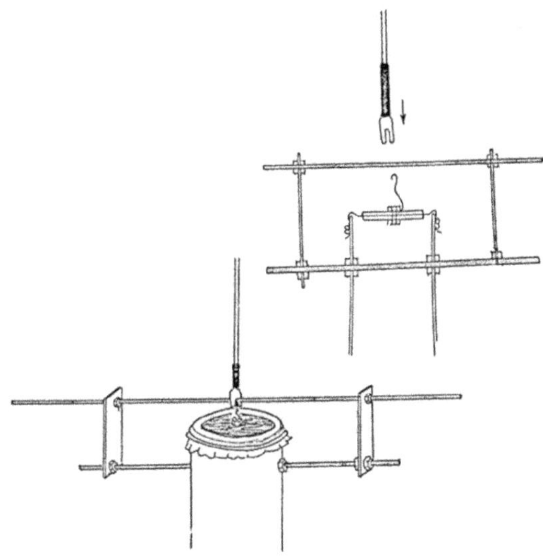

Figure 10.8 Two representations of the Piston Pipe Resonator, shown with a friction rod vibration source.

Reverberant Devices

Yet another element frequently contributes to the Baschets' sound mix: acoustic reverberation. In common parlance, this word refers to the effect that occurs when a vibration from some source is somehow picked up in some other medium and sustained along with the original vibration and beyond. Reverberation is usually distinguished from "echo" in that it is not a distinct repetition of the original sound, rather a lingering and slow dying away of its overall pitch and timbre. We are quite used to hearing natural room reverberation (the lingering of sound in a room as it bounces of the walls after the original sound source has ceased), and, of course, we are all awash in various forms of electric and electroacoustic reverb these days. Used judiciously, the effect is quite natural and adds warmth and richness to most sounds.

The Baschets have built acoustic reverberation into many of their instruments. They do this by incorporating materials and appendages which pick up the vibration from the initial vibrator and then sustain it independently for a time. In concept, these materials are independent from the main resonator and feed into it to enrich the sound of the initial vibration. In some cases, in practice, the reverberant device may to some extent take on the function of the main resonator, transmitting on its own a goodly amount of vibration to the air directly.

The devices the Baschets have used are not made to retain an unbiased replica of the original vibration. Distinctive character and idiosyncratic response all are part of the plan. They create both echo and overtones. Some can even be tuned – that is, their response modified for overall effect or for specific frequencies. They are usually mounted with screws so they can be added or removed according to the kind of sound desired.

Whiskers

The reverberation device they have most commonly used is wonderfully simple. It has been a part of their work almost from the start. They call it "whiskers." It consists of a large number of spring steel wires affixed at one end to the body of the instrument and free at the other. They are generally between 60 and 100 cm, and within this range, they are all different lengths, giving them a large range of resonating frequencies. They are quite free to bend and sway and occasionally contact on another. They pick up vibration from the body of the instrument and sustain it in their characteristically free spring steel style. The wire itself has little enough surface area that it conducts only minimally to the surrounding air; for the most part, it feeds back into the instrument and the main resonator.

A great many of the Baschet instruments have whiskers. In a few cases, they have cast off their supportive role and actually function as the main sounding element.

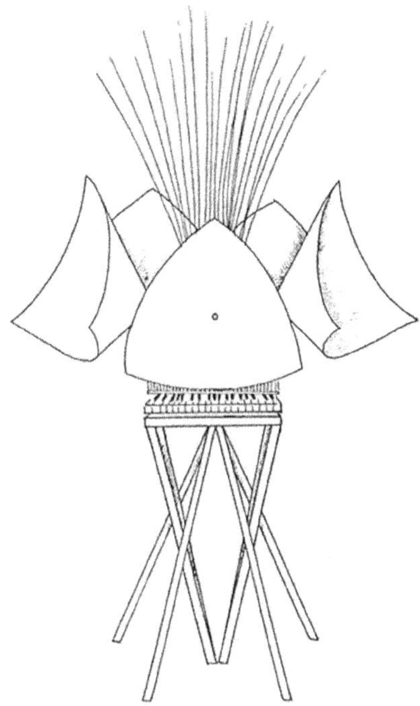

Figure 10.9 Aluminum Piano, with whiskers and cones.

In addition to adding their own peculiar resonance to the sound of any instrument, they contribute yet another strange and intriguing visual element (Figures 10.9 and 10.10).

Free Sheet Metal

A second approach to reverberation has been the addition of free sheet metal resonators. These are usually fairly large (as much as 150 by 100 cm), and they are mounted near their middles so that the edges are free to sag, sway, flop, and vibrate as they will. They have been used in structures with balloons, but having such large radiating surfaces of their own, they may serve in practice as the main resonator for the instrument. In their work with free metal sheets, the Baschets found that they tend to have very irregular response curves, responding like glorious thunder when fed a frequency they are partial to, and rather weakly for others. The good news is, it was these explorations that led the Baschets to try increasing the rigidity of the sheets by folding and bending, and ultimately to the creation of the beautiful cones and flowers that distinguish most of their later instruments (Figure 10.11).

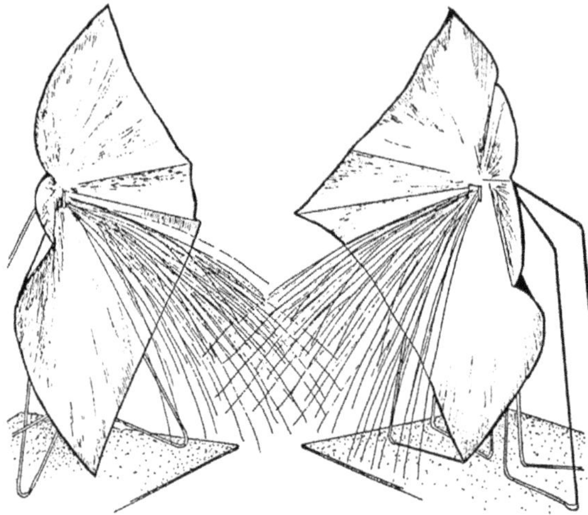

Figure 10.10 Ballet Mache, created for a ballet. Dancers would dance or jump between the whiskers.

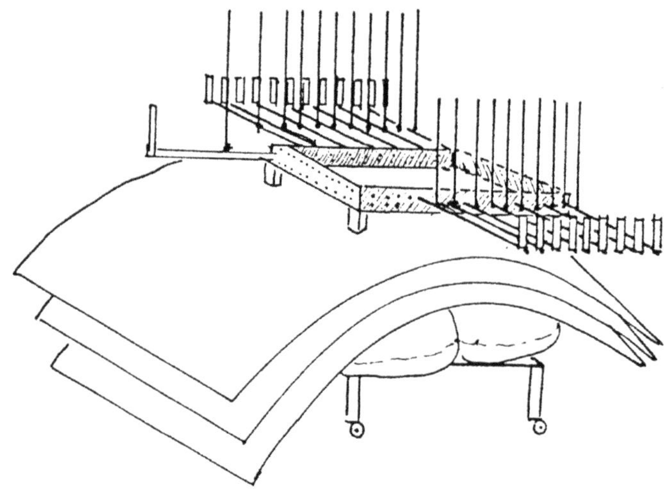

Figure 10.11 Glass Rod Structure with Three Steel Sheets.

Springs, Disks, Rods, and Sympathetic Strings

The Baschets have experimented with a number of other reverberant devices to provide spiced (that is to say, not unbiased) reverberation to the basic sounds of their instruments, including various arrangements of metal springs and disks. They have also occasionally employed sympathetic strings – strings not sounded directly by the player but which enrich the sound by responding to vibrations initiated elsewhere in the instrument. We should mention here too that even without the addition of dedicated reverberation devices, the Baschet instruments are full of sympathetic vibrations due to the nature of their construction. The rigid metal assembly means that any vibrations that arise will be conducted throughout the instrument. If, for instance, one of a set of tuned rods is set in vibration, that vibration will be communicated to all the other rods, and those which have natural resonances at the same frequency will respond and add their part.

Conclusion

The Brothers Baschet led a remarkably fertile lives of ongoing musical acoustic exploration. This article provides only the broadest overview.

Notes

1 This passage is taken from *Sound Sculpture: The Baschet Experience – Shapes, Sounds and People, 1945-1965*, an unpublished manuscript by François Baschet (p.27). Since this article was originally written, two published versions of material from this manuscript have appeared: *The Sound Sculptures of Bernard and François Baschet*, edited by Martí Ruiz and published by Universitat de Barcelona Edicions in 2017, and *Les Sculptures Sonores: The Sound Sculptures of Bernard and François Baschet* published by Soundworld in 1999.

2 This scheme is laid out in the *Sound Sculpture* manuscript (see the previous note) and also appears in the article "Sound Sculpture: Art, Music, Education and Recreation," by Bernard and François Baschet, which appeared in *Leonardo*, Volume 20 #2, 1987. Most of the description given here is lifted directly from these sources.

Related articles

The most important source for information on the work of the Baschets is François' own writing. See endnote 1 for bibliographical details.

Other articles and letters pertaining to the Baschet brothers which appeared in *Experimental Musical Instruments* are listed below. These articles can be accessed through the Internet Archive at https://archive.org/details/emi_archive/.

"More Baschet Sounds: A Mostly Pictorial Presentation of Architectural works, Museum Installations and Educational Instruments Built by the Baschet Brothers"by Bart Hopkin and François Baschet, appeared in *Experimental Musical Instruments* Vol. 4 #1, pages 13-16

Notes and comments from François Baschet appeared in the letters section of the following issues of *Experimental Musical Instruments*: Vol 3 #1, Vol 3 #6, Vol 5 #5, Vol 6 #1, Vol 6 #3, Vol 7 #4, Vol 7 #5, Vol 9 #1, Vol 10 #1, and Vol 10 #2

11

Tata and his Kamakshi Veena

David Courtney

This article originally appeared in *Experimental Musical Instruments* Volume 3 #4, December 1987.

Introduction

Experimentation in the construction of new musical instruments is a fundamental aspect of human culture. Wherever there are humans, there will be those who strive to create new music with new musical instruments. Previous articles in this journal have tended to concentrate on western experimentation, with particular emphasis on work presently underway on the west coast. It will be very interesting to show the work of one maker who is far removed both culturally as well as geographically from the west coast. This article will deal with the work of an East Indian who is known as *Tata*, and his *Kamakshi Veena* (Figure 11.1).

Preface

The subject of experimentation in the evolution of musical instruments is complex. Concentration on the mere technical aspects of new instruments is certainly interesting but incomplete. It is incomplete because the development of new instruments should be considered along with the cultural environment in which they develop. The cultural environment is important because 1) it subtly lays down a general direction for experimentation and 2) it determines whether the new instrument will be amalgamated into the total musical/cultural environment.

The western approach is largely Aquarian in its outlook. The artists and craftsmen who fashion new musical instruments tend to do so from a sincere desire to transcend their present musical environment. The instruments themselves tend to be a social statement.

The Indian environment for experimentation is largely economic in its motivation. We see (especially at the village level) that musicians tend to make their own musical instruments because the commercially available ones are cost-prohibitive. Most Indian musical instruments cost several hundred rupees (approximately twenty

Figure 11.1 Kamakshi Vina & bow.

dollars) and extend into the thousands (a few hundred dollars). This to us may seem very inexpensive but to an Indian villager it might represent several months' earnings. Making one's own instrument therefore becomes a matter of necessity. The availability of materials then enters the picture as a significant factor in the development of instruments. In general, the villager experimenting with new musical instruments is not making a "statement" but merely making variations upon established themes to be readily absorbed into the larger fabric of Indian culture.

The diversity of Indian culture has not been surpassed by any country in the world. It has over a dozen languages with over 800 dialects. The religions are mixtures of Hindu, Islam, Sikh, as well as Christian, Buddhist, and variations on tribal animism too numerous to mention. The economic disparities are equally varied. Many of the industrial families have combined wealth that is comparable to the Rockefellers. At the same time, there are villagers whose annual earnings may be less than $20. The musical environment is composed of two major classical schools and hundreds of regional styles.

It is this diverse musical heritage that interests us greatly. It would be impossible for us to fully explore the entire topic of the continuing evolution of Indian instruments, so we will concentrate on one man.

Tata and his Kamakshi Veena

It was about January 1980 when communal violence interrupted my studies in the Southern city of Hyderabad. It seemed a reasonable time to visit some of my in-laws, so my wife and I hastily made travel arrangements and departed for Akividu. Akividu (pronounced Ah-Ke-We-Do) is a very small town (hardly more than a village) whose only points of interest are its train station and its missionary hospital. It is located in coastal Andhara just a few hours by train from Vijaywada. The usual round of eat/sleep/gossip quickly got boring so I decided to see if anything musical was going on.

I had been attracted to a strange musical instrument that I had seen being played. I had studied Indian music all of my adult life and had spent years in India, but this was something new to me. It was a bowed instrument similar to a violin but had features in its construction which differed from anything I had seen before. So I decided to find out more about it.

I was informed that there was a wandering minstrel who made these and sold them at fairs and sometimes in the local bazaar. Nobody knew much about him. They just knew he was an elderly villager whom everyone affectionately called "Tata." *Tata* is the Telugu word for grandfather. The feelings of community are so strong in India that it is very common to hear people being referred to as Father, Mother, Sister, and so on, who are of no apparent relationship (Figure 11.2).

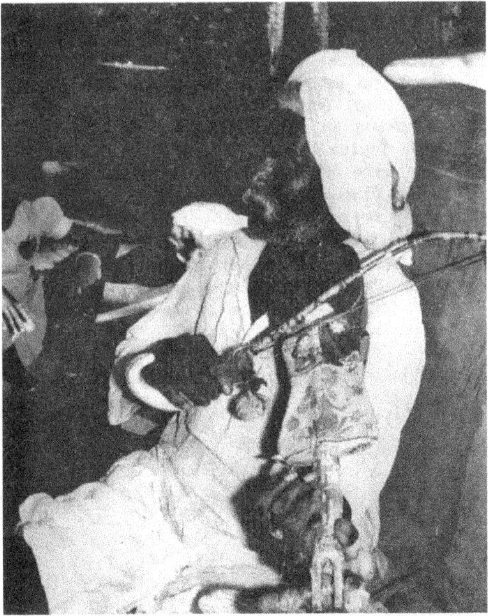

Figure 11.2 The maker plays.

Inquiries were made as to his whereabouts. It was an easy job to find out where he lived because Akividu is such a small place. A boy was fetched and sent to see if Tata could be persuaded to show me how to make his Kamakshi Veena. A short time later the boy returned to say that Tata did not want to come. I then resolved to go and see him.

The lane leading to Tata's dwelling was like so many I had seen in India, very narrow and unpaved, with vastly different levels of building construction existing together. One could see the so called *paka* house made of cement and stone and adjacent to it would be a *jhompadi* which might be nothing more than a grass mat covering some bamboo. Typical for all the houses would be a section on the ground in front of the door which is carefully coated with cow dung. On top of the dung are white patterns drawn with a mixture of chalk dust and rice flour. This *mugu* or *rangoli* is almost a morning ritual and is considered a symbol of the cleanliness and propriety of the householder.

It was a little difficult to find Tata's house. The main problem was communication. I was used to moving about freely in Hyderabad. It was a large city and its local language was a dialect of Urdu in which I had become comfortable. Akividu, on the other hand, was purely a Telugu area, a language of which I could understand only a little and speak none. This problem was solved through one of the inhabitants who spoke Urdu as his mother tongue. I was thus able to use him as my interpreter, though one can imagine the mental gyrations one must go through to first formulate ideas and mentally translate them into Urdu, and then have another person interpret them and translate them into Telugu.

We found the house where Tata lived. It was a low broad structure with mud walls and a thatched roof supported by bamboo and logs. Traces of whitewash still remained on the walls outside. Tata did not seem surprised that I had come but he did not want to see me and was visibly afraid of my presence. It seems that he had got the idea from the boy that I wanted to take him away with me to America, and he didn't want to go. His attitude changed when he found out that I was not going to take him from his home and family and that I just wanted him to show me how he made his instrument. I made it quite clear that I would pay him for his help. He agreed to come the next day to my house.

The next day he arrived and was very cheerful and proud that a *Dora* (white man) had taken interest in his work. Through the rest of that day I learned that he must have been in his late 60s (he himself did not really know how old he was). His real name was Nageshwar Rao, and he was married and had grown children. Telugu was the only language he spoke and his cast was *Adi Andhara* (scheduled class, what used to be called "untouchable").

He brought with him a bag, out of which he proceeded to pull his materials and tools. These materials included sticks, horse hair, membrane covered bowls, bamboo, resin, colored paper, and string. The tools were also very crude by the standards of western craftsmen.

He explained that he was going to show me how to make a Kamakshi Veena. The word *Kamakshi* is derived from a goddess of the same name and the connection is obviously in reference to the narrow waist which both the statue and the musical instrument have. The word *Veena* is a generic term which can be applied to any and all stringed instruments.

Kamakshi Veena is essentially a piece of bamboo of about two feet in length which is made of three sections (two nodes and the shafts at both ends). The bamboo is cut about four inches above one node and about ten inches below the other node. The ten-inch section is cut longways and carved to make a spike. This spike will pass through the resonator. The four-inch section has four holes burned into it for the two tuning pegs (see the diagram).

The heart of the instrument is a bowl or container which has skin or membrane stretched over its face. The method of attachment varies but in the case of this instrument, it appears that a membrane (peritoneum) was simply stretched over the face very soon after the animal was slaughtered. Although Tata did not show me this part, my previous work with tabla makers in Hyderabad leads me to believe that fresh peritoneum is so sticky that it does not have to be glued. The bowl has two holes in it. These holes are so that the bamboo neck can pass through.

The bamboo neck is then cut and holes are made by taking a hot iron rod and burning the holes in. (I can tell you from personal experience that holes which are burned are less likely to split at a later date than holes which are drilled.)

A strong strip of bamboo of about one centimeter in width and forty-five in length is then bound firmly at its midpoint to the base of the spike with string. It is then bent and lashed with more string in the middle. The other ends are bent and passed into holes on the side of the bamboo. The whole thing is now glued (Figures 11.3 and 11.4).

Cardboard pieces are now cut and placed in the open area. The entire center section is now built up with papier-mache. Thereafter, a very thin layer of colored paper is applied to the bamboo. The whole is now allowed to dry.

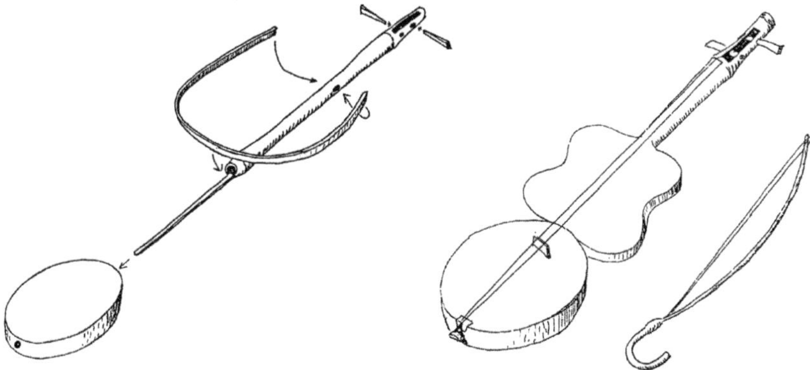

Figure 11.3 The structural elements on the left come together to create the basic form at right.

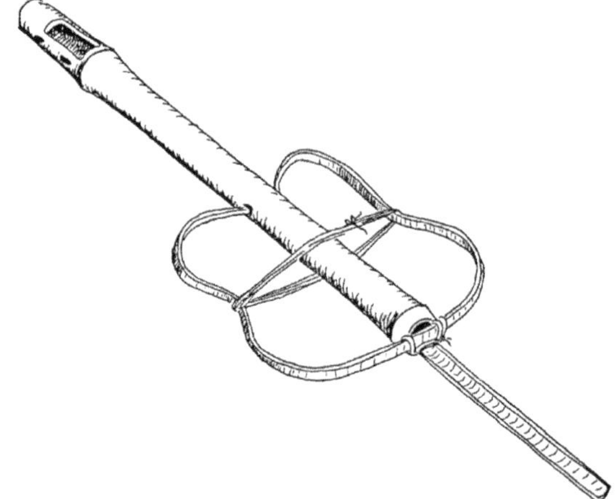

Figure 11.4 The finished bamboo framework.

The tuning pegs can be made while it is drying. A small section of large diameter bamboo is taken and whittled to the shape shown in the diagram. One might also mention that there is no effort to remove the outer "skin" of the bamboo. On a previous occasion while I was studying under a veena maker, I was told that the outer layer of bamboo, though very thin, is the strongest part of the plant.

We can also build the *Kaman* or bow while the glue is drying. The bow is of incredibly simple construction. Horse hair is simply fixed at both ends of a stick. There does not even appear to be any effort to alternate the hairs as is done with violin bows in the West. They are simply fixed as is. There is however one interesting embellishment. A curved article such as a stick or similar item is attached to the end of the bow which is held. I have seen such an embellishment on many bows for folk music around India. It appears that this functions to give the bow a more comfortable balance and feel as well as cosmetic value.

The final assembly is performed after the whole thing has dried. The bamboo/papier-mache assembly is inserted into the resonator. There is no need to glue or otherwise bind this assembly because the holes were cut so that it is already a tight fit (Figure 11.5).

The strings are attached in an incredibly simple manner. Both strings are of equal gauge and have loops made on one end. These loops should be large enough to pass around the end of the spike of bamboo protruding from the end of the resonator. They then pass along the upper surface of the resonator, over the neck and down into the open space where the tuning pegs are. The strings are simply tied to the pegs. This itself is interesting because Tata has no way of drilling holes small enough to allow a string to pass, since he does not own a drill. The strings are now tightened, but as they are tightened, a small piece of cardboard is jammed in between the string and the rim of the resonator to protect the skin. Otherwise the string would cut into the membrane at that point, creating a weak spot which would

Figure 11.5 Tata at work.

ultimately tear. Also, during this tightening process, a small square coin, or a piece of coconut shell or a piece of plastic, or just about anything, is inserted between the string and the membrane. This functions as the bridge (Figure 11.6).

The tightening of the strings proceeds until they reach the correct pitch. The tuning is not critical from the standpoint of key or interval. Since this instrument is intended to be used as an accompaniment to the voice, it will be tuned to match the key of the singer. The interval between the strings may vary, but Tata seemed

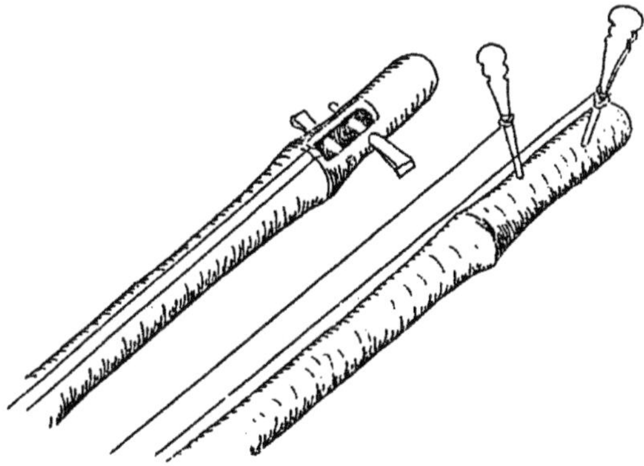

Figure 11.6 On the left in the drawing is the stringing arrangement on the Kamakshi Vina. Notice that the strings lie flat against the bamboo. On the right is the method used for most Indian tribal/folk instruments. Notice the strings "float" in the air.

to prefer a tuning which corresponded to the tonic for the innermost string and a minor third down for the second.

The Kamakshi Veena is now finished.

Commentary

The question of what this instrument is can be quite a question indeed. When Tata himself was asked about it, he indicated that what he was doing was making a violin. (Indeed a few people have attempted to call violin Kamakshi Veena, though this is admittedly quite rare.) Our first impression of such a statement is of course astonishment, but it is equally astonishing to Indians to find sitar being called "lute." It is a basic human process to take something new and strange and translate it into something familiar. This process of translating the unfamiliar into the familiar is exactly what Tata did with his instrument. He saw a violin, which is common in southern India but rare in his community, and fused elements of its construction with indigenous folk instruments along with some embellishments of his own.

The folk influences from which Tata drew as a base are readily apparent, especially in the membrane-covered resonator. This type of resonator is a very common element of Indian stringed instruments. The only western instrument of the class would be the banjo (of African origin). Apparently there are no such bowed instruments in the west. For anyone who has never heard such an instrument, it is impossible to describe the richness of the tone.

Another element which is definitely of Indian origin is the extensive use of bamboo. Westerners tend not to realize the desirable properties that bamboo has. In a nutshell, we could say that it is cheap, replenishable, surprisingly consistent (within each strain), and more importantly, has a strength/weight ratio which is unsurpassed by any of the woods western civilization uses.

Tata's Kamakshi Veena has certain elements of the violin in its construction which stand out from its Indian elements. The most important of these is the manner in which the strings attach at the tuning pegs. Most Indian folk instruments have just a single bridge at the resonator. This makes it necessary that the tuning pegs stand up vertically, serving to hold the strings up at the end opposite the bridge. The strings in a sense just float above the neck. The Kamakshi Veenaa, in contrast, shows some western influence (or perhaps influence from the more developed instruments like Saraswati Veena), in that the strings pass over a second very small bridge at the end of the fingerboard and down from there to attach to tuning pegs. These pegs, in contrast, have a horizontal orientation. Of course, this second bridge is nothing more than the node of the bamboo.

Another element is the shape of the papier-mache assembly. Tata himself declared that it was there for the sound, but by looking at its construction it obviously doesn't function as a resonator like we see in a violin. However, that is not to say that it doesn't influence the sound. Even a cursory glance at the structure of bamboo makes it evident that bamboo makes a very good natural Helmholz resonator. This resonator would tend to give a coloration to the sound of the instrument. The act of burning a hole into the side and padding with glue, string, and papiermache would certainly dampen these natural resonances. My own personal feeling, though, is that the influences of the bamboo as compared to other factors would be negligible anyway. Nevertheless, we should be open to the possibility that this dampening could be a significant influence.

There are a number of innovations in the Kamakshi Veena which are neither Western nor traditionally Indian in nature. These appear to have sprung spontaneously from Tata's own creativity. I feel that the most important innovation is the use of peritoneum for the resonator. Most Indian instruments that use membranes use the rawhide, usually goatskin, but on occasion I have seen buffalo, or even snake or lizard (on the *kanjira*, a southern Indian tambourine, for example). I have encountered the use of this type of internal membrane only once before, and that was on an obscure south Indian drum called *Uddaku*. The characteristics of the skin are very important in determining the final tonal quality of the instrument. The general thinness of the peritoneum, coupled with the its high tension when it is mounted on the instrument, would tend to yield a sound rich in overtones, especially compared to a resonator made of rawhide.

More evidence of Tata's creativity is seen in the fact that all of his instruments are different. Variation is especially evident in his choice of resonators. I have seen tin cans, serving bowls, and a variety of other containers utilized for his instruments. All of them were probably chosen with availability as an important consideration.

It is doubtful that Tata's contributions to Indian folk music will ever receive any more attention than this one article. It is most likely that his innovations will quietly be absorbed into the fabric of Andhra Pradesh's village culture as a natural part of the evolution of Indian instruments. But we can get a perspective on our own musical evolution by noticing the parallels between Tata's world and ours.

Related articles

Additional articles by David Courtney pertaining to Indian instruments and music which appeared in the *Experimental Musical Instruments* journal are listed below. These articles can be accessed through the Internet Archive at https://archive.org/details/emi_archive/.

"A Comparative Tuning Chart" by several authors, in EMI Vol. 6 #2, pages 11–17. This article is also included in the current collection

"The Tabla Puddi" by David Courtney in EMI Vol. 4 #4 pages 12–16

"Bridges: An Indian Perspective" by David R. Courtney, Ph.D, in EMI Vol 7 #5, pages 8–11

12

Bamboo is Sound Magic

Darrell DeVore (1939-2005)

This article originally appeared in *Experimental Musical Instruments* Volume 3 #4, December 1987.

Bamboo is simply the most significant plant on Earth. Its utilitarian purposes are unsurpassed by any living thing. A sacred plant that has always influenced world culture, bamboo has special meaning to the world of music and instrument making. It is a prime source of Sound-Magic (the essential musical voice of any sounding material), giving the world more music and musical possibilities than any other natural material.

Bamboo was playing beautiful music 100 million years before there were humans around to hear it. All this accumulated ancestral Sound-Magic still grows in bamboo stands everywhere, waiting to be heard, to be discovered and shaped into instruments for realizing music.

The pure sound of bamboo goes directly to the center of human spiritual consciousness as a life affirming signal: distinct, familiar, recognizable, universal.

Bamboo Experience

Eighteen years ago, bamboo planted its roots deep in my consciousness and gradually grew until it changed my life. It became a major voice in a continuing dialogue I've had with music throughout my life. The pursuit of music led me to bamboo. Bamboo led me to instrument making. Instrument making led me to new music. Shaping bamboo became a way to survive economically and spiritually.

I became a "fluteman," making, playing, and selling bamboo flutes on the streets, on college campuses, in parks, stores, wherever. In a period of a dozen years I made and distributed more than 5,000 flutes to thousands of people. I learned to make several different types: open-end (Ney), slit-end (Shakuhachi), notched-end (Quena), noseflutes, water flutes, membrano flutes, multi-flutes (for multi-players), panpipes, whistles, self-standing, self-playing wind flutes, as well as other aerophones such as trumpets, didjeridus, and saxoboos (bamboo with fingerholes and a saxophone mouthpiece).

Percussion developed parallel to flutemaking. Bootoos (described later in this article), boobams, a variety of hand drums, shakers, rasping sticks, batutus (tongue idiophones), water percussion, hanging marimbas and windchimes, bull roarers and new outer-air inventions (O-trads, windwands) of bamboo and rubber bands.

New chordophones evolved from ancient musical bows in the form of "Spirit Catchers" (bamboo bows strung with sacred microtonal sound objects and mounted on walls or resonating boxes).

I eventually utilized elements from all these forms to structure environmental sound sculptures for listening and participating, for dance and ritual.

Dealing with that much bamboo… sawing, cutting, splitting, shaving, filing, sanding, burning thousands upon thousands of sound holes, rubbing, polishing, binding… the repetition of work becomes a ritual. One tries to perfect motions with tools in order to shape bamboo efficiently without harming the magic of the material.

Working with bamboo put me in touch with shaping spirit – the same shaping spirit that helped our ancestors form musical instruments thousands of years ago.

The primitive instrument maker engendered the instrument he made with life, giving it great significance. Such an instrument was approached with reverence because it was capable of music that was necessary to further life.

Bamboo has led me to hear sound phenomena that are new and fresh, beyond human imagination. It is in this bamboo sound phenomena that new music forms will grow and flourish, continuing a tradition of Sound-Magic well into the future.

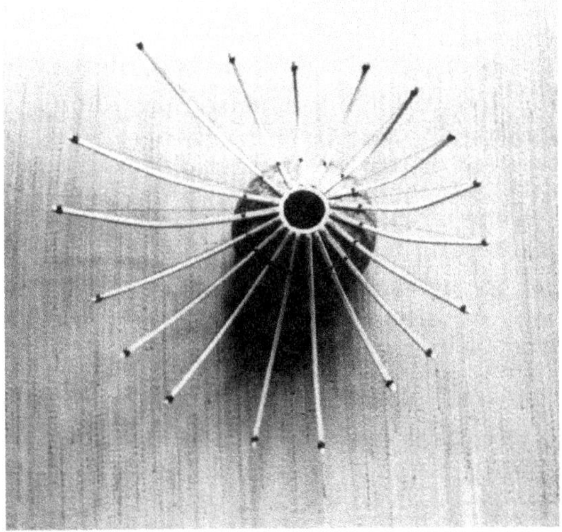

Figure 12.1 Circular chordophone. Bamboo bows, nylon strings mounted on a gourd resonator. The strings are beaten with sticks.

Figure 12.2 Self-playing wind flutes.

Bootoo (Stamped Ideaphones)

Bootoo (Two Bamboos) are the perfect example of the power of Sound-Magic in bamboo. These simple-looking primitive objects are capable of wide-ranging musical ends, from simple native percussion music to advanced psycho-acoustic sound phenomena.

Bootoo are two bamboo tubes, open at one end and closed the bottom by the node. Two finger holes are burned or drilled in each tube to allow tuning of pitches. (I use only two holes because of the difficulty of gripping the tubes for striking while at the same time independently fingering the tuning holes. This is trickier than it looks.) A row of notches is burned or filed perpendicular to the length and centered directly above the holes to facilitate rasping. The tube ends are rounded by filing and bound by waxed nylon to prevent cracking.

Tuning can be random or in harmonic relationship to the fundamental tone of the tube. It's fairly easy to tune simple triadic scales such as: C, Eb, F, in one tube, and F, Ab, Bb in the other. Disregarding tuning systems, the two finger holes placed anywhere within a practical zone of balance and finger comfort will produce magical sounds that are musically related. The voice of bamboo sounds good singing any scale.

Bootoos are used in a variety of ways as idiophones (percussion), aerophones (flutes and singing tubes), earphones (listening tubes), and telephones (for long distance calls to ancestors) (Figure 12.3).

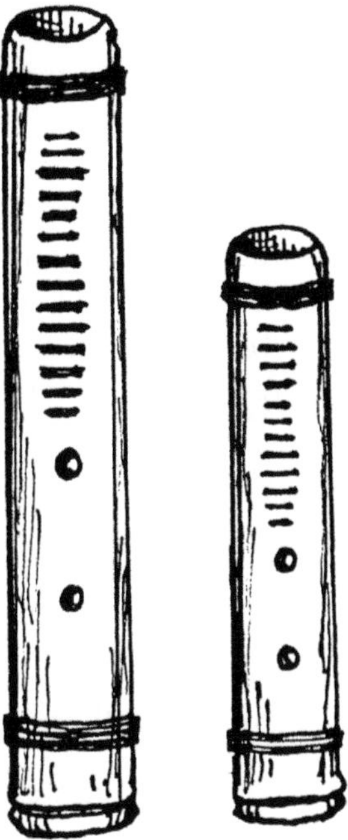

Figure 12.3 Bootoo.

Bootoo Percussion

Bootoo are played holding the tubes vertically, open end up, while gripping firmly (not tight) with fingers in position to stop the tone holes. With a downward motion, strike sharply a point on the bottom edge below the closed bottom node against any dense surface (hard ground, concrete, rocks, bricks, tree trunks, knee bones). This will bring out the pure sound of bamboo. Striking both tubes rhythmically while at the same time closing and opening the finger holes produces a wonderful music of swooping microtonal Sound-Magic. Bootoo can also be played singly by holding a tube in one hand while striking the closed end edge with a mallet.

Interesting sustained tones can be achieved by slowing drawing the bottoms of the tubes along a rough, hard surface (concrete, asphalt) in a continuous motion.

A multiphonic rasping sound can be made by holding one tube rigid while rubbing the bottom edge of the other tube against the notched grooves of the rigidly held tube in rhythmic patterns.

Bootoo Flutes

Bootoo become aerophones when played like a flute. A tube is held horizontally to the mouth with one hand (first finger stopping the hole nearest the open end), while sound is produced by blowing across the open tone hole nearest to the closed end (as in any transverse flute). Working the finger hole produces two distinct tones. Place the palm of the free hand against the open end of the tube and move the palm back and forth while sounding the flute tone. This action creates a wide variable pitch-range of the sound coming from the open tube end. Two or more Bootoo sounded together in this manner create magical new music (Figure 12.4).

Singing Tubes

Place the open end of Bootoo so it completely covers your mouth and sing into it while manipulating the finger holes for some very unusual vocal effects (Figure 12.5).

Bootoo as Listening Tubes (Bamboo Walkman)

One of the most delightful and surprising sound experiences occurred when I first put the open ends of a pair of Bootoo snugly to my ears and rhythmically closed and opened the two finger holes of each while listening to external sound sources. My ears were amazed. I was hearing other-worldly music bouncing and echoing off the walls of the inner chamber of bamboo. Each time I lifted a finger, opening the hole, all perceivable external sounds rushed into the chamber, along with a complex of wind voices stacked upon the tonal frequency of the bamboo interior fundamental. Precise coordination of left- and right-handed finger stopping resulted in mind-boggling stereo psycho-acoustical effects. Instant new music.

It is an especially powerful experience to listen to and play these tubes with live acoustic instruments and live singing voices, including your own. Recorded music,

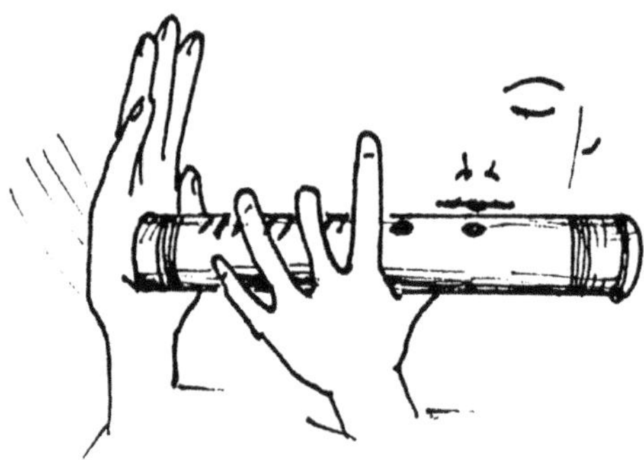

Figure 12.4 Bootoo Flute.

Figure 12.5 Bootoo as singing tube.

Figure 12.6 Bootoo as listening tubes.

industrial noise, birdsongs, any sounds are transformed by this process into an infra new music form that only you can hear and that cannot be reproduced. Bootoo may be the first bamboo personal stereo system (Figure 12.6).

I have enjoyed much music with these instruments over the years. I came upon this design about 15 years ago as an evolutionary offshoot of my bamboo flute-making. I have composed and performed with tuned sets of Bootoo that resulted in music probably very similar to the Haitian Ganbo music (described in *EMI* Vol. III #2) and other bamboo music of the world.

The beauty of Bootoo is in their simplicity and versatility. They can be made easily with few tools from any size available bamboo.

Instruments like Bootoo embody the essence of a Sound-Magic that comes from ancient times to enlighten and instruct the present. A primitive tradition of new music is passed on to the future, as contemporary musicians and instrument makers

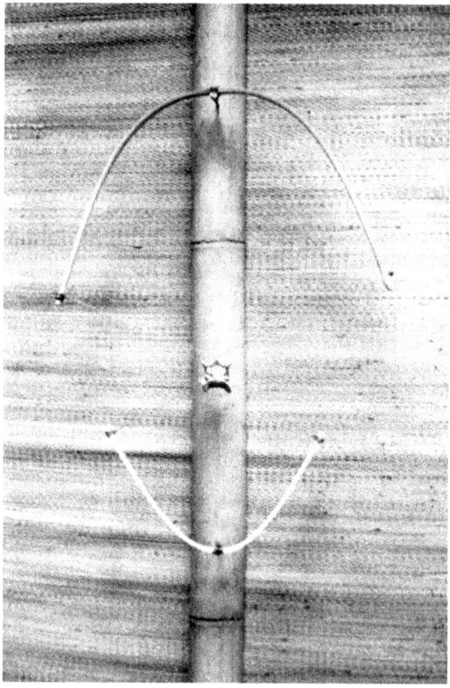

Figure 12.7 Bambow, a spirit catcher of bamboo, nylon, and bone.

Figure 12.8 Membranoflutes, made of bamboo and goatskin.

continue to explore the many sounds of bamboo. There is a belief among some primitive people that spirits live in the inner chambers of bamboo. Listening to Bootoo, that is clearly true.

Related articles

Other articles which appeared in *Experimental Musical Instruments* by or about Darrell DeVore, and articles about bamboo as an instrument-making resource, are listed below. These articles can be accessed through the Internet Archive at https://archive.org/details/emi_archive/.

"Spirit Catchers and Windwands (Music in Circular Motions)" by Darrell DeVore in EMI Vol. 5 #4

"The Mallet Kalimba" by Robert Rich in EMI Vol. 2 #3

"Percussion Aerophones" by Bart Hopkin in EMI Vol. 6 #3

"Bamboo: The Giant Musical Grass", parts 1-3, by Richard Waters, in EMI Vol. 10 #3, Vol. 10 #4, and Vol. 11 #1

"Bamboo and Music", parts 1 &2, by Richard Waters, in EMI Vol. 14 #2 and Vol. 14 #3

13

Dachsophon

Hans Reichel (1949-2011)

This article originally appeared in *Experimental Musical Instruments* Volume 4 #3, October 1988.

In the following I'll try to talk about a quite complex, though rather unpretentious looking musical instrument which I made in 1986. I call it *Dachsophon* (*Dachs* is the German word for badger) and have played it in concerts of improvised music since then, alongside my "usual" guitar work. The sounds it emits have been described as "eerily vocal" – in fact, it's not exaggerated to say that you can make this instrument talk, sing, cry, grunt, whistle, grumble, shriek, yawn, yodel, and whatever – but very often these noises are colored by a strange, sick , unearthly expression. It can sound pretty melodic too, sometimes not unlike a flute or other wind instruments. It can remind you of soundtracks for weird, unheard of animation films, or of "tormented animals in a queer veterinarian hospital" as a reviewer wrote somewhere. Another one fancied to hear some "pissed off cats," as well as the croak of a bullfrog. Be that as it may, when you see the sound source of all this, you might think it's a joke.

It is nothing but a flat stick of wood (a ruler, for example) clamped to an edge of a table. If you take a bow, preferably a strong double bass bow, it's not big news that you can squeeze some more or less interesting noises out of this arrangement, and even alter the pitch of the sounds, just by pressing your fingers or something else on various points of the stick while bowing it. This way of doing things is not very efficient, but not unpromising either. To be brief, one fine day I took the curved block of wood which you can see in the photo below. Originally I had made this object to fool around with on guitar strings – but when I started pressing it here and there on the stick clamped to the table, moving it like a seesaw while operating the bow with the other hand, I was quite stunned by the clarity and flexibility of the notes coming out of this dull stick. It seemed to be suddenly animated by that block, which was obviously functioning like a kind of "oppressive" mobile fingerboard.

As you see, both sides of the dax (that's how I simply call the wood block now) are slightly curved. One of them is plain, for playing slide notes; the other one has guitar frets which enable you to play scale-like passages or distinct melodies, so to speak.

The position of the frets results from a logarithmic succession chosen at random, and the hole in the middle of the block was made for better handiness. Moving the dax near the clamp will produce very low notes, because the major part of the stick can vibrate freely. Moving the dax closer to the far end or closer to the point of bowing will evoke higher notes, for obvious reasons – actually you can get sounds going up to the limit of the hearing range. However, you can't generalize this behavior of the stick, as you can do with that of a vibrating string: there are always a few points where a scale is interrupted, giving way to a lower or higher tone. It depends on the bowing technique too – needless to say, doing it fiercely or tenderly will make the ruler react in very different ways (Figures 13.1 and 13.2).

Of course, the sound outcome of a stick depends on its material, length, thickness, shape, and position on the edge. If you make yourself a little collection of those pieces, each one with more or less different features, you can enlarge the range and the variety of sound colors considerably.

As for the material, anything which is sufficiently rigid will do unless it is too thick to vibrate audibly. It works very well with metal, glass, etc., but wood turned out to be the most versatile stuff, as usual. Appropriate thickness is about ¼ inch, length about 11-13 inches, but there is much room for variation (Figure 13.3).

The shape matters a lot: it is useful to make pieces with more complex outlines, because it makes a difference at which point on the stick the bow is "busy" (if you

Figure 13.1 The Dax (all photos by Hans Reichel).

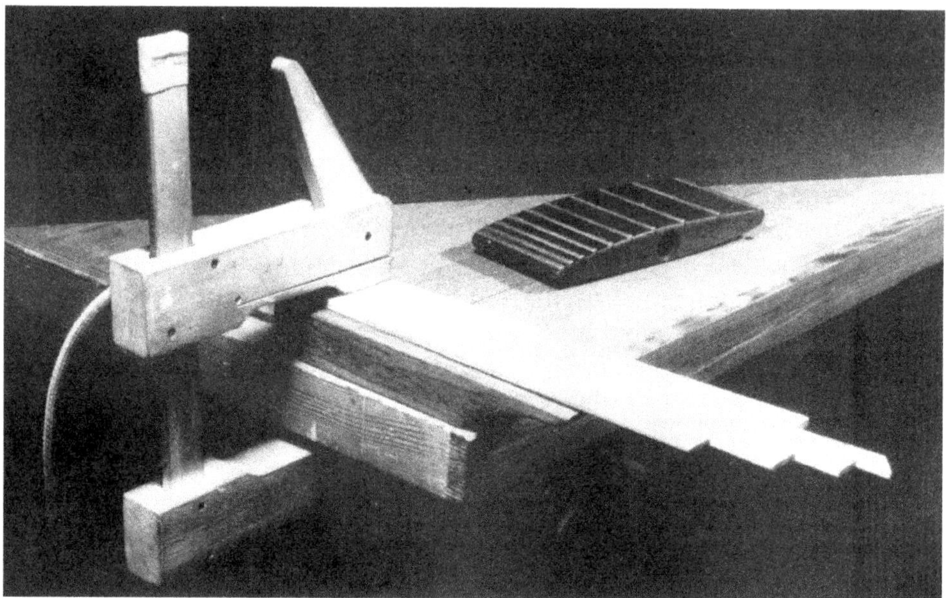

Figure 13.2 The Dachsophone, ready to play.

stroke the tip of the tongue, you will get other sound colors and scales than if you do it at the side edge, or in a corner, for example). You can make pieces with sharp or round edges, with wings, with holes or rifts, or with a thick frame and a thin "soundboard" – these are actually louder than the other ones.

Talking about loudness: extreme amplification of all this doesn't pose a problem, because there can't be any disturbing feedback. Just attach one or two contact mics to a separate small but solid sound box, and clamp this together with the stick, as shown in the drawing. Thus, you can easily match even with the loudest tenor sax player without losing one hair of your bow. On the other hand, playing the Dachsophon on stage doesn't seem to look very intelligent – people can hardly see what you are doing over there, unless they are seated right in front of the thing. Some have even thought I was trying to make a fire, because of the clouds of rosin emitted by the bow. Anyway, it's a delightful and charming contraption to play with, and it goes very well together with string and wind instruments. What I especially like about it: it's sometimes unpredictable, and always good for a surprise (Figures 13.4 and 13.5).

Now and then people have asked me why I don't make a more sophisticated device – a turntable, for example, to the rim of which a large number of sticks is fastened. Maybe the clamp made it look to them like something still in the experimental stage, not yet ready to be introduced as a finished instrument. In fact, one could imagine some more elaborate constructions based on this principle, maybe

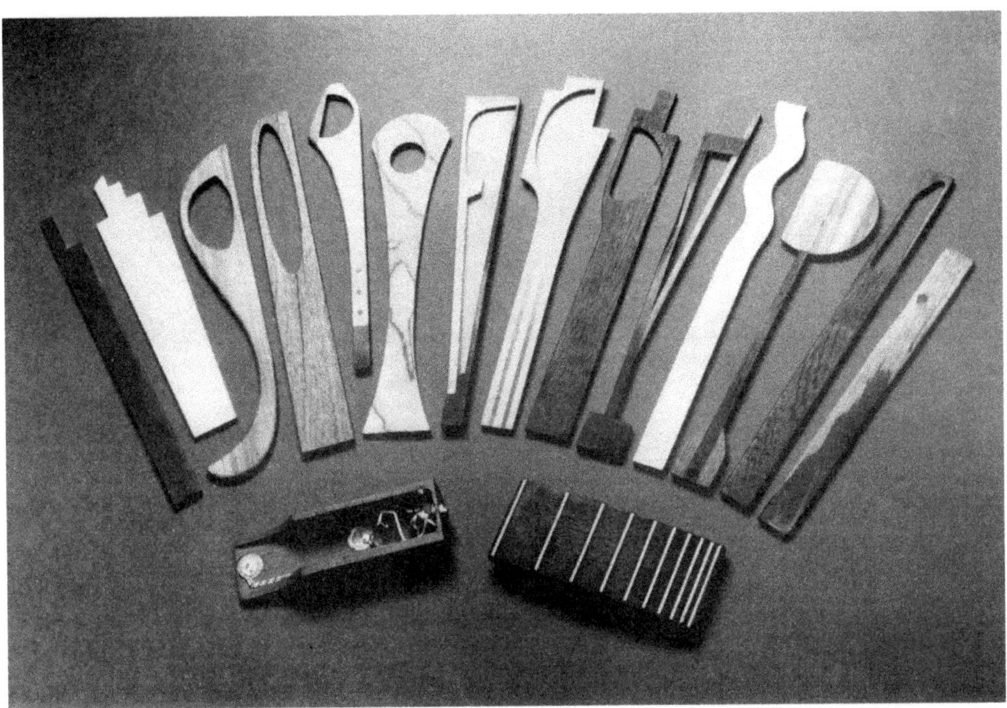

Figure 13.3 Fourteen sticks, each of which makes a Dachsophon. They are all made of wood: ebony, spruce, Brazil pine, mahogany, cedar, plywood, maple, rosewood, sandalwood, persimmon-wood, and African wenge (partly in combinations). I shouldn't say they all sound completely different – but each has a few "personal" features not obtainable in the other ones. For example, the ebony stick on the left has a quite mellow, mild personality, good for playing melodies in the bass- and mid-range. The light piece of spruce next to it is decidedly nasty, and you can imitate a lot of poultry with it (thanks to the "teeth" at the tip it has a very wide range of sound colors). When you play it at a certain point, you will be surprised to hear a quite untalented trumpet student in his first lesson, driving the neighbors mad. The sticks with the holes are very good for "singing" when bowed at the thin edges (the dax can easily produce a nice tremolo on its plain side). In a certain range, the triangle-shaped piece in the center can sound very much like a clarinet, the way it is played in the folk music of southeast Europe (but only when you use the fretted side of the dax very fast). The dark stick of wenge (second on the right) is one of the craziest in this collection: you can make a police siren with it, as well as a hysterical Mickey Mouse, and a lot of other strange things which it would be going too far to describe here. I know that all this will sound hard to believe anyway… what all these sticks have in common, more or less, is a certain quality of a ventriloquist – both funny and uncanny, in a way.

even in combination with strings. On the other hand, as the Dachsophon is now, it can easily be packed up in a few moments, and be carried away in a small plastic bag – that's not bad either. And it doesn't need to be a table that it is fixed to. You can clamp it to the frame of a grand piano, or a dog-kennel; whatever the case may be, something will happen... For example, you can create a fine natural reverb by

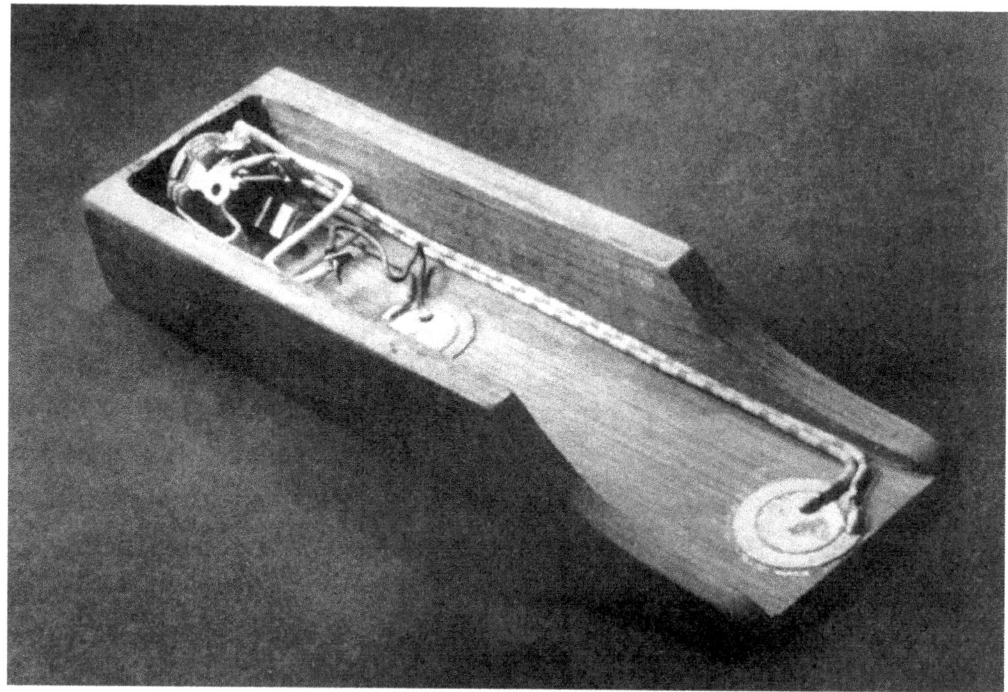

Figure 13.4 The sound box (with two contact microphones) shown upside down.

fastening a stick on the body of an acoustic steel-string guitar which has a contact microphone close to the bridge. That mic will pick up the sound of the stick directly (through the wood of the guitar), as well as the sounds of the strings vibrating sympathetically with those tones emitted by the stick which match the fundamentals and harmonics inherent in each string. This is a well-known thing, of course – but, since the Dachsophon is an extremely "fast" instrument, it is able to produce a big amount of tones and overtones covering almost the whole audible range within a few seconds, if desired – so the strings can't help but keep on ringing all the time. The result (as sensed by the contact mic) is an impressively dense, sustained, reverberated soundscape.

I think, to introduce "the sticks," that's it for the moment. In case somebody wants to try it out, the only problem would probably be to get the fretted block. But you don't have to be a carpenter to make it, and as for the frets, somebody in a guitar repair shop could certainly help.

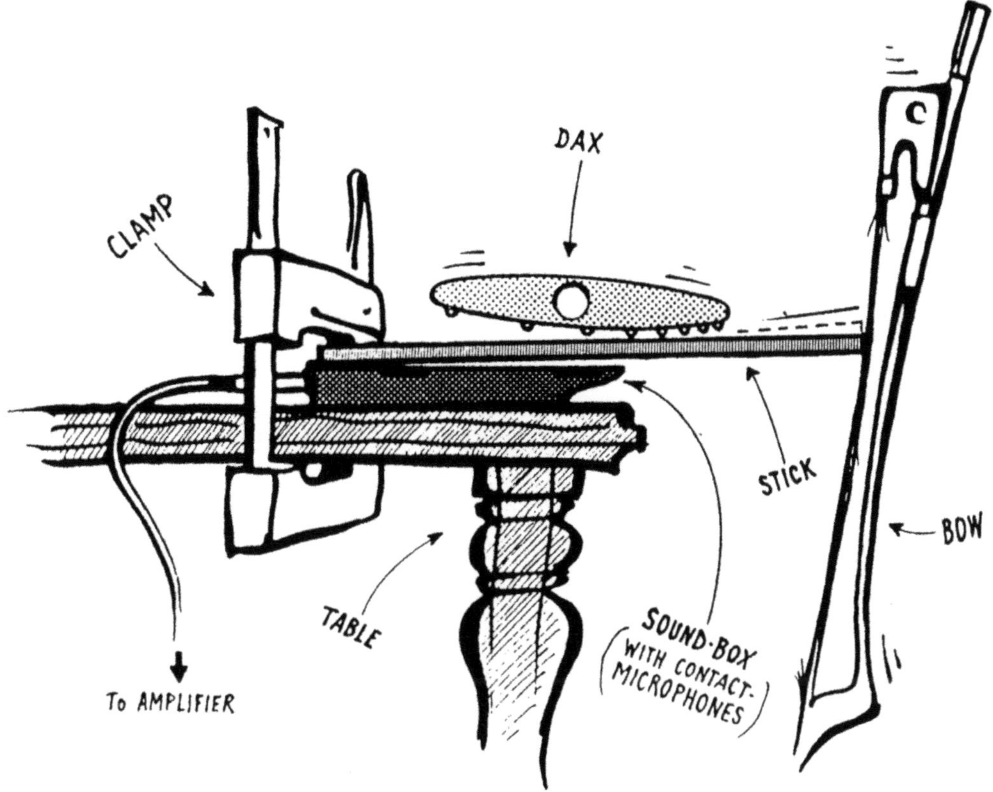

Figure 13.5 Dachsophon clamped over the sound box.

Related articles

Another of Hans Reichel's instruments is described in "Hans Reichel's Pick-Behind-the-Bridge Harmonic Guitar" by Bart Hopkin in *Experimental Musical Instruments* Vol. 5 #1. This article can be accessed through the Internet Archive at https://archive.org/details/emi_archive/.

14

The Pikasso Guitar

Linda Manzer

This article first appeared in *Experimental Musical Instruments* Volume 4 #6, April 1989.

I became interested in making musical instruments in 1969 when I searched for a dulcimer and found a "kit" for half the price of an already built one. The sheer pleasure I felt in listening to an instrument that I had assembled with my own hands is still a vivid memory. The bug had bit, and little did I know that, wiggle as I might, all paths in my future would lead right back to instrument making.

In 1974, I began an apprenticeship with Canadian guitar maker Jean-Claude Larrivee. I stayed with Larrivee for the next 3½ years. I once calculated during the years I was there we made 1500 steel string and classical guitars. Jean, as well as being an excellent teacher, had a knack for hiring the nicest, most dedicated, and skilled employees. This made for a very supportive atmosphere, conducive to learning and also a lot of fun.

In 1978, I set up a shop with lute maker Michael Schriener in Toronto above a pool hall. I had little money and no tools. This is where the school of hard knocks kicked in and I started struggling. I skipped many meals and recall doing nothing but guitar making for the next few years. In 1982, I met Pat Metheny after one of his concerts here in Toronto, and he ordered a six-string guitar. Over the next five years, he ordered a series of diverse and exotic instruments from me. With a few exceptions, I had never before built, or in some cases even envisioned, these instruments, so I was required to research or design them myself. Pat would tell me what he wanted and I would do my best.

The Pikasso Guitar began when Pat Metheny asked me, with a smile, how many strings could I put on a guitar? Our resulting collaboration yielded the instrument shown here. It has 42 strings in three groups of 12 strings and one rather normal neck with 6 strings. The 6-string has a hexaphonic pickup to trigger his Synclavier as well as a Fishman piezo under its saddle. The other three sets of strings also have piezo pick-ups. The second neck sounds somewhat like a koto (usually tuned chromatically); the third neck sounds slightly like an autoharp due to its duplication of strings; the fourth neck has an extremely wide range and is used primarily for accent or color. There are also two brass fittings inside the body so it can be mounted on a stand thereby leaving the hands free when performing. There are two access doors so that adjustments can be made to the interior electronics without loosening any strings. Each of the pickups runs through a custom-designed preamp en route to the master onboard control panel (designed by Mark Herbert, Boston, MA). The "Manzer Wedge" concept was invented for the Pikasso guitar and was first used on this instrument. The lower bass side is thinner under the arm and the treble side is fatter on the knee side therefore leaning the top back and enabling the player a better aerial view of the strings. The added benefit of this feature is that the guitar becomes more ergonomic and comfortable for the player. This instrument can be heard on the album *Song X* by Pat Metheny in the song "Mob Job" (Figure 14.1).

As I worked on this series of instruments over the course of a few years, word was circulating that I made "unusual instruments." I received a call from Bruce Cockburn asking me to help him out with a problem. He was going on a world tour in a few months. He had a charango from Nicaragua that he loved, but it was very delicate and he feared it would not stand up to the rigors of the road. Together we came up with the idea to build an electric semi-solid body charango

Figure 14.1 The Pikasso Guitar.

Figure 14.2 Eight-string Drone Guitar made by Linda Manzer for Pat Metheny, with a sitar-style buzzing bridge. The individual bone saddles under each string can be adjusted to create different qualities of buzz.

with a custom-made Fishman pickup. It had hollow chambers inside to keep the warmth of an acoustic instrument, but it was stable because it was largely solid and it was not made from an armadillo shell. I colored it blue to match his other instruments – why not, eh?

In 1988, I was commissioned to build a 20-stringed instrument for Ángel Parra. He had had such an instrument built in Chile by a carpenter and it was in pretty rough shape. He told me that this was a Chilean instrument, but I have been unable to find any information about one anywhere. I suspect that the instrument I used as a reference was one of a kind. I presented it to him before a concert in Toronto and to my surprise and delight he played it during his concert that night.

Most recently I was commissioned to build an arch-top 6-string bass guitar. It had an 86 cm scale length with German spruce top and curly maple back and sides, with a solid core down the center of the body. This had a twofold purpose: First to stabilize the top from the tremendous down pressure of the strings, and secondly to prevent feedback problems, as the instruments was to be amplified in a concert situation. The result was an acoustic bass sound in a smaller more manageable size.

I find it impossible to begin working on an instrument until I have a clear picture in my mind of what I want. Otherwise it's like driving in downtown Boston without a roadmap. You don't get where you are going and you use up a lot of gas, but hey, maybe you get to see a lot more of Boston that way! I guess it is all how you look at it.

Related articles

Linda Manzer's article appeared as part of a collection of articles on unusual instruments from contemporary luthiers under the title "Shape and Form, Contemporary Strings," Parts 1 and 2, appearing in *Experimental Musical Instruments* volume 4 #5 and 4 #6. Other featured makers included Fred Carlson, Susan Norris, Francis Kosheleff, Clif Wayland, Steve Klein, and William Eaton. These articles can be accessed through the Internet Archive at https://archive.org/details/emi_archive/.

15

The Piatarbajo

Its History and Development

Hal Rammel

This article appeared in *Experimental Musical Instruments* Volume 5 #2, August 1989.

Among ensembles of invented and reinvented musical instruments, none may perhaps be more unique than the orchestras played and led by a single musician. The history of the one-man band is a study in uniqueness, and among the most unusual is the *piatarbajo*, designed and built by Joe Barrick. At the hands (and feet) of Joe Barrick, the piatarbajo performs as a five-piece band, providing all the rhythmic and harmonic accompaniment needed for Barrick's lead guitar, mandolin, fiddle, harmonica, or vocal. But, while the piatarbajo was built some 15 years ago, the idea, as Joe Barrick expresses it, "to do it all,"[1] has been around much longer and has provoked a wide range of responses and solutions. Borne of this impulse to play it all, alone, have been an extraordinary variety of musical instruments performed upon by an equally diverse, inventive, and determined array of musicians.

The one-man band exists, in all its uniqueness and independence, as a most elusive yet persistent musical tradition. Defined simply as a single musician playing more than one instrument at the same time, this ensemble is limited only by the mechanical capabilities and imaginative inventiveness of its creator. Despite its generally accepted status as an isolated novelty, it is a phenomenon with some identifiable historic continuity.

The historical survey that follows confines itself to one area of one-man band invention: instruments that place the musician sitting before a contraption facilitating ensemble performance by both hands and feel. Excluded here is the most familiar form of one-man band, the guitar, harmonica, vocal ensemble, and its many variations (e.g., Henry "Ragtime Texas" Thomas' guitar and quills, or Stovepipe No. l's guitar and stovepipe, played jug fashion).[2] Also omitted are the far older combinations of pipe and tabor (from medieval jongleur to rural English "whittle and dubb," and their transplanted forms in the native music of South America and the Caribbean). A third category falling outside the present discussion are the immensely various permutations of washboard and miscellaneous homemade percussion.[3]

Sitting or standing, the feet have always been an important resource for expansion of the one-man ensemble sound. Whether providing foot-stomping time-keeping (e.g., Chicago's Maxwell Street regular Daddy Stovepipe)[4] or more delicate timbral accompaniment as in the foot-played triangle of North Carolina's Will Blankenship in the 1930s (added to harmonica and autoharp), refining the role of the feet was often the impetus to further mechanical development. Percussion historian James Blades suggests that the foot-pedal operated pair of cymbals may have had its inspiration in the one-man band's use of a drum or cymbal operated by a cord attached to the performer's foot. While a foot-pedal bass drum and cymbal became a significant addition to the trap-drummer's kit in the early part of this century, it also played an exceptional role in the expanding capabilities of the one-man band.

Undoubtedly the most well-known and well-recorded one-man band is that of Jesse Fuller, whose invention of the foot-operated bass greatly expanded the sound of his music. Fuller called his invention the *fotdella* (a combination of the words "foot" with "killer-diller"). It was an upright string bass with six piano strings hit by individual hammers attached to six pedals he played with the big toe of his right foot. Because it was limited to only six bass notes, he sometimes played tunes without the fotdella. The sock cymbal he played with his left foot was also his own construction (except for the topmost cymbal), as was his neck harness (holding harmonica, kazoo, and microphone plugged into the amplifier of his 12-string guitar), and the stool he sat on to play. With his vast repertoire of blues, spirituals, rags, and pop tunes (his "San Francisco Bay Blues" became a folk standard), Fuller is probably the only one-man band so fully recorded. The high-spirited, well-integrated playing of his various instruments makes his accomplishment well worth the attention and praise.

Like the Musiker in L. Frank Baum's *The Road to Oz* (1909), whose lungs were full of reeds and who had only to breathe to squeeze out his music, the key to elaboration of the one-man band lies in making the instrument an extension of the musician's body. Exemplary of such imaginative mechanical experimentation is the work of Fate Norris. Norris is best known for playing banjo in the hillbilly string band The Skillet Lickers, although poorly recorded and hardly audible in any of their many recordings. Born somewhere in northern Georgia in the 1890s, Norris played with the Skillet Lickers throughout their recording career from 1924 to 1931. He apparently had experience in the medicine show circuit and had a flare for comedy that included a "talking doll" and a complex one-man band of six instruments that he took to fairs and fiddler's contests throughout the south. An attendee at the Skillet Lickers' appearance in Nashville in 1927 reports:

> Fate Norris, of Dalton, Georgia, the one-man wonder, who plays six individual instruments in an individual band, will also furnish entertainment. Mr.

> Norris has in his band two guitars, bells, bass fiddle, fiddle, and mouth harp. He devoted seventeen years to the mastery of his art.[5]

Fiddler Bill Helms recalls seeing Norris at a fiddler's convention in Chattanooga:

> Fate Norris was there too, had a musical soap-box – made out of soap boxes with a pocket knife, and strings from mandolins, guitars, fiddles, autoharps. Had pedals and knee pads. Played two instruments with his feet, played a mouth harp.[6]

There are no recordings of Fate Norris' one-man band, only these descriptions and a few photographs. Another newspaper account of Norris' appearance in Nashville adds, "It required fifteen years, Mr. Norris says, for him to perfect his performance."[7] What took place over those 15 years? How did such a band develop? No such information about Fate Norris exists, but there is a description preserved about another one-man band of similar complexity and construction heard and seen almost one hundred years earlier in London. Henry Mayhew's Survey of London Street life in the 1840s and 50s, *London Labor and the London Poor (A Cyclopedia of the Condition and Earnings of Those that Will Work, Those that Cannot Work, and Those that will not Work)*, includes this testimony from a street musician known only as "blind performer on the bells" (Figure 15.1).

> I started the bells that I play now, as near as I can recollect, some ten years ago. When I first played them, I had my fourteen bells arranged on a rail, and tapped them with my two leather hammers held in my hands in the usual way. I thought next I could introduce some novelty into the performance. The novelty I speak of was to play the violin with the bells. I had hammers fixed on a rail, so as each bell had its particular hammer; these hammers were connected with cords to a pedal acting with a spring to bring itself up, and so, by playing the pedal with my feet, I had full command of the bells, and made them accompany the violin, so that I could give any tune almost with the power of a band. It was always my delight in my leisure moments, and is a good deal so still, to study improvements such as I have described. The bells and violin together brought me in about the same as the piano. I played the violoncello with my feet also, on a plan of my own, and the violin in my hand. I had the violoncello on a frame on the ground, so arranged that I could move the bow with my foot in harmony with the violin in my hand. The last thing I have introduced is the playing of four accordions with my feet. The accordions are fixed in a frame, and I make them accompany the violin. Of all my plans, the piano, and the bells and violin, did the best, and are still best for a standard.[8]

Figure 15.1 Fate Norris' One-man Band (Photo courtesy of Rounder Records).

Except for this statement, no other personally related account of the gradual development of a one-man band exists. In the careful addition of each instrument and the technique and mechanical means to play it, there is a unique sort of virtuosity, a triumph of imagination and invention over physical limitation, not the least being blindness since one month of birth. Such inspired resilience is not uncommon among one-man bands and recalls this itinerant musician seen around Harlan County, Kentucky in the early part of this century:

> Back before I can remember a travelling minstrel by the name of Charlie Page or Paige came through the mountains putting on little shows with puppets at the local schoolhouses. He was a one-armed man, his arm being off right up to his shoulder, but he played a fiddle, blew a mouth harp and rang a bell at the same time. My father said he held the bow between his legs, and had a harp holder around his neck and tapped the bell on the floor with his foot, and he said he was a fairly good fiddler.[9]

Knowledge of the existence of the one-man bands of Fate Norris and Charlie Page would be lost without the preservation of a few photographs and sketchy recollections of people who saw them. Such testimony illustrates the obscurity of this tradition. Its history has many such great gaps and sudden flashes of detail, demonstrating how incomplete and arbitrary the historical record can be in discussing almost any musical tradition.

Mayhew's "blind performer on the bells" was fifty-five years old when he related his story. He had been playing on the streets most of his life, for the previous twenty-three years alone, playing a piano he had rigged onto a wheeled cart. "I was one of a street-band in my youth, and could make my 15s. a week at it. I didn't like the band, for if you are steady yourself you can't get others to be steady, and so no good can be done." Jesse fuller (whose calling card read "Lone Cat") had a similar experience, finding other musicians "all too busy – running around, drinking, and gambling."[10] Daddy Stovepipe, Dr. Isaiah Ross (still performing blues with his guitar, harmonica, cymbal ensemble), and many others echo these sentiments, all affirming the independence of their enterprise.

Practical considerations like these are the most often expressed motivations of one-man band performers, but if everyone who experienced such frustrations felt moved to create a one-man band, they would be far more in number. There is something deeper at work in this extraordinary impulse to play it all, alone, at one time, with all the requisite physical agility, and to play it so joyfully. There is a radical independence at work here, an urge to confront and explore human capabilities and possibilities to their limits, an urge to realize a unique and playful thought. In other words, even if the musical results may at times have been rough and musically limited, what is significant is that the attempt was made, that the idea was considered and acted upon, often with a lifetime of devotion.

Figure 15.2 Unidentified one-man band (from a promotional postcard with no written information).

Joe Barrick was born of Choctaw parents in Pauls Valley, Oklahoma in 1922. His first musical instrument, at age 15, was the mandolin; he recalls wanting something light he could play as he walked down the road. From a musician friend, he learned his first three chords, but it wasn't long before those three chords were too limited and he taught himself a more complicated style of mandolin playing and moved on to fiddle and guitar. He walked and hitchhiked all over southeastern Oklahoma playing free at dances (or, rather, parties; dances were frowned upon). He learned tunes from other musicians and off the radio; few books or records were available to him. "You'd hear a tune on the radio, then go off and learn it right quick. Didn't have records. Seemed like you learned quite a bit that way. You remembered it!" One of Joe Barrick's earliest influences was Bob Wills. He greatly admired his fiddle playing and still plays a lot of tunes associated with Wills and his Texas Playboys.

Joe Barrick settled in California when he got out of the armed services and began playing in western bands, mostly on fiddle and mandolin. During this thirty-year period, before moving back to Oklahoma fifteen years ago, he began, utilizing skills gained in his regular employment as a carpenter, to design and build his own instruments. One of his first used the skull of a cow as the body. "It was western music. Everything was western and everything western you'd always see a cow's skull laying around. I just got the idea of making one so it'd be western music."

The cow's skull guitar and mandolin, the Oklahoma guitar (with its body shaped like the state of Oklahoma), and the toilet seat guitar ("set on it") were all made around this time when he began playing by himself. "I used to play these just for show. That's when I started to want to play by myself. It was hard to keep anybody together to play with anybody."

Thinking about how to play a backup rhythm for himself (especially for his fiddle playing) without having to rely on less dependable musicians led to ideas about how a rhythm guitar could be played without the hands but with the feet. His solution was the invention of the *piatar*, a guitar played with hammers like a piano and operated with the feet. Pedals at his right foot control hammers that strike the strings of the guitar, while pedals worked by the left foot operate moveable frets that push up on the strings determining the chords that will be played. It began as an old board with a guitar neck and pedals mounted on it. He "pictured it" long before actually trying to build the device, and when he did, "I surprised myself with how good it sounded. It was just simple. That's where I learned how to play it … on that one flat piece of board. So it worked out good. Then I made another one that had all this other stuff on it. That's the way it worked out. I wanted to see how it worked."

Joe Barrick's next version of the piatar, built about 15 years ago, added bass guitar and banjo, and later a snare drum. These sit in shelf-like arrangement in a box at his feet. As with the piatar, hammers operated with the right foot strike the drum and strings of the banjo, bass, and guitar on the downbeat; treadles worked by the left foot operate the moveable frets and play the chords. Hence, the *piatarbajo*; its name derived from the instruments utilized in its invention. Along with the piatarbajo, he plays a double-necked guitar/mandolin that also has a cow's skull

Figure 15.3 Joe Barrick with his Piatarbajo.

as its body. To this, he sometimes adds harmonica (held in a neck harness), and, occasionally, fiddle. This arrangement is designed so that he can move smoothly from one instrument to the next: "to switch to it and keep going." The fiddle is not part of the apparatus yet. "I have to stop and pick it up. I play the guitar and mandolin mostly, because it's built together." However, the piatarbajo was designed so that more could be added to it; he has not gotten around to a final "dressed up version." "I try to do it all! Almost have to … by myself … Well, might as well do it all. I do a little of everything."

Each instrument on the piatarbajo has its own pick-up mike and comes out of a separate speaker. There is a regular bass amp for the bass part so that it sounds like a bass guitar. There are separate amps for the guitar and banjo and they sound out on the downbeat with the snare drum, so that it all blends together. The bass is on one side and the other amps on the other so that it will sound like individual musicians playing. "It's not all just one lump. It don't come out like that. It sounds like several pieces." This balancing of the individual instruments is of crucial importance and Joe Barrick takes great care in achieving that full, balanced sound. "You can tell when it comes out good. Boy, you can hit it and it rings out! Well, you know you got it!"

Joe Barrick's repertoire draws on the music he has listened to all his life, mostly country songs, lots of tunes associated with Bob Wills: "San Antonio Rose," "Spanish Two-Step," "Joe Turner Blues," "Over the Waves." When asked to name favorites, he mentions "Sally Goodin," "Arkansas Traveller," "Eighth of January," "Mocking bird," "The Waltz You Saved for Me," "Faded Love." He plays clubs much less frequently today than he once did and much more often for special community events; for openings, music festivals in southeastern Oklahoma, parties where he is invited to play, fundraisers, VFW get-togethers, events sponsored by Choctaw organizations (in 1977, he received an award of honor from the Choctaw Nation of Oklahoma "in recognition of outstanding work designing musical instruments"). He regularly appears at the annual World Series of Fiddling in Langley, Oklahoma (organizers of this event awarded him a Citation of Appreciation last year for his efforts "to preserve the music of past decades and his performance on the piatarbajo"). Barrick also takes the piatarbajo to schools and nursing homes and it's always a special event for him when people want to get up and dance. "I play for some of these rest homes and they like to dance. Some of the bands that come there didn't want them to dance, so they'd ask me if I'd mind if they'd dance. 'Well, that's what I want you to do! I can play better when you're dancing!' It's hard to play for someone that just sets there." Music is a social event for Joe Barrick: "As long as I can make music and have a few laughs I'm happy."

As a young man, Barrick had the feeling that there was more to music than three mandolin chords, and although he has greatly expanded his array of instruments and is carrying much more than a single mandolin, he has creatively preserved

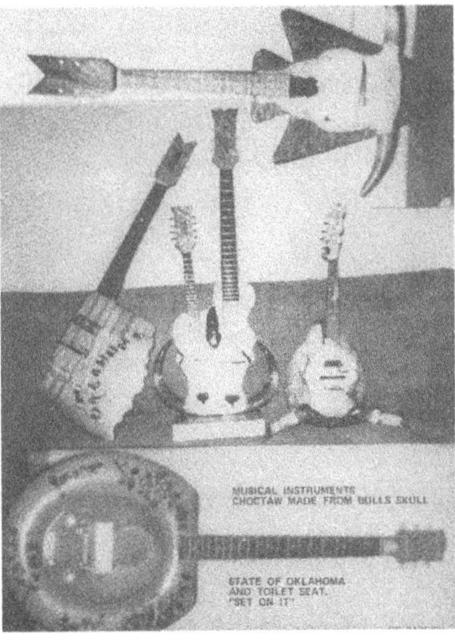

Figure 15.4 Electric guitars made by Joe Barrick.

Figure 15.5 "Amigo/Joe Barrick" – A signed promotional photograph.

his independence and self-determination. "No one tells me when to practice and I can play any song I want without having to hope the rest of the band likes it." With a firm sense of the sound he seeks to achieve, a sound that rings out and brings joy to others, Joe Barrick sets music to the dance of life. He plays it all himself, and the elusive tradition of the one-man band is living and thriving in his life and music.

As a category of musicianship, the one-man band transcends cultural and geographic boundaries, spans stylistic limits, and defies conventional notions of technique and instrumentation. From the musicking of centuries of street performers and itinerant musicians to the elaborate inventions of Fate Norris, "blind performer on the bells," and Joe Barrick, and the further-out improvising of Rahsann Roland Kirk, Jerome Cooper, and Jon Rose, the link uniting these diverse ensembles can be found in their close relationship to the beginnings of all musicking, to the playfulness that inspires all the earliest of musical experiences, of imagining and inventing, and of realizing what was once only imagined.

Notes

1 Quotations from Joe Barrick are drawn from interviews conducted by the author in December 1986 and February 1987.
2 Henry "Ragtime Texas" Thomas' guitar and quills – an early African American panpipe – can be heard performing "Old Country Stomp" recorded in Chicago in 1929, presently reissued on *Songsters and Saints*, Vol. 2 (Matchbox MSEX 2003/2004). Stovepipe No. 1 (Sam Jones) played guitar and harmonica, occasionally adding a piece of stovepipe played in similar fashion to a jug for bass as well as melodic lines. A familiar figure on the streets of Cincinnati in the 1920s, his "A Chicken can Waltz the Gravy Around" (with David Crockett, guitar) has also been reissued on *Songsters and Saints* Vol. 2.
3 Further detail about these other varieties of one-man band can be found in an earlier version of this article, "Joe Barrick's One-Man Band: a History of the Piatarbajo and other One-Man Bands," *Musical Traditions* 7 (London, 1988).
4 Born in 1867, Daddy Stovepipe (Johnny Watson) spent his last years performing on Chicago's Maxwell Street. No relation to Stovepipe No. 1, Daddy Stovepipe played guitar and harmonica, and made a point, in an interview with Paul Oliver, of identifying his foot-stomping as his third instrument. Daddy Stovepipe here is mentioned here as only one of many guitar/harmonica/percussion one-man bands that could be cited.
5 Charles K Wolfe, "When the Skillet Lickers Came to Nashville," *The Grand Old Opry* (London: Old Time Music Booklet 2, 1975), Page 102.
6 "Interview with Bill Helms," *John Edwards Memorial Foundation Quarterly* Volume 2, part 3 (June, 1967), page 57.
7 Wolf, page 103.
8 Henry Mayhew, *London Labor and the London Poor* (New York: Dover publications, 1968, reprint of 18 61–6 62 addition), Vol. 3, p. 161.
9 Edward Ward, "Music in Harlan County, Kentucky," John Edwards Memorial Foundation Quarterly, Vol. 15 (Spring 1979), page 21.
10 C.H. Garrigues, "Jesse Fuller," liner notes to *Jesse Fuller, "The Lone Cat"*, Goodtime Jazz M–12039.

Related articles

Several other articles by Hal Rammel appeared in *Experimental Musical Instruments*. In addition, one other article on one-man bands by another author appeared. These articles, listed below, can be accessed through the Internet Archive at https://archive.org/details/emi_archive/.

"Spotlite on William Roof": JoAnn Jones in EMI Vol. 5 #2

"The Triolin" by Hal Rammel in Vol. 3 #4

"The Devil's Fiddle: Past and Present", Part One and Two, by Hal Rammel, in EMI Vol. 7 #3 and #4

"Hal Rammel's Sound Palette" by Mike Hovanksek, in EMI Vol. 8 #4

"The Experimental Sound Studio Invented Instruments Ensemble" by Hal Rammel, in EMI Vol. 10 #2

"The Alfalfa Viola" by Hal Rammel, in EMI Vol. 10 #3

"New Discoveries from the Cloud Eight Archive of Musical Instruments and Fortean Musicology: The Prehistoric Brass/Woodwind Connection" by Davey Williams with illustrations by Hal Rammel, in Vol. 11 #1

In addition to these articles, letters from Hal Rammel appeared frequently in the *Experimental Musical Instruments* letters column

16

A Comparative Tunings Chart

Bart Hopkin, David Courtney, and Larry Polansky

This article originally appeared in *Experimental Musical Instruments* Volume 6 #2.

Experimental Musical Instruments thanks the many people who helped in the creation of this chart and accompanying notes, including writers, proofreaders, and referees, as well as composers & theoreticians who have allowed their work to appear.

In *EMI*'s February 1989 issue, we published a sound spectrum chart, laying out values for frequency, pitch name, and wavelength, over the entire audible range. We follow now with another approach to mapping the sonic continuum: a comparative chart of tuning systems. It shows how pitch sets from various sources compare to the most basic just intervals, to the familiar 12-tone equal temperament scale system, and to one another. Tunings from diverse music cultures, historical European scale systems, and scales created by contemporary composers and theorists are represented on the chart (Figure 16.1).

Preliminary Notes

To make sense of the chart, anyone who has not already done so should abandon the idea that there are exactly 12 possible tones per octave. There are 12 tones per octave on the piano and most other western instruments but there is an infinity of possible tonal gradations between any two piano notes, and our ears are capable of hearing far finer pitch distinctions than 12-per octave calls for. As the chart shows, different people at different times have approached the division of the octave's tonal continuum in many different ways and continue to do so.

A tuning, as we're using the word here, is a set of pitches available, by convention, for use in a given musical style or tradition. The word may imply a set of absolute pitches – that is, a specific set of frequencies, as when in western tradition one speaks of a C Major scale – or it may be meant to indicate only a set of relative interval relationships, as when one simply speaks of a major scale without specifying a key. In the latter case, the tuning retains its identity as long as its relative pitch

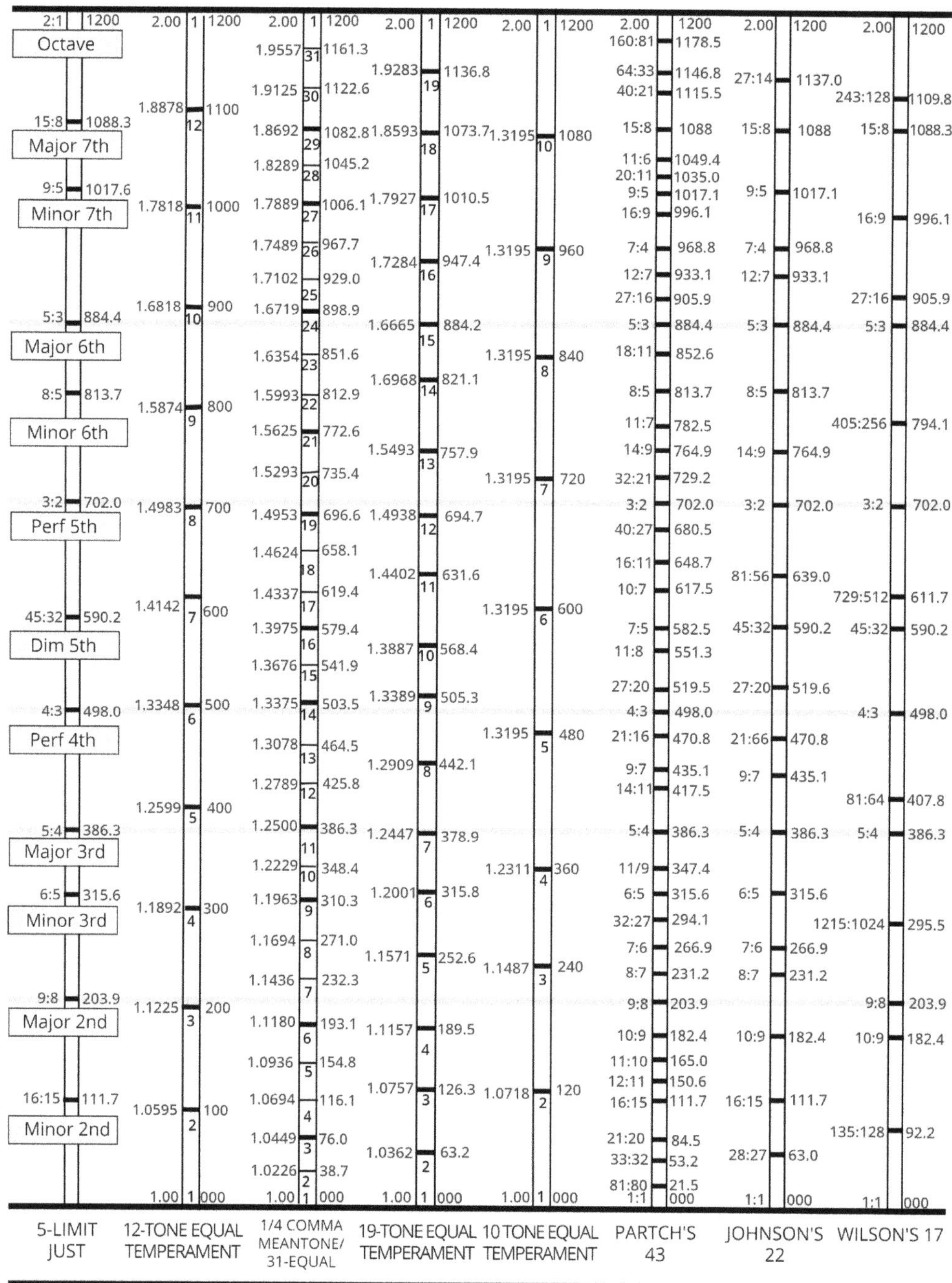

Figure 16.1 Comparative Tunings Chart.

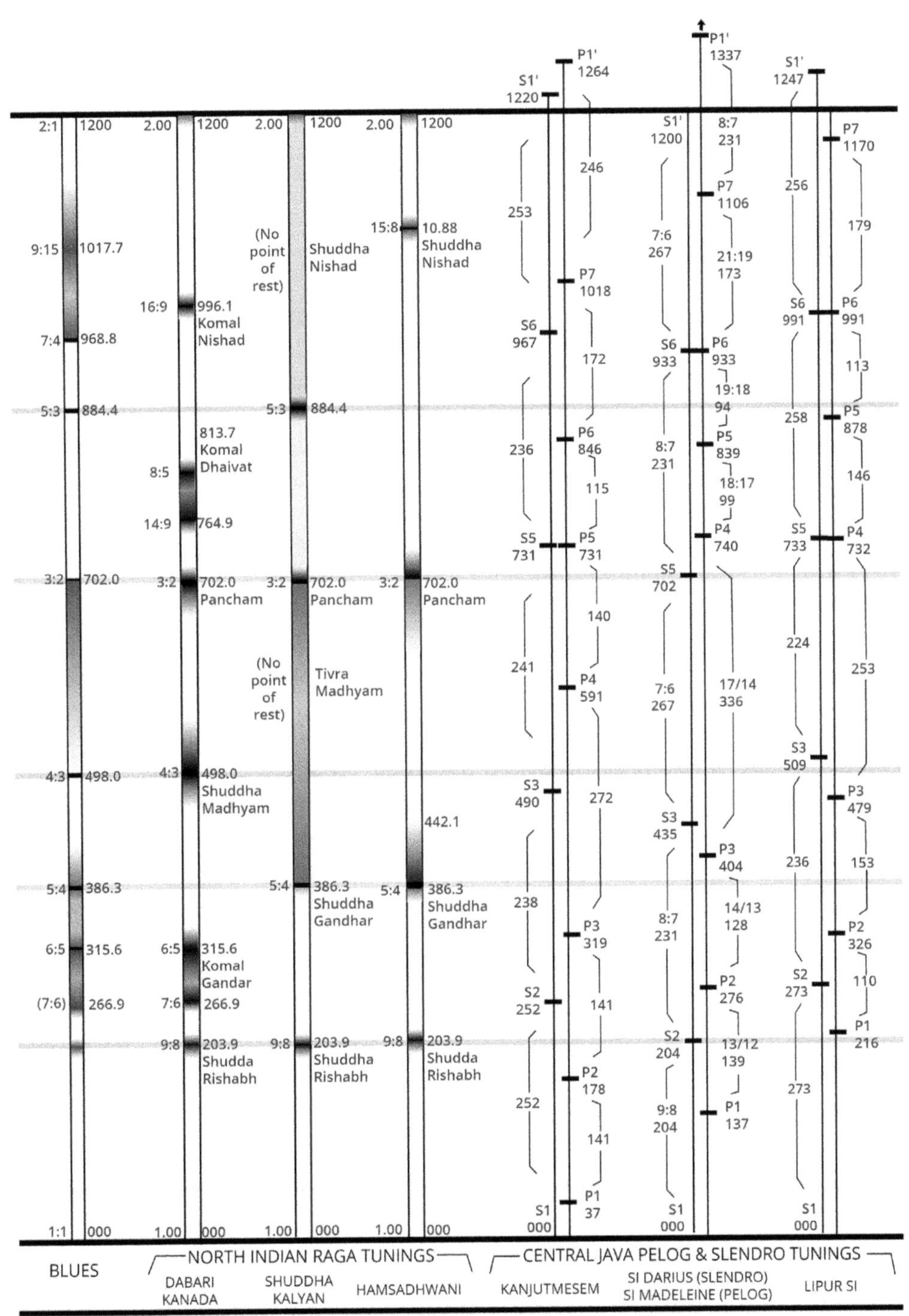

Figure 16.1 (Continued)

relationships are retained, no matter how it might be transposed from one absolute pitch level to another.

In the chart given here, all is relative. For comparative purposes, all the tuning systems are built over a common, unspecified fundamental. Relative pitch relationships can be seen, but no specific frequencies are given.

Reading the Chart

Here is how things are laid out on the chart. Each tuning appears in the familiar vertical ladder arrangement, starting at the bottom of the page, with the ascending pitches spaced out above according to the sizes of the intervals within the set of pitches. Vertical distances are calibrated for uniform interval size, so that a given interval always corresponds to the same vertical distance.[1] Each tuning is given over a range of one octave or a bit more.[2] To help demarcate the tonal territory, horizontal reference lines cross the entire chart at heights corresponding to certain basic just intervals: the major second 9:8, the major third 5:4, the perfect fourth 4:3, the perfect fifth 3:2, and the major 6th 5:3.

The degrees of each scale are indicated by short bold horizontal marks, except for ambiguous or variable scale degrees, which are represented by dappled areas of varying density. Below the mark, where applicable, is a name or number for the particular degree of the scale. To the right of the mark a cents value appears, and to the left will be seen a ratio or decimal number. Both the cents value and the ratio or decimal are given in relation to the first degree of the pitch set. (The three Javanese tunings are handled slightly differently; for full information, see the notes below on those tunings.)

Let's review the meaning of the relative pitch indicators just mentioned.

Identifying Intervals by Ratios: Just Intonation

Musical intervals can be described as ratios of rates of vibration. For example, we hear the interval of an octave between two pitches whose vibrational frequencies have the ratio 2:1, meaning that one tone is vibrating at twice as many cycles per second as the other. It matters not what the two frequencies actually are; as long as they form a 2:1 ratio, we will hear the octave. Similarly, 3:2 corresponds to the interval of a 5th, 5:3 to a major 6th, and so forth. The ratios appearing on the chart give the frequency of the given scale degree over the frequency of the starting tone of the tuning, reduced to the simplest terms.

The idea of expressing intervals as integral ratios forms the basis of just intonation theory. Tunings and intervals that are expressible in these terms can be called just.

Identifying Intervals by Decimals: Tempered Scales

European tunings are generally said to be based, in the ideal, in simple just intervals. But over the last couple of centuries, European musicians have seen fit, in tuning their instruments, to slightly detune, or "temper," those intervals. This is done to create scales in which interval sizes are more uniform than they would be in a pure just system. There are certain practical advantages to this.

To create tempered scales, a different sort of mathematic is called for. Ratios are out. A logarithmic approach is in.[3] Alongside each degree of the tempered scales on this chart, where the ratios appeared for the just scales, you will see a decimal number between 1 and 2. Like the ratios, the decimal is the number by which the frequency of, the first degree must be multiplied to give the frequency of the degree in question.[4] The decimals are, like the ratios, a way of expressing an interval.

The Cents System for Measuring Musical Intervals

One of the primary methods for precise measurement of musical intervals uses the unit called the cent, defined as 1/100th of a semitone in 12-tone equal temperament. The octave thus comprises 1200 cents. Aside from being useful in its own right as a calibrator, this cents value provides a quick and easy comparison to nearby pitches in 12-TET. For example, you can quickly recognize that a tone in some exotic tuning standing at 270 cents above the tonic is 30 cents (3/10 of a semitone) below the minor third in 12-TET, since by definition the 12-TET minor third is 300 cents.

Worth Bearing in Mind: Descriptive vs. Prescriptive Tunings

Tunings can be determined either descriptively or prescriptively. Tuning information is descriptive when it amounts to description of existing musical practice: someone observes a musical tradition, analyzes the music to determine what pitches are being used, and lines them up in ascending order to create a scale. Tuning information is prescriptive when it is designed not to describe what someone has done, but to tell someone what to do, as when a musician devises a tuning for a particular musical purpose and then proceeds to use it. This chart contains both prescriptive scales – deliberately created by composers or theorists for one purpose or another – and descriptive scales, representing existing musical practice in various cultures. Prescriptive scales are theoretical in nature and can usually be presented neatly in black and white. Descriptive scales inevitably entail more difficulties. Intonation in actual musical practice generally turns out to be subtle and highly nuanced, and not easily pinned down with accuracy and confidence. It may vary from moment to moment and musician to musician. For this reason, we have included more extensive notes on two of the traditions presented here – the Indian and the Indonesian – written by musicians and scholars with extensive experience in the areas.

What do All these Tunings Sound Like?

Are there ways you can actually hear these tunings? Different readers will have different resources available to them, and those with computers and software set up for this kind of thing have a great advantage. But for the rest of us, the easiest approach might be to construct a monochord, or perhaps a harmonic canon (which can be thought of as a many-stringed monochord zither). The scale degree ratios or decimals correspond inversely to string lengths, and so it is not difficult to calibrate a rule alongside or beneath the string(s) and mark the string stopping points for the relevant pitches. Alternatively, those with digital tuners (which are getting less expensive all the time, and which usually have meters calibrated in cents) can work from the cents values on the chart to tune zithers, metal conduit marimbas (sometimes called tubulons), or any other easily made instruments of dependable pitch.

The Tunings

5-Limit Just

5-limit just intonation is normally considered to be the theoretically ideal intonational basis for music in the European tradition, and some other music traditions as well. The designation "5-limit" refers to the fact that 5 is the largest prime number required in either numerators or denominators to build the ratios of the tuning. (Roughly speaking, the larger the limit number, the more harmonically complex and potentially dissonant will the intervals of the tuning be perceived.) Although this scale has 12 tones per octave, nominally analogous to those of standard 12-TET, it functions quite differently, since it includes only those tones needed to play in a few closely related keys. The sound or mood of an accurately tuned 5-limit is often described as sweet and restful. A common form of 5-limit is presented here; other variations are possible.

12-Tone Equal Temperament

12-TET, dividing the octave into 12 equal steps, is the standard tuning in most current western music; it is what we hear every day and it is what has been taught as correct in mainstream practice for a couple of centuries now. 12-TET has found favor over other equal temperaments (equally spaced divisions of the octave with different numbers of steps) because it manages to do a fairly good job of approximating the intervals of five-limit just – that is, it happens to have intervals reasonably close to the important five-limit intervals. Other equal temperaments that do comparably well in this regard have inconveniently large numbers of tones per octave.

Quarter-Comma Meantone/31-Tone Equal Temperament

Quarter-comma meantone, an unequal temperament, was one of the widely used temperaments prior to the ascendance of 12-equal in the 18th century. The early unequal temperaments sought to achieve excellent approximations of just intervals in some keys, at the cost of poor approximations in some other keys (which were then avoided). Meantone temperaments were designed to produce good 3rds and 6ths (which are the poorest intervals in 12-TET) in the chosen keys. Quarter-comma meantone is generated by adding series of fifths, slightly detuned such that a series of four such fifths will add up to two octaves plus a flawless just major third. Continuing around the cycle (with octave displacements) until the fifth has been added 11 times completes the scale.

It happens that if, instead of stopping after 11, you continue the cycle of additions of the same fifth, you come very close to duplicating your original starting tone after 31 additions. In doing so, you have effectively generated a 31-tone equal temperament. Thus, quarter comma meantone can be considered, with just a little fudging, to be a subset of 31-tone equal temperament. Accordingly, 31 ET and Quarter-Comma Meantone are presented on a single axis on the chart here.

The cents values and scale degree multipliers given are correct for 31-TET except on the 12 meantone degrees, where the correct meantone values are given. The two differ by no more than a little over a cent.

More Equal Temperaments: 10 and 19

19-tone equal temperament has been cited as a practical short-term option for moving to higher order ETs, since it approximates important just intervals nicely, and could be accommodated using keyboard layouts and notational systems based in the familiar forms used for 12-TET.

10-tone equal temperament is included here for opposite reasons: it's an example of a blatantly noncomformist tonality. There are no intervals comfortably close to normal 3rds, 4ths or 5ths; in short, the listener is deprived of familiar touchpoints, and the tonality must simply be heard on its own pungent terms. For more on 10-TET, see Gary Morrison's article in *Interval* Vol. 2 #1, Fall 1979.

Monophonic Fabric (Partch's 43)

Harry Partch, whose just intonation work from the 1940s to the early 70s proves more prophetic with each passing day, set forth this 11-limit scale as his most comprehensive tonal resource. For some of the reasoning behind his choice of intervals, and for some insight into one of our century's exceptional minds (and a serious instrument maker as well), read Partch's *Genesis of a Music* (New York: Da Capo Press, 1979).

Ben Johnston's 22-Tone Microtonal

This is one of several scales appearing in the chart which were prescriptively devised by a contemporary composer for specific musical purposes. Ben Johnston writes, "It is the chromatic scale in a system using prime numbers 2, 3, 5 and 7 to generate its ratios. It results from combining major and minor scales generated using rules derived from the traditional Pythagorean 7-tone diatonic scale. It was used in my *String Quartet No. 4 (Amazing Grace)* as the basis of the closing section, and its proportions also govern rhythmic subdivisions in some parts of that work. Cf. Randall Shinn's "Ben Johnston's String Quartet No.4," *Perspectives of New Music*. The scale derivation is discussed in *Rational structure in Music*, reprinted in *1/1*."

["Rational Structure in Music" has since appeared in *Maximum Clarity and Other Writings on Music* by Ben Johnston, University of Illinois Press, 2006.]

Erv Wilson's Just 17-Tone Genus

Erv Wilson writes: "A chain of diatonic tetrachords (16/15, 9/8, 10/9), repeated seven times, is the origin of this 17-tone scale. This corresponds to a chain of alternating major thirds (5/4) and minor thirds (6/5). A synthesis of tetrachordal melody and triadic harmony is thus achieved. This is similar to a 13th-century Persian scale tuned as a chain of fourths (4/3)." For more on this scale, see Erv Wilson's *Some Basic Patterns Underlying Genus 12 & 17* [now available online at http://www.anaphoria.com/genus.pdf].

Blues

Blues is surely one of the most exotic, subtle, complex, rich, and beautiful approaches to intonation imaginable. It has provided a major counterbalance to the predominance of 12-TET in European and American music of this century. Blue tonalities, of course, are highly flexible and ambiguous. They make extensive use of sliding pitches; minute intonational inflections have musical meaning. Blues has a peculiar veiled quality in which pitches can be hinted at without actually being sounded, or seemingly reached for in such a way that the reaching, rather than the reached-for, becomes the musically significant thing. To represent all this graphically would be impossible. On the chart, the blue tonalities appear, inevitably, primarily as a series of gradations.

North Indian Ragas

Notes by David Courtney, Ph.D.

The discussion that follows is relevant only to the North Indian musical system. This system, known as *Hindustani Sangeet*, is the system that most Americans are familiar with. There is another system called *Carnatic Sangeet*, for which the information given here does not apply.

Debate as to the exact intervals used in Indian ragas has been going on for at least half a century. I do not claim to be the final word on the subject. Indeed, variations from one performer to the next may actually preclude the possibility of a resolution.

The method for determining the musical intervals appearing on the chart was quite simple. I have been working with a computer program which allows me to specify musical intervals from a table in the software. This approach allowed me to program small compositions in the various ragas (modal forms) and manipulate the intonation. These variations were then evaluated by Indian musicians. It was then possible to fine-tune the tables to the required values. The initial values for the table came from two sources: Standard ratios derived from Helmholtz, and some very fine theoretical work by Dr. Jayant Kirtani. The conclusions of both Helmholtz and Kirtani were basically in agreement and validated by the computer experiment. However, a few minor points relevant to the raga required modification.

Raga (or Rag) is an aesthetic concept embodying both musical and non-musical components. For the present purposes, however, I will confine my discussion to modality, because this is the area where intonation has the greatest significance.

A raga may be composed of 5, 6, or 7 notes. These notes are derived from a 12-note scale much as Western scales are derived from the 12-note chromatic scale. The augmented fourth (*tivra madhyam*) has been subject to considerable variation. The value of 53/37 was derived entirely theoretically by Dr. Kirtani. I have placed this value in my computer and have found it to be most acceptable.

It must be pointed out that these values do not have the same significance as they would in a Western system. The reason is that these "notes" are not considered discrete building blocks for the music. In practice, the entire tonal continuum is used, and the specific pitches indicated here are simply considered convenient resting places. This totally microtonal approach to the music produces some very interesting variations in the utilization of the pitches. The concept of "notes" being mere resting points has some interesting twists to it. There appear to be three possible ways of regarding a given "note." In the simplest approach, a single pitch serves as the resting point. This approach is illustrated in the chart under Rag *Hamsadhwani*. It is also possible to have notes in the rag which do not have any resting point at all. This is demonstrated by the Shuddha Nishad (major 7th) and the Tivra Madhyam (augmented 4th) of Rag *Shuddha Kalyan*. These notes exist only in the form of an elongated slide from Sa (octave) down to the Dhaivat (major 6th) and another slide from Pancham (5th) down to the Gandhar (major 3rd). Another concept is to have oscillating resting points. This is illustrated by the two values for Komal Dhaivat (minor 6th) in Darbari

Kanada. This approach is characterized by alternating between the two values in a slow shake or vibrato.

Even the supposedly "simple" resting point turns out to be not quite so simple. As described earlier, the "resting point" concept presupposes that the pitch can "wander" around. If such a process were random, we could expect to see what is referred to as a normal distribution. In reality, however, it is common to see a skew in the distribution around the stated values. The most common examples are the use of Shuddha Nishad (major 7th) and the Komal Rishabh (minor 2nd). It is a very common practice to raise the Shuddha Nishad in ascending passages and to flatten the Komal Rishabh in descending passages. Other notes may also show a skew, but these tend to display themselves only in particular ragas.

Data for three Indian scales appear on the chart: Darbari Kanada, Shuddha Kalyan, and Hamsadhwani. Many many more ragas exist. Remember that what is presented here is only pitch information, which is only a skeleton; each raga has many other associated musical and extra-musical qualities.

Pelog & Slendro

The gamelan ensemble of Central Java often takes the form of twin instrument sets tuned to two different sets of intersecting pitches. The two tunings, called *slendro* and *pelog*, are not universally fixed, but vary, within limits, from gamelan to gamelan. Tunings from three different gamelan appear on the chart. The data was gathered by Larry Polansky, who has provided the following comments.

Notes on the Tunings of Three Central Javanese Slendro/Pelog Pairs

by Larry Polansky

In the accompanying chart, three slendro/pelog pairs are given. The first is the measurement of the famous slendro/pelog gamelan at the Mangkunegaran Palace in Solo, Central Java, called *Kanjutmesem*. The second is the tuning for Lou Harrison and Bill Colvig's gamelan *Si Darius/Si Madeleine*, an aluminum gamelan built at Mills College. The third is my own measurement of the pelog/slendro pair of the two slenthem (a large, resonated bronze mallet instrument) in gamelan *Lipur Sih*, built and tuned by Tentrem Sarwanto of Semanggi, Solo, Central Java, and owned by myself and Jody Diamond.

In my notation for these tunings, I have started each pair on slendro 1 (S1), but not because this is an any way a "root," tonic, or fundamental. There is nothing to suggest that S1 is any more primary than other notes in slendro. Pitches are labeled simply as S1-S6, P1-P7, with S1' and P1' denoting a higher octave of that degree. The absolute cents intervals from S1 are written next to the pitch name, and the cents interval to adjacent pitches is written between the pitches.

I have included these three *pairs* of slendro/ pelog because I am interested in the ways that the two gamut relate, a subject which has not received as much attention as the absolute interval widths of each individual gamut. I hope that these simple graphs will give the reader some idea of the complex system of tuning in Central Javanese "double" gamelan.

Some notes on the slendro/pelog pairs described here:

1) These tunings, including the Harrison/Colvig gamelan, represent intonational ideas from Central Javanese pelog and slendro in the court gamelan tradition. Quite different tunings will be found in villages, other ensembles (like street *Siteran* and so on), and other parts of Indonesia (Bali, Sunda, East Java, Cirebon, etc.).
2) Certain important aspects of Central Javanese tuning practice are not fully represented in these charts, notably: stretched octaves (generally 10-20 cents), the wide variation in interval size in different gamelan, the relationships of tuning to *pathet* (mode), and the variation in tuning when different *tumbuk* (a note that is the same in slendro and pelog) are used. Note that two out of the three gamelan described here are tumbuk 6 (meaning that slendro 6 is equal to pelog 6). Tumbuk 6, 5, and to a lesser extent, tumbuk 2, are the most common tumbuk. These ideas, and others, deserve discussion at length but are beyond the scope of these brief notes.
3) A common theoretical model for gamelan tunings is the *Idealized ET Tuning*, which presents slendro as five-tone equal, and pelog as a subset of nine-tone equal. For tumbuk 6, the pitches in ascending sequence would be:

   S1 (nominally at 0 cents), P1 (160¢), S2 (240¢), P2 (293.3¢), S3 (480¢), P3 (560¢), P4 (693.3¢), S5 (720¢), P5 (826.6¢), S6/P6 (960¢), P7 (1093¢,), S1' (1200¢), P1' (1360¢).

   This model is useful and appears frequently in the writings of both western and Javanese theorists. In James Tenney's recent work, *The Road to Ubud*, for prepared piano and gamelan, the piano is tuned to a pelog based on nine-tone equal. As far as I know, no gamelan has ever been tuned exactly this way.
4) The measurements for Kanjutmesem are taken from an important article by Surjodiningrat, Sudarjana, and Susanto, of Gadjah Mada University, in Jogjakarta (1972) entitled "Tone Measurements of Outstanding Javanese Gamelans in Jogjakarta and Surakarta." I have used their spelling for the name of the gamelan (rather than the more modern Kanyutmesem). Their measurements, which correct some important flaws in Kunst's earlier measurements (*Music in Java*, Volume II, edited by E.L. Heins, The Hague, Martinus Nijhoff, 1973, 3rd edition) are in hertz (hz.). There is a 1 hz. difference between S5 and P5 here that I have "rounded off" for simplicity and my interest in showing a tumbuk, but the reader is advised that whether a "real" tumbuk is perceived in this

tuning is open to question. In the computations of cents values from the actual frequencies measured, I have rounded off to the nearest cent. This gamelan tuning is included because it is one of the most recorded gamelan in the world and has been influential in the study of classical Central Javanese gamelan in the west. Its current tuning is significantly different than the one measured by these authors in 1972, since it has been "renumbered." Kanjutmesem, like Lipur Sih, has significantly stretched octaves, so that, for example, P1 is not a 2/1 to P1'. Note also the unusually wide (for measured Central Javanese tunings) P3-P4 interval – likely the result of a renumbering in the past, meaning that this interval appeared between two different pelog tones at an earlier time. But, as Carter Scholz has pointed out, this interval (P3-P4) is smaller than two steps of nine-tone equal, and considerably smaller than Si Madeleine's 336¢. The slendro is remarkably close to the much-maligned five-tone equal model of slendro.

5) Lou Harrison and William Colvig have built several gamelan, each with a different tuning. The Si Darius/Si Madeleine tuning, I believe, represents a fairly early model of Harrison's just intonation interpretation of slendro and pelog and is influenced by, but not a copy of, the tuning of gamelan *Kyai Udan Mas* (a tumbuk 6 gamelan which has been in residence at the University of California at Berkeley since 1976). Harrison's more recent tunings have avoided the perfect 3/2 fifth of these earlier tunings, and he is now working with stretched octaves as well. However, this pair is an important model for the study of American tuning systems in gamelan, and it is interesting to compare it to some actual Javanese tunings. Joan Bell Cowan's M.A. thesis from Mills College, "Gamelan Range of Light: The Influence of Instrument Building in Composition" (1985), presents Harrison's tuning in comparison with two other American tunings for aluminum gamelan, her own *Range of Light* and Daniel Schmidt's *Berkeley Gamelan*.
6) The tuning for the slendro/pelog pair of gamelan *Lipur Sih* was measured within a few days of its construction and first tuning (in December, 1988), using some tuning measurement and analysis software I wrote for the Amiga computer using the computer music language HMSL. For these measurements, the steady-state (middle) section of the tone was used. There is usually no low 6 or high 1 (P1') on a pelog slenthem, and the low 6 of the slendro slenthem, not given in the table, is 234 cents below the low S1.
7) It is interesting to compare this data with two "charts" (Figure 16.2). These are constructed from descriptions and drawings made by Tentrem Sarwanto as an explanation of the possible archetypes of relationships between pelog and slendro. The first chart is the possible and common relationships between slendro/pelog pairs with tumbuk 6, and the second of those with tumbuk 5. These two charts are quite general as intervals may vary significantly, and pelog/slendro relationships are not consistent from gamelan to gamelan, even with the same tumbuk. In this chart, I have tried to show: the range that each pelog interval

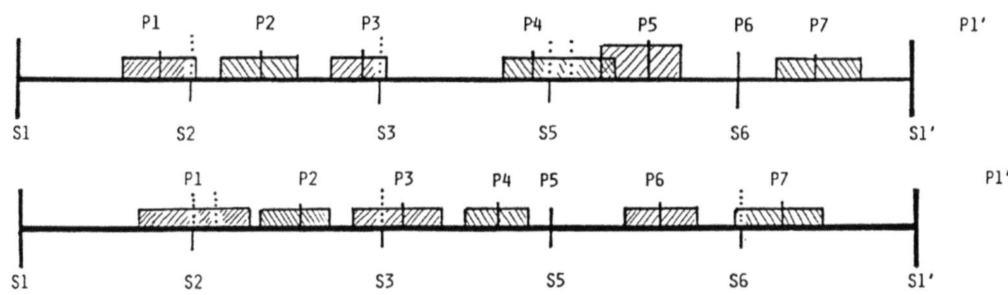

Figure 16.2 Tentrem Sarwanto's generalized chart showing possible relationships between pelog and slendro, with timbuk 6 (top) and timbuk 5 (bottom).

may assume in relation to a given slendro interval; the "most likely" value that each pelog interval might assume in general proximity to its slendro neighbor (indicated by a heavy vertical line); and possible equivalences between slendro and pelog, besides the tumbuk itself (indicated by a dotted line). For example, with a tumbuk 6, slendro 5 can usually replace pelog 4 "in a pinch" (which is economically more feasible in the case of the larger bronze instruments: it means that only ten pitches are needed instead of eleven!). Other equivalences are possible and common, such as the near equality of slendro and pelog three in Lipur Sih. I have described these pairs in terms of the relationship of pelog to slendro because I believe that slendro is most often tuned first, and pelog is "fitted" to the slendro. While this is not always the case, there is theoretical and empirical evidence that this is a common occurrence.

Notes

1 Calibrating for uniform interval size in this way means that increments in frequency are not uniformly spaced arithmetically, but logarithmically. This makes sense from the ear's point of view.
2 The one-octave range is adequate if you assume that the same set of intervals would in practice be duplicated in other octaves. That's not true of all tuning systems, and for that reason, the range is exceeded on some tunings given here.
3 For more on the mathematical basis of equal tempered scales, see Chrisotpher Banta's "Scales and their Mathematical Factors" in *Experimental Musical Instruments* Volume 1 #5, February 1986.
4 These decimals in most cases are irrational numbers, extending infinitely to the right of the decimal point. On the chart, they are rounded off to five significant digits.

Related articles

Many articles that appeared in *Experimental Musical Instruments* touched on tuning systems to a greater or lesser degree. Following are articles in which tuning was the primary focus. These articles can be accessed through the Internet Archive at https://archive.org/details/emi_archive/.

"Scales and their Mathematical Factors" by Christopher Banta in EMI Vol 1 #5

"Some Introductory Words on Just Intonation" by Bart Hopkin in EMI Vol. 2 #4

"Alternative tunings on Fretted Instruments – Refretting and Other Approaches" along with several related articles by various authors, in EMI Vol. 3 #6

"The Sound Spectrum: Pitch Names, Frequencies, and Wavelengths" by Bart Hopkin, in EMI Vol. 4 #5

"Relating Timbre and Tuning": Bill Sethares in EMI Vol. 9 #2, also included in this collection

In addition, reviews of books and organizations relating to scales and scale systems appeared in several issues including EMI Vol. 1#4, Vol. 2 #4 and #4, Vol 8 #4, and Vol 10 #3 and #4

17

Famous Color Instruments of the 19th and 20th Centuries

Kenneth Peacock

New York University

This article appeared in two parts in *Experimental Musical Instruments* Volume 7 #2 and Volume 7 #3, September and November 1991.

Introduction

Many readers of *Experimental Musical Instruments* would agree that musical connotations are intended by painters who have referred to their paintings as "compositions." And multi-media artists today commonly use the term as well. Specific connections to music can be seen in such paintings as Mondrian's "Broadway Boogie-Woogie," Kupka's "Fugue in Red and Blue," or "Symphony in Violet" by Albert Gleizes. Actually, the issue of musical means of expression employed by artists in other fields is much older than one might at first suspect. Music has frequently been associated with the concept of color since before the time of Aristotle and the ancient philosophers believed that harmony could best be described as the union of varied colored things.

"Color in music" has had many meanings. At different times, the term has described purity of tone, instrumental quality, melodic ornamentation, or even literal color in manuscripts. For example, in the 11th century, red and yellow lines indicated the pitches F and C before the development of musical staves. Comparisons between music and color have seemed a most natural human activity, and the topic has been of interest to writers in many fields.

In 1690, the English philosopher John Locke mentioned that a blind man had claimed the sound of a trumpet was like the color scarlet. This prompted heated international discussions during the 18th and 19th centuries concerning the possibility of correspondence between light and sound. Understanding the nature of light was perhaps as important to the debate as the development of an art-form which unified different modes of expression.

Both sides of the question were articulated by the poet Goethe who first embraced but later rejected the idea of an analogy between musical tone and colored light. Such analogies were encouraged by the new theory that both sound and light were the result of similar vibrations of a medium. Air provided a medium for sound, and "luminiferous ether," thought to pervade the universe, was the substance through which light could travel. It was postulated that only the rate of vibration was responsible for the nature of the effects produced in each case. This idea gained support because coincidental mathematical similarities in vibration ratios are seen if one end of the visible spectrum is compared to the other, and in turn compared to vibrational relationships within a musical octave. For example, the vibrational frequency of violet is approximately twice that of red. And some believed that if the sound of middle-C could be raised by 40 octaves, one would see red light.

Sir Isaac Newton had been the first to observe a correspondence between the proportionate width of the seven prismatic rays and the string lengths required to produce the musical scale D, E, F, G, A, B, C. Several 19th-century scientists cautioned against over-simplifying the analogy, but the belief that light and sound were physically similar persisted in textbooks published after the first third of the 20th century. The existence of luminiferous and electro-magnetic ether was disproven by the Michelson-Morley experiment of 1887; yet over 20 years later, one writer interested in the possibilities of combining color and music attempted to "sum up the scientific side" by stating:

> In a general way it seems to be indicated that harmonic colours are the results of vibratory effects upon the eye of multiples of like measurements, thus fulfilling exactly the analogy according to which harmonious effects are produced upon the ear.

There always has been a considerable lag time before artists have incorporated new scientific findings into their work, but in the case of what was termed "color-music," scientific agreement seemed an unnecessary prerequisite to the development of the new art-form. One consequence of the lively debate over correspondence between colored light and sound was widespread interest in a viable color-transmission instrument which could be operated from a musical keyboard. Early proposals date from shortly after the initiation of the controversy.

The Eighteenth-Century *Clavecin Oculaire*

A French Jesuit named Louis-Bertrand Castel (1688-1757) was the first to respond with a proposal for the performance of color-music. Well known during his lifetime, Castel was considered an eminent mathematician, but his writings also show an interest in aesthetics. His work attracted attention in England where he was made a Fellow of the Royal Society – an honor probably bestowed because of his reputation in mathematics, not because of his *clavecin oculaire*.

Castel wrote two major essays concerning a "harpsichord for eyes." The first was in the form of an extended letter (dated 20 February 1725) published in the November 1725 issue of *Mercure de France*. A development of those ideas appeared in *Memoires de Trevoux* (1735) under the title, "Nouvelles experiences d'optique et d'acoustique." It was translated into German and annotated by the composer Georg Philipp Telemann who had it published in quarto. This second essay was also translated into English in 1757 by an anonymous student who had apparently assisted Castel with some of his experiments. The idea for a *clavecin oculaire* was stimulated by writings of Athanasius Kircher (1601–1680), who had experimented with the *magic lantern* – an invention which became the slide projector. In his first article on the subject, Castel asked:

> Why not make ocular as well as auriculor harpsichords? It is again to our good friend [Kircher] that I owe the birth of such a delightful idea. Two years ago I was reading his *Musurgia*: there I found that if during a beautiful concert we could see the air which is agitated by all the various tremors of the voice and instruments, we would be astonished to see it sown with the most vivid colors.

Apparently Castel considered himself a philosopher rather than a technician, and he originally intended only to theorize about the feasibility of making color-instruments, but his ideas encountered skepticism. This prompted him to construct a model which was completed in 1734. It is not clear whether he wanted his *clavecin oculaire* to accompany sound or to substitute colors for sound according to his particular system of correspondence which was influenced by the discoveries of Sir Isaac Newton.

Newton had described important prismatic experiments in papers read before the Royal Society, and this information was later published in his *Opticks* (1704). Reflecting a preoccupation with systems of order which characterized philosophical thinking during the Age of Reason, Newton considered the fundamental order of the spectrum, that is, red through violet, to be equivalent to the "natural" order of tones from C to B. Castel, however, believed that the color blue was analogous to C, and he modified Newton's distribution of the visible spectrum. The harpsichord keyboard of Castel's *clavecin ocullaire* controlled colors indicated in Table 17.1.

Describing how one should play his instrument, Castel wrote:

> Do you want blue? Put your finger on the first key to the left. Do you want the same only 1 degree lighter? Touch the 8th note. If you want it 2 degrees, or 3 degrees..., touch the 15th, or 22nd, or 29th, or the last to the right. If you want blue-green, touch the first block to the left. Do you want red, and which red? Crimson-red? That is the 4th black. You have only to know your clavier and know that blue is C and red is G etc. This you can acquire with three days practice.

This description clearly implies an instrument of five octaves. In his *Optique des couleurs*, Castel proposed to implement his color system via harpsichords constructed with 12 octaves! He believed the limits of aural perception encompassed

Table 17.1 Keyboard colors proposed by Castel for his *clavecin oculaire* (18th century)

C	Blue	F-sharp	Orange
C-sharp	Caledon (blue-green)	G	Red
D	Green	A-flat	Crimson
E-flat	Olive Green	A	Violet
E	Yellow	B-flat	Agate
F	Apricot (yellow-orange)	B	Indigo

12 octaves (from 16 to over 65,000 cycles per second), and since colors were analogous to sounds, arrangement of color tints should follow a similar pattern. Thus, by mixing various amounts of white and black into each of the 12 pigments, 144 different colors would be obtained for the *clavecin oculaire.* Castel had first attempted to use prisms for his instrument, but the colors obtained by refraction of light were probably not of sufficient luminosity. He abandoned this method. Later experiments were conducted with candles, mirrors, and colored papers. Each key operated one of the 144 cylindrical candle covers, allowing light to shine through colored paper when the flame was exposed.

Telemann reported that Castel was encouraged by his friends to seek practical realization of his plans, and "without the aid of craftsmen he [had] nearly finished [the project]." This statement was apparently based on Castel's own remarks, but it is misleading because no full-sized *clavecin pour les yeux* was ever built. Albert Wellek concluded over a hundred years later, after investigating all of the documents pertaining to the issue, "Above all there can be no doubt that Castel's construction was certainly begun, but by no means did this lead to a fortunate termination." There also seems to be little evidence that the model of Castel's instrument performed according to his expectations. A copy of the 1757 brochure entitled *Explanation of the Ocular Harpsichord Upon Shew to the Public* (the anonymous translation of Castel's essay of 1735) is located in the British Museum. The envelope which contains the pamphlet has a handwritten note signed by M. Low, the first owner:

> I was admitted among a select party to a sight of [this instrument] at the Great Concert Room in Soho Square; but to a sight of the instrument only, for nothing was then performed nor afterwards, as ever I heard, neither did I ever know why.

In spite of this, there is no question that the interesting experiments of Father Castel were directly responsible for development of other theories and instruments in the first half of the 19th century. Although these yielded no lasting results, they led to later innovations which initiated our own audio-visual age.

Nineteenth-Century Instruments

Castel had proposed a color-music technology employing diffuse reflection of light from pigment. In 1789, Erasmus Darwin (grandfather of the renowned naturalist) suggested that the newly invented Argand oil lamps might be used to produce visible music by projecting strong light through "coloured glasses." This is probably the basis for the instrument described by D.D. Jameson in his pamphlet, *Colour-Music* (1844).

Jameson specified a system of notation for the new art form and also described his apparatus in some detail. A darkened room in which the walls were lined with reflecting tin plates provided the setting for his "colourific exhibition." In one wall, 12 round apertures revealed glass containers filled with liquids of various colors corresponding to the chromatic scale – "the bottles seen in the windows of druggists' shops can be used for this purpose." These acted as filters for light projected from behind the wall. Moveable covers were activated by a seven-octave keyboard, and each was raised a specific height depending on which octave was chosen.

Another intriguing instrument was constructed between the years 1869 and 1873 by Frederic Kastner. This was a type of gas organ which the inventor called a *Pyrophone. EMI* readers will recall letters to the editor on Kastner's instrument written by François Baschet, with a technical diagram (Vol. VI #5); Michael Meadows (Vol. III #4); and others (Vol. III # 6). His idea was undoubtedly developed after hearing the sound made by gas jets which where commonly used for interior lighting before electricity. Supposedly, the apparatus produced sounds like the human voice, the piano, and the full orchestra. Cylindrical fillers covered ignited gas jets (Figure 17.1). Kastner later extended possibilities for the visual portion of his experiments after electricity became available. A device he termed the "singing lamp" was essentially,

> a sort of pyrophone with thirteen branches, all decorated with foliage and furnished with burners containing several gas jets, which opened into crystal tubes. These burners were brought into play electrically, through an invisible wire that connected to a keyboard in a neighbouring room or street – or indeed another part of the town.

In 1877 Bainbridge Bishop, who had been interested in the concept of "painting music," constructed a machine in the USA which was to be placed on top of a home organ (Figure 17.2). A system of levers and shutters allowed colored light to be blended on a small screen at the same time a piece of music was performed on the organ. Sunlight was first used as the source of illumination, but later an electric arc was placed behind the colored glass. Bishop's invention attracted P.T. Barnum's attention, and he had one of the instruments installed in his Connecticut residence. A bold proposal was advanced by William Schooling in an article published in the

Figure 17.1 Frederick Kastner's Pyrophone (ca. 1870). Ignited gas jets opening into Crystal tubes were controlled from the keyboard. In addition to a visual display, the instrument was reputed to produce sounds like the human voice or a full orchestra.

19th century (July 1895). Although there is no evidence that an instrument was actually built, his suggestion to use vacuum tubes of various shapes is remarkable. Contacts on a keyboard would electrically activate individual tubes, and the intensity of illumination could be varied with a pedal to alter current levels. It is not difficult to see this concept as a forerunner of today's computer-controlled artistic or commercial lighting systems. They are so prevalent that we now take these displays completely for granted.

The best-known color instrument of the last century was patented in 1893 by Alexander Wallace Rimington (1854–1918). The inventor, a Professor of Fine Arts at Queen's College in London, called his apparatus the *Colour-Organ* and this name has become the generic term for all such devices designed to project colored light. Rimington described his instrument and the color theories upon which it was based in a book entitled *Colour Music: The Art of Mobile Colour* (1911).

Figure 17.2 Bainbridge Bishop's device for "painting music" (1877). The light-producing apparatus was placed on top of a normal organ, and a system of levers and shutters allowed colored light to be blended on the screen while music was performed. P.T. Barnum had one of these at home. It could be considered similar to today's "color-organs" which are attached to home stereos.

Rimington was convinced that physical analogies of some kind existed between sound and color. In his book, he repeatedly compared the two phenomena, claiming both "are due to vibrations which stimulate the optic and aural nerve respectively." Much space was devoted in the appendix for supporting viewpoints. He did admit, however, that the analogy was along broad lines and that the correctness of his own theory was open to question. Rather than attempting to show an exact parallel between vibration frequencies of light and sound, he divided the spectrum into intervals of the same proportions as occurred in a musical octave. Thus, the ratio between two light waves approximated that for a corresponding interval in sound. Rimington's color scale is shown as Table 17.2. Each octave contained the same colors, and registral placement of colors was directly proportional to saturation, that is, higher octaves contained more white light.

Table 17.2 Colors produced by the keyboard of Rimington's *Colour-Organ* (1893)

C	Deep red	F-sharp	Green
C-sharp	Crimson	G	Bluish green
D	Orange-crimson	A-flat	Blue-green
E-flat	Orange	A	Indigo
E	Yellow	B-flat	Deep blue
F	Yellow-green	B	Violet

The Colour-Organ stood over ten feet high (Figure 17.3). It was a very complex apparatus, employing many filters varnished with aniline dye and 14 arc lamps, and it required a power supply capable of providing 150 Amps. The five-octave keyboard resembled that of an ordinary organ and was connected by a series of trackers to a corresponding set of diaphragms in front of special lenses. Stops were furnished to control the three variables of color perception: hue, luminosity, and chroma or color purity. One stop allowed the performer to spread the spectrum band over the

Figure 17.3 Alexander Rimington with his *Colour-Organ* (1893). This instrument produced no sound, but trackers from the octave keyboard were connected to lens diaphragms and filters for fourteen arc lamps.

Figure 17.4 A page from the autograph manuscript of Alexander Scriabin's *Prometheus the Poem of Fire* (1911). The color part for a *Tastiera per Lucee* appears at the top of the orchestral score. An additional voice was added to this part in the published version.

entire keyboard instead of one octave – proof of Rimington's flexible attitude concerning the analogy between particular colors and tones. Like all earlier mechanisms (with the exception of Kastner's pyrophone), Rimington's instrument was not capable of producing any sounds. He did recommend, however, that compositions played in color be performed simultaneously on sound-producing instruments because this added to the enjoyment of the color. No new notational system was needed because musical compositions were played on the keyboard in a normal manner and thereby "translated" into colored light.

Rimington's efforts attracted considerable attention. According to a report in the *Musical Courier* (8 June 1895), Sir Arthur Sullivan improvised on the Colour-Organ shortly after its completion. He played with his eyes closed however. On 6 June 1895, Rimington presented a private lecture-demonstration in London which was attended by over a thousand people. His Colour-Organ was accompanied by piano, a normal sound-producing organ, and a full orchestra. It is interesting that this is the same instrumentation called for in Alexander Scriabin's famous 1911 color-symphony, *Prometheus* (Figure 17.4). Scriabin probably knew of Rimington's work, and he was the first composer to include a part for projected light – his *Tastiera per luce* – in a score for orchestra. Four subsequent public concerts were given by Rimington in 1895 in which compositions by Wagner, Chopin, Bach, and Dvorak were performed. There were apparently no other presentations until after Rimington published his book in 1911. Sir Henry Wood contacted the inventor in 1914 concerning a proposed realization of the color part in Scriabin's *Prometheus*, but World War I prevented implementation of the performance, and another attempt to produce this work with colored lighting was not undertaken in England until 1972.

Color Instruments Developed After Scriabin's *Prometheus*

Many reviewers of Scriabin's color-symphony have remarked that the first performance of *Prometheus, the Poem of Fire*, given in Moscow on 15 March 1911, did not include color realization because the machine to perform the lighting part would not operate. Additional information has not been located, but it seems reasonable to assume that if such a machine did exist, it was probably a larger version of the model constructed by Scriabin's friend Mozer. Apparently no expense was spared for the original production of *Prometheus* as Serge Koussevitsky gave the work an unprecedented nine rehearsals. It is puzzling, if such an instrument was available, that no attempt was made to use this *Tastiera per luce* in subsequent performances.

The first public presentation of Scriabin's symphony accompanied by colored lighting (20 March 1915) was in Carnegie Hall. According to a report published in the *New York Times* the day before the performance, Modest Altschuler had contacted the president of the Electrical Testing Laboratories for assistance in realizing the color portion of *Prometheus*. Preston S. Millar, a specialist in electrical lighting, had been assigned to supervise construction of a color-projection instrument later named the *Chromola.*

Two versions of the Chromola were built over a period of three months, and "thousands of dollars were spent before the project was completed." Special lamps were manufactured by General Electric as the instrument projected 12 separate colors. It was operated from a keyboard with 15 keys – the extra keys repeated the first three colors of the scale. When key contacts were closed, a low-voltage DC circuit activated the 110-volt AC circuit to one of the projecting lamps. Unlike previous devices, this machine was not built to demonstrate a particular association between color and sound. It was intended solely for performances of *Prometheus*.

A problem was quickly encountered because there is no indication of appropriate colors for the luce part given in Scriabin's score. Six years after the first effort to present *Prometheus* with coordinated light, colorist Mathew Luckiesh wrote:

> Some of those responsible for the rendition of this music, with color accompaniment, had, at different times previous to the final presentation, accepted both the Rimington scale and Scriabine's code (the latter having been discovered later in a musical journal published ... in London) as being properly related to the music.

Modest Altschuler's Carnegie Hall production of *Prometheus, the Poem of Fire* generally met with disfavor. One critic dismissed the colored lights, which were flashed onto a small white screen, as a "pretty poppy show." For various reasons, the Chromola was considered one of the instruments of the orchestra rather than equal in effect to the combined instrumental and choral forces as Scriabin intended. The audience also evidently expected more. Technical problems contributed to difficulties, and not enough time was allowed for an artistic setting. Inappropriate theater facilities further diminished the possibility of a successful color realization.

A report in the *New York Times Magazine* (28 March 1915) indicated that an apparently successful, private presentation of the color symphony accompanied by the Chromola took place at the Century Theater "about February 10th." Members of the distinguished audience included Isadora Duncan, Anna Pavlova, and Mischa Elman. On that occasion, Millar (the inventor of the color-instrument) was quoted as saying:

> It was my dream to utilize an entire theatrical stage, hanging parallel curtains of thin diaphanous gauze from the proscenium, back to the rear wall of the theater, thus giving the light depth and sufficient space to expand and create atmosphere.

According to Luckiesh, others suggested that colors be projected onto loose folds of material, "kept moving gently by electric fans placed at a considerable distance." Had such solutions been adopted, the world premiere of Scriabin's *Prometheus* with color realization would have undoubtedly received a better press.

In the years following Altschuler's rendition of *Prometheus* with colored light, a great number of "color-organs" appeared. It may be difficult to prove a direct link to the color symphony, but Scriabin's composition must have encouraged these developments.

While Castel's 18th-century *clavecin oculaire* and 19th-century innovations such as Rimington's Colour-Organ had been conceived to reveal physical connections between light and sound, most instruments built during the early decades of this century were not intended to express direct association. One exception, however, was a device invented in 1912 by an Australian named Alexander Hector. On his instrument yellow corresponded to middle C – a pitch associated with red by Rimington and blue by Castel. This frequent difference of opinion concerning "correct" color associations prevented the establishment of a consistent aesthetic for performances of color-music. If the same musical composition was performed on separate instruments (Rimington's or Hector's for example), the resulting translations would yield entirely different colors.

By the early 1920s, it became apparent that there was no indisputable correspondence pattern between colors and sounds. For this reason, many predicted the evolution of a new and independent art form – pure light manipulation which had no connection to sound. Experimenters attempted to resolve technical difficulties, and most seem not to have been aware of the work of others in this very old field. Nearly every color-organ inventor in the nineteenth and early twentieth centuries was under the delusion that he was the first to conceive of color-music. Mary Hallock-Greenewalt is perhaps the extreme example. Her book is a self-panegyric in which she claimed in the opening pages, "It is I who have conceived it [color-music], originated it, exploited it, developed it, and patented it." She concluded (over 400 pages later) that the art she named "Nourathar" – an arbitrary combination of two Arabic roots – is an aid to better health. Thus, her conception of a medicinal use for art anticipated by over half a century the philosophy of today's New Age practitioners.

Mrs. Greenewalt's apparatus for the performance of color-music was named the *Sarabet* (after her mother, Sara Tabet). In 1919, the machine was demonstrated for the first time. Her elaborate instrument was operated from a small table-like console (Figure 17.5A). A sliding rheostat controlled reflection of seven colored lights onto a monochromatic background. In her concerts of light, Mrs. Greenewalt emphasized variation of luminosity which she considered parallel to nuances in musical expression. Particular colors were treated as subordinate to diverse intensities of color. A new notational system for performance on the Sarabet was patented by the inventor (see Figure 17.5B).

Many color-projection instruments appeared shortly after 1920 – the year generally considered to mark the birth of kinetic art. For example, in 1920 the English painter Adrian Klein designed a color projector for stage lighting. His instrument, which demonstrated a color theory involving logarithmic division of the visible spectrum, was operated from a two-octave keyboard (Figure 17.6). Leonard Taylor, another English experimenter, built a device whereby 12 colored lights were activated from a 13-note keyboard. Although no relay switches were used, various "organ stops"

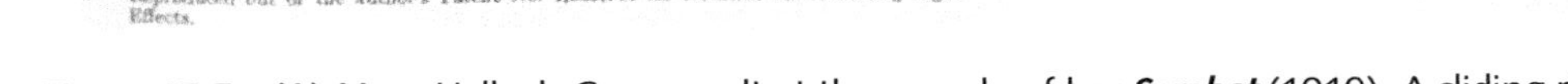

Figure 17.5 (A) Mary Hallock-Greenawalt at the console of her ***Sarabet*** (1919). A sliding rheostat controlled reflection of seven colored lights onto a monochromatic background. (B) Mrs. Greenewalt's patented system of light notation indicated subtle variations of luminosity which were parallel to nuances in musical expression. This Beethoven composition was originally for solo piano.

Figure 17.6 The 2-octave controller of Adrian Klein's color projector for stage lighting (1920). This instrument demonstrated Klein's color theories in which the visible spectrum was logarithmically divided.

controlled individual colors which could then be diluted with a variable-intensity daylight lamp (the thirteenth note). Similar color experiments were carried out between 1920 and 1925 by Achille Ricciardo who built a colored-light instrument for the *Teatro del Colore* in Rome, and Richard Lovstrom who during the same period in the USA patented an apparatus to perform color-music. The Czech artist Zdenek Pasanek worked with a color-keyboard as did Alexander Laszlo, who introduced his device (called a *Sonchromatoscope* – Figure 17.7) in 1925 at the Music-Art Festival at Kiel. Laszlo's book, *Die Farblichtmusik*, was published the same year. His preludes for piano and colored light employed a special system of notation.

From 1920 to 1925, Ludwig Hirschfeld-Mack studied at the *Weimar Bauhaus.* During the summer of 1922, he and others were rehearsing one of the shadow plays which were often presented at the Bauhaus. When one of the acetylene bulbs they were using needed replacement, Hirschfeld-Mack accidentally discovered that shadows on a transparent paper screen were doubled. By using acetylene bulbs of different color, "cold" and "warm" shadows appeared simultaneously. The principle

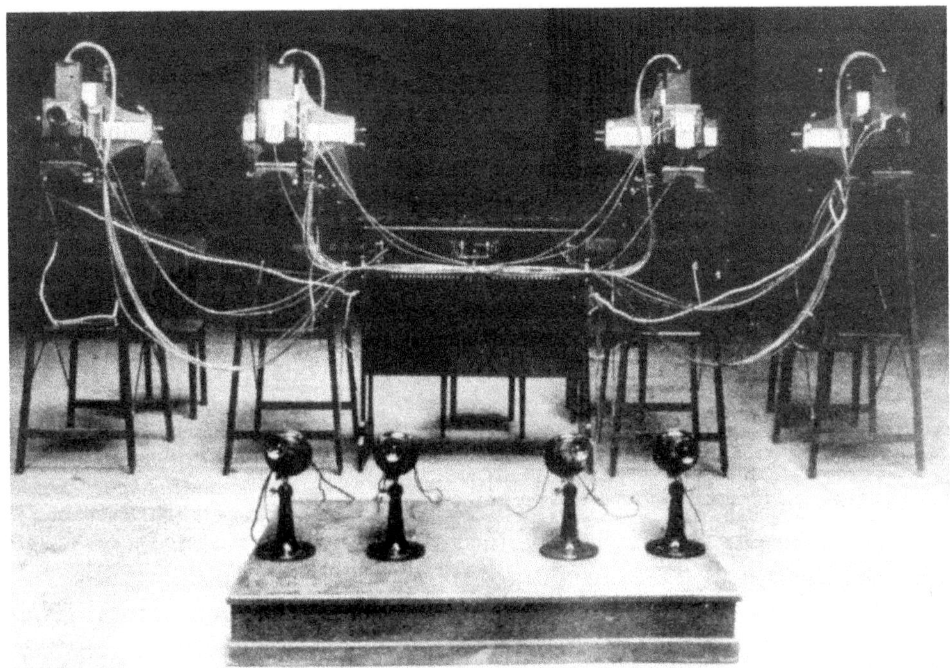

Figure 17.7 Alexander Laszlo's *Sonchromatoscope* was introduced at the 1925 Music-Art Festival al Kiel. Lazlo published a series of preludes for piano and colored light which employed his own system of notation.

was refined in subsequent years by using a type of color-organ. This device enabled Hirschfeld-Mack to present reflected-light compositions with his own music. The lighting technique was introduced to the public in 1923 at a film matinee at the Berlin Volksbühne and later in Vienna with Fernand Léger's experimental films. After 1960, Hirschfeld-Mack lived in Australia where he continued his activities, dispensing with the keyboard of his apparatus.

Wilfred's *Clavilux*

Most famous of the experimental color-instruments was the *Clavilux,* developed in 1922 by Thomas Wilfred at a cost of over $16,000. Wilfred completely rejected theories which presumed correspondence between light and sound. Light alone was the principal feature of a new art-form which he named "Lumia." It is true that Van Deering Perrine, the noted American painter and friend of Isadora Duncan, had experimented with various color-instruments around 1912 – and he may have been the first to reject the direct allusion to music – but Wilfred was able to develop the full implications of pure light manipulation. He considered the term "color-music" a metaphor; yet his art resembled music by including factors of time and rhythm in live performance.

Wilfred first used light in a purely abstract manner, but later decided form and movement were essential. These he achieved via filters which permitted the projection of moving geometrical shapes onto a screen. Vasily Kandinsky's theory that geometrical patterns supplement non-objective use of color possibly influenced Wilfred's work.

The *Clavilux* was introduced to the public on 10 January 1922 in New York although the first of several instruments had been partially completed in 1919. This was after more than a decade of experimentation. Wilfred's main instrument, employing six projectors, was controlled from a "keyboard" consisting of banks of sliders (Figure 17.8A). An elaborate arrangement of prisms could be inclined or twisted in any plane in front of each light source. Color intensity was varied by six separate rheostats which the performer operated delicately with his fingers. Selection of geometric patterns was effected via an ingenious system of counterbalanced disks. Wilfred's shifting light performances have been compared by many to the beautiful display of the Aurora Borealis (see Figure 17.8B).

During the years 1924 and 1925, Wilfred gave an extensive recital tour throughout the United States, Canada, and Europe. The late Percival Price once told the author about a *Clavilux* recital (5 January 1925) he attended in Toronto. "Before the concert there seemed to be an attitude of snobbery toward the new art, but after Wilfred began to perform everyone was spellbound." Nearly all of the published reviews substantiate this conclusion, and the critics' difficulty in finding the right words to describe the effect of the performance is evident. Deems Taylor, for example, wrote:

> The fact that Thomas Wilfred's *Clavilux* is commonly known as the color-organ is not the only reason why a music reviewer should have attended his recital last night in Aeolian Hall. For this new color-art might very aptly be called music for the eye. ...it is color and light and form and motion, but it is not painting, nor sculpture, nor pantomime. It is difficult to convey in words. Describing the Clavilux to one who has not seen it is like describing an orange to an Esquimo.

Wilfred notated his compositions, and they were given opus numbers like musical works. The most enthusiasm seems to have been generated by *A Fairy Tale of the Orient* (op. 30). A writer for the *Louisville Times* (20 November 1924) described the work as "an Arabian night of color, gorgeous, raging, rioting color yet not rioting either. ... Jewels were poured out of invisible cornucopias; lances of light darted across the screen to penetrate shields of scarlet or green or purple." Another reviewer described Wilfred's work in the *Manchester Guardian* (18 May 1925) as like

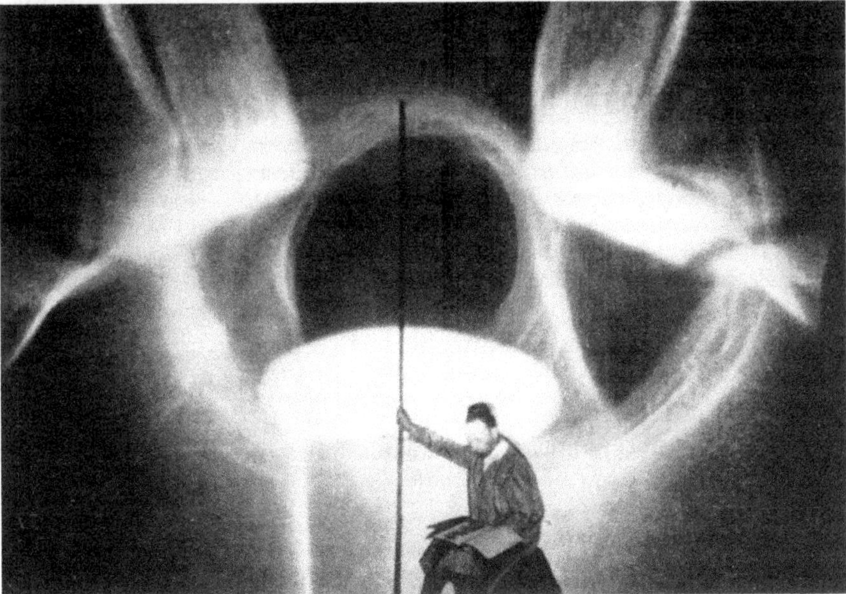

Figure 17.8 (A) Thomas Wilfred at the console of his *Clavilux* (1922). Moving geometrical shapes were projected onto a screen as the performer controlled the color intensity of various light sources. An elaborate arrangement of prisms could be twisted or inclined in any plane.
(B) Wilfred preparing one of his "Lumia" compositions for the world tour of 1924–1925. Critics compared his performances of kinetic light to the magnificent display of the Aurora Borealis.

a dream of "some unearthly aquarium where strange creatures float and writhe, and where a vegetation of supernatural loveliness grows visibly before the spectator." From the reports, it seems that Wilfred's art probably surpassed the dramatic effects produced by today's Laser performances – especially since his audiences had never witnessed anything like it before. Although most of Wilfred's recitals were presented in complete silence, there were also collaborative performances where music was interpreted in colored light. One such rendition of Rimsky-Korsakav's Scheherazade was presented in 1926 with the Philadelphia Orchestra directed by Leopold Stokowski.

Conclusion

In the decades following Wilfred's introduction of the *Clavilux,* many artists experimented with the technique of interpreting music in colored light. George Hall, for example, built a device in the 1930s which he called the *Musichrome.* It was equipped with eight keys to control two sets of four colors each. In a brochure about his instrument, Hall indicated no set rules to follow when interpreting musical compositions. "The accompanist must follow his own color reactions to the music played. Generally speaking, heavy, loud, thunderous music calls for the use of red, although there are times when an intense blue is desirable."

Frederick Bentham gave performances on a *Light Console* at the Strand Electric Demonstration Theatre in London shortly before the Second World War (Figure 17.9). At his concerts, he accompanied phonograph recordings of many works including *Pictures at an Exhibition*, *The Firebird*, and *Prometheus*. Describing a presentation of Scriabin's color-symphony on 31 March 1937, Bentham wrote:

> The first snag was that I could not feel in accord with the music (I wonder what my present day impression would be) and the second was the difficulty of looking at the score at the same time as keeping my eyes glued to the stage. A colour organist who does not do the latter is equivalent to the musician who ploys without listening to himself. ... After some rehearsal the ordered sequence of colour changes was extracted but ... in the end I took the colours as on organist might take a simple theme for improvisation and let myself go to the music.

Extensive technical innovation after World War II made possible the permanent installation of a large number of "color-organs" in theaters and galleries all over the world. These instruments were operated either by live performance or they have been programmed to present light sculptures. One such work, Wilfred's *Lumia Suite* (opus 158), was displayed during the late-1960s in the New York Museum of Modern Art. For two decades (and until recently), the engineer and lighting designer Christian

Figure 17.9 Frederick Bentham adjusting the *Light Console* before a performance at the Strand Electric Demonstration Theatre in London (1937). He presented many concerts with orchestral recordings and improvised lighting.

Sidenius gave performances of colored light with music at his private installation in Sandy Hook, Connecticut. His elaborate equipment included stereopticon color-projectors, and he called his concerts, "Lumia, the Theatre of Light" in honor of Wilfred's original Lumia Theatre of New York. And today, similar multi-media experimentation is evident. For example, Jan Gjessing and his artistic collaborators in Norway have presented many concerts with their stereoscopic projectors. Gjessing works with what he has termed "sound-to-light liquid cells" which modulate visual patterns in real time.

Within the past 15 years, the decreasing cost of technology has fired a revival of interest in the practical development of instruments to perform color-music. One result has been that today's consumers of both art and of entertainment events have come to expect that their aesthetic experiences will be generated by mixed-media – often including colored light and sound. Annual summer concerts of sound with lighting are presented in Paris; and in the Soviet Union, a large organization under the direction of Bulat Galeev has constructed color-instruments which have been used to present huge outdoor spectacles of sound and light attended by thousands. In the USA and Europe, many color-music concerts of more modest attendance have been presented by various groups. And audiences in our multimedia age have responded enthusiastically to this veritable explosion of activity. Enterprises such as Laser Images Incorporated, for example, have toured colleges with portable color instruments for live performances. In addition to selections in rock style, music by Carelli, Strauss, Holst, and Copland has been included in the repertoire accompanied by colored light. The Laser Arts Society holds monthly meetings in California to discuss creative applications of Lasers for kinetic sculpture. Nor have commercial applications of color-music been neglected. For the recent holiday season, Macys provided its "gift to New York City – a one-of-a-kind outdoor extravaganza with lasers, lights and holiday music to delight one and all." Shoppers could purchase an inexpensive "color-organ" inside the store (the well-known devices are attached to a home stereo and different audio frequencies trigger various colored lights), then they could step outside to witness the display of lasers with music. Their show was indeed impressive. Huge colored-laser patterns moving in time to music were projected onto the side of their tall building which occupies an entire city block.

Although experimenters during the past two centuries could hardly have anticipated today's widespread use of laser light in combination with electronic computers, these marvelous inventions are in some ways refinements of earlier technological proposals for a viable color-music instrument. Every generation, it seems, must re-discover and re-define the art of color-music for itself. And rarely does there appear to be awareness that previous activity has occurred. The current catalog of one major video company, for example, informs its clients that with their product, "a new art form was born. Blending color, music and movement, this new medium is a marriage of sight and sound."

A version of this article was originally published in *Leonardo* Vol. 21, no. 4, pp. 397–406 (1988).
Fuller referencing and footnoting appears in the 1988 *Leonardo* version of the article.

Related articles

Other articles pertaining to color organs which appeared in *Experimental Musical Instruments* are listed below. These articles can be accessed through the Internet Archive at https://archive.org/details/emi_archive/.

"The Flame Componium and Reflections on the Pyrophone" by Qubais Reed Ghazala in *Experimental Musical Instruments* Vol. 10 #3

"Building a Color Organ: The Harmonicophone Shows Notes and Harmonics Sounded" by Manuel Comulada in *Experimental Musical Instruments* Vol. 7 #6

In addition, this current collection contains a group of articles on pyrophones by Michel Moglia, Etiye Dimma Poulsen, Norman Anderson, and Bart Hopkin

18

Membrane Reeds

Indonesia and Nicasio

Bart Hopkin

This article originally appeared in *Experimental Musical Instruments* Volume 7 #3, November 1991.

This article describes an unusual sort of reed for use in wind instruments. When I began experimenting with this reed type, I thought I was doing original work. I later discovered that there is nothing new under the sun and that instruments using the same principle already existed, albeit in very different form, on the other side of the globe in Indonesia. And – who knows? – perhaps they can be found elsewhere as well. In this article, we'll look at both the Indonesian forms and my own work.

I have been calling the reed type a *membrane reed,* because the most important element is a small, stretched membrane of some kind covering the mouthpiece end of a wind instrument tube. When air is forced under this membrane and into the tube, it starts a standing wave in the tube at the tube's resonant frequency. If the tube happens to be equipped with properly spaced toneholes, the thing will play like any other wind instrument. I'll describe all this in detail soon. But first, since the word 'reed' may be confusing in connection with something so un-reedlike as a stretched membrane, let me clarify the sense of my usage.

The term *reed instrument,* as it is commonly used, covers a subset of the larger category of wind instruments. Reed instruments usually make use of a tuned air column in the form of a tube of some sort. The tube is equipped with something that serves as an open-and-closable gateway at the mouth, so that an air flow into the tube can be alternately restricted and allowed to pass more freely. Whatever serves as the gate is flexible enough to be responsive to pressure variations occurring within the tube, so that a standing wave set up in the tube will cause the gate to open at a moment of maximum internal pressure, and close at minimum pressure. Thus, when air is forced through the gate and into the tube, it tends to enter not in a steady stream, but in a series of pulses. The frequency of the pulses is determined by the open-and-shut frequency of the gateway, which (if all goes well) is controlled primarily by the standing wave frequency in the tube. By these means are the reed instrument's vibrations instigated and perpetuated, and its sounding pitch determined.

Whatever serves as the gate in this gating arrangement is the reed. Many reed instruments actually use pieces of natural reed (usually carved from the cane reed species known as *arundo donax*). But the term is applied inclusively to all instruments using such gating systems, and many of them use non-reed reeds. We are familiar with several sorts of reeds, including the cane reeds used in the air-gating arrangements of clarinets and oboes; the free reeds, usually of metal, used in harmonicas, accordions, and most organ reed pipes; and the 'lip reeds' used in trumpets and other brass (while this last usage is unconventional, several writers on musical instrument acoustics have found the phrase useful).

It was thinking about these several common types that led me to the simple question, *what other ways might there be to create pressure-responsive gating of an air stream as it enters a tube?* In other words, could one come up with other musically useful reed types?

Nicasio

I had the seemingly promising idea that it should be possible to create a sort of double reed in reverse. Rather than closing under pressure from the outside air source, then reopening in response to a combination of its own internal stresses and pressure from within the tube (which is what an oboe reed does), it would open under pressure from the outside source, and then reclose under its own internal mechanical forces and pressure from within the tube. This is what the trumpeter's lip reed does, but my thought was to create a separate mechanism to do it. And I had an already familiar and functioning model, for this is also what the mouth of a balloon does when you inflate the balloon, then pinch and pull tight the opening to release the air with a protracted squeal. In an attempt to lend some dignity to the situation, I will refer to this type of air-flow gating system as a *labial reed.*

My idea was to introduce the output of a labial reed into a tube and hope that its opening and closing frequency would conform to the resonant frequency of the tube. This didn't seem an unreasonable expectation, given that human lips as well as other sorts of reeds do so, and all of them are heavier and less compliant than

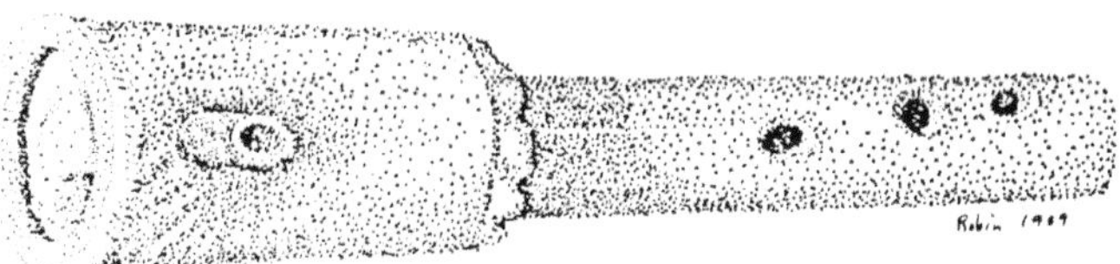

Figure 18.1 Membrane reed instrument from Sumatra (Drawing by Robin Goodfellow).

balloon lips. As a practical matter, it is easy to attach a labial reed to an air tube: one need only cut off the body of the balloon near the base of the neck and stretch the opening of the cut-off neck portion over the end of the tube. One can then pinch the neck with both hands somewhere near its midpoint and pull it apart just as you would to make a balloon squeal, while blowing through the mouth of the balloon (Figure 18.2). In this way, it is possible at least to tryout the general principle; later on, one could consider how to create an adjustable no-hands balloon pincher mechanism, leaving the hands free to cover tone holes or whatever.

Now, I still think that this labial reed idea has potential. However, I must report that in the limited work I did with it, the results were not very rewarding. There are many possible reasons for this. So why didn't I pursue it farther at the time? Because I kept getting distracted by something else – a balloon reed configuration that immediately produced much more musical results.

I found that if the balloon neck is pulled over the rim of the tube (as shown in Figure 18.3), a different sort of gating system is created that can take the place of the pinching of the neck.

When the player blows into the balloon mouth, the air must squeeze its way under the taut balloon rubber and over the tube rim, thence into the tube – it has nowhere else to go. This squeezing-under business it does, it turns out, in pulses: the balloon

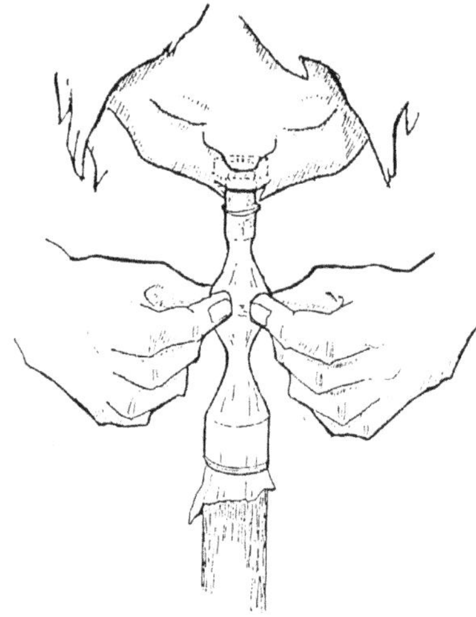

Figure 18.2 Labial reed in a crude and simple arrangement. The cut-off neck of a balloon is fastened over the end of an open tube, while the mouth of the balloon is held in the player's mouth. It is shown here connected to an inexpensive plastic plumbing fixture whose shape allows the player to hold it behind the teeth, preventing it from slipping out of the player's mouth. Initial experiments with this arrangement didn't prove especially effective from a musical point of view. Should we conclude that it is an inherently flawed idea, or does it yet hold some potential?.

Figure 18.3 Membrane reed in a simple but effective configuration. Again using the plumbing fixture to make it easy to hold the mouth of the balloon behind the teeth, the player blows through the neck of a balloon stretched over the end of a tube. The neck is held at an angle such that, to enter the tube, the air must squeeze between the balloon membrane and the rim of the tube. This it does in pulses, setting up a standing wave in the tube.

rubber lifts to allow a burst of air through, then comes back down under its own elastic pull. The pressure wave created by that original pulse travels the length of the tube, is partially reflected back, and returns to the balloon end of the tube, where it helps to lift the balloon again, allowing another pulse through. As long as the air stream continues through the mouth hole the cycle is repeated, at a frequency corresponding to the resonance frequency of the tube. In short, the system behaves exactly as a wind instrument reed should!

This general arrangement is my version of the membrane reed. It is similar in principle but quite different in configuration from the Indonesian form. Since I initially happened upon this system, I have done some further experimentation, arrived at some refinements and built a couple of instruments. The basic configuration speaks wonderfully easily and can produce a very good tone. It generally does not sound unique or in any way strange; in fact people often compare it to existing wind instruments – but with no consistency: listeners have likened it to virtually every orchestral woodwind instrument. The tone quality very much depends, of course, upon what sort of air column shape is being used, as well as a host of factors affecting the reed itself. The sound emanates primarily from two distinct points on the instrument: 1) the highest open tone holes (or, if none are open, the far end of the tube) – which is typical for reed instruments – and 2) the surface of the membrane itself. The tone coming directly from the membrane tends to be brighter than that coming from the tonehole, and sometimes buzzy.

Despite the system's initial cooperativeness, it has proven difficult to domesticate for musical purposes. This is not surprising – after all, people have been refining the designs of the standard reed instruments for a thousand years. The biggest difficulties for membrane reeds have been in the area of pitch control. The sounding pitch, while primarily determined by tube length, is also subject to several factors affecting the reed. This is true of all reed instruments, of course: the player can force the sounding pitch up or down by techniques which alter the effective stiffness of the reed, or by varying vocal cavity resonances. With these stretchy membrane reeds, however, the effects are multiplied to a point where pitch stability can become a problem.

There are two main sources of intonational instability: 1) the player can greatly increase the stiffness of the membrane by making it more taut, so that slightly pulling on the neck may drastically raise the pitch. (This can variously be done during playing by stretching the mouth piece on the balloon neck farther away from the tube edge, or by altering its angle of approach). In addition, the length and rigidity of the section of balloon neck leading up to the rim of the tube – factors which can be altered quite a bit by stretching or slackening during playing – affect the resonant properties of that enclosed air space, which in turn has its effect on the membrane reed behavior. 2) Blowing harder usually has the disconcerting effect of flattening the pitch considerably: apparently restoring force on the stretchy membrane does not vary in a linear fashion with amplitude, so that as the membrane vibrates increasingly vigorously it fails to spring back quickly enough to maintain a constant frequency.

Either of these two factors (the sharpening effect of increased tension or the flattening effect of forceful blowing) may cause the pitch to bend as much as a minor third. Fortunately the player can use the two effects to offset one another. One can stay in tune while playing fortissimo by stretching the reed enough to counteract the flattening effect of over-forceful vibration. Were someone to devote the amount of time to this instrument that professionals normally devote to established instruments, this compensation might become second nature; not only that, but the freedom afforded by such a wobbly instrument would surely become a great expressive virtue. As an added advantage, the pitch change associated with changes in reed tension means that the instrument can easily be tuned by adjusting the reed and its mounting. Still, it must be said that at this point in my explorations with membrane reeds, the problem of pitch instability remains a serious one and has prevented me from making an instrument that I find fully satisfactory.

Membrane reeds can be designed to work over a large pitch range, from the contra-bass region to mezzo-soprano. The size of the membrane itself becomes a factor here. Normally, if only as a matter of convenience, the diameter of the membrane will correspond to the diameter of the sounding tube for the pitch range in question. For typical wind instruments in the soprano range, internal tube diameters near the mouthpiece end range (very roughly) from perhaps ¼ to ⅝ inches, since these small diameters are what work well with the relatively short soprano tube lengths. Problems may arise for membrane reeds in the highest musical ranges because it becomes difficult to create an effective membrane over these small diameter openings. On the other hand, where longer tube lengths allow it, one can create membrane reeds of very large diameter – like two, three, four or more inches. A membrane reed over a six foot length of four-inch diameter sewer pipe produces an impressive sound indeed. It's quite something visually as well, since the large membrane can be seen vibrating with amplitudes of a half inch and more above and below the rest point. Not surprisingly, relatively large, heavy weight balloons work best in such applications. They tend to be more stable than very thin balloons in their patterns of oscillation (less prone to jump to the higher overtones), and of course they don't break as easily.

It's also quite possible to make short fat membrane reed instruments, with three- or four-inch diameter membranes over tubes of less than a foot in length. With these, pitch control becomes more difficult and one loses the option of creating tone holes for different pitches. Such instruments will speak though, and often with a nice tone. With them, and to some extent with instruments of longer air columns, one can try a different approach to pitch control: instead of using tone holes or other means to change the effective length of the air column, one can focus on pitch and timbral variations achieved by control of the reed. Reed control may come through the stretching effects described above, plus the player can experiment with using the free hand to press the large vibrating membrane itself at different points and with different pressures. The range is limited, but the possibilities are still quite enjoyable for something so easily made.

We should take note that the presence of a membrane reed affects the resonant frequency of a given tube length, producing a result that is different from either a rigidly stopped end or an open end. The softness of the membrane lowers the sounding pitch as compared to what it would be with a rigidly stopped tube of the same length. This is true of all reed instruments, but the effect is exaggerated by the greater compliance (softness) and larger surface area of the membrane reed as compared with other reeds types. Tube length calculations and tone hole spacings that work with other instruments will have to be modified for membrane reeds.

Descriptions of membrane reed instruments I have made, plus diagrams of some of the other workable reed configurations, appear in Figures 18.4, 18.5, and 18.6.

And now, with a basic understanding of the workings of membrane reeds under our belt, we turn to the Indonesian instrument.

Indonesia

The Indonesian instrument actually came to my attention before I began my own experiments with balloon reeds. Robin Goodfellow, a frequent contributor to *Experimental Musical Instruments* who always seems to come up with such things,

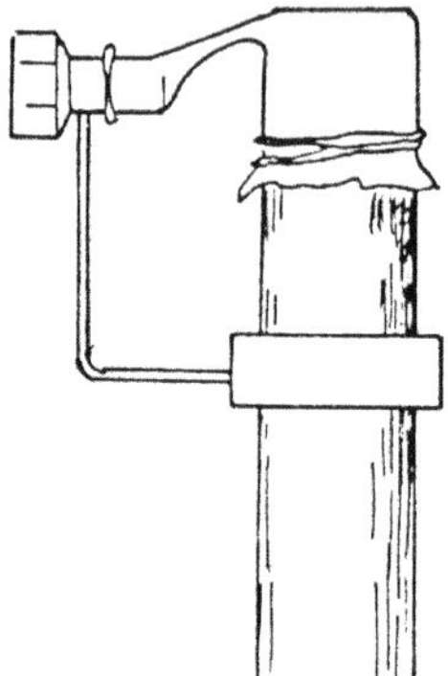

Figure 18.4 A fixed position system for the simple membrane reed set up shown in Figure 18.3. Because changes in tension at the balloon's neck cause considerable detuning of the sounding pitch, it helps to stabilize the position of the mouth piece. An easy way to do this is through a flexible steel support. Using a non-rigid spring-tempered steel rod or bar allows the player to flex the position for pitch bending and vibrato, confident that it will return to an unchanging default position to assure a consistent pitch basis.

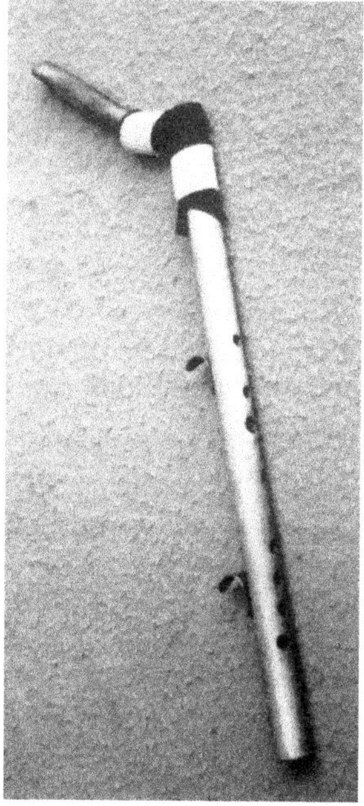

Figure 18.5 Alto membrane reed instrument. A fourteen-inch sounding tube with eight tone holes gives this instrument a range of an octave, starting at G below middle C. The tone holes alone produce something less than a chromatic scale, but by flexing the reed one can fill in the spaces between. The tone is warm, not bright, and somewhat reminiscent of the upper register of a clarinet.

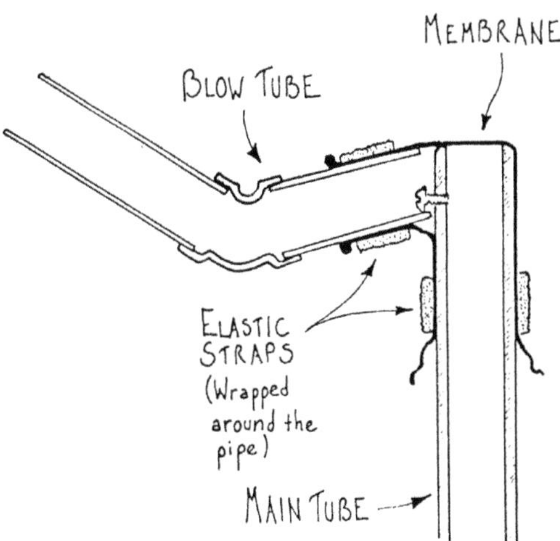

Figure 18.6 The blow tube/membrane reed arrangement used in the alto membrane reed instrument shown in Figure 18.5. In this arrangement, the balloon membrane serves the additional function of holding the blow tube in place, with its position checked by a screw protruding from the main tube. Elastic straps with velcro fasteners wrap around to hold the balloon on both the blow tube and the main tube. Tension on the membrane can be altered during playing by rocking the blow tube down a bit, which has the effect of increasing the space between the top part of the rim of the blow tube and that of the main tube. This turns out to be a natural and easily-mastered performance technique.

showed me a couple of them. She had gotten them from Saul Robbins, a friend who had recently returned from that part of the world. Here is Saul's description of his coming upon the instrument:

> Last year my companion Erin and I travelled around the world. Indonesia was our first country and Sumatra our third island to visit. Part of the time there was during the month of Ramadan, the fasting month, which is a time to test one's devotion to Allah by fasting from sun up to sun down. Needless to say it is a tough test, mode even more difficult by the need to continue working, driving vehicles, and so on. There were many stories told about buses careening off the road from the driver's lack of stamina.
>
> Having survived most of the month's excitement we settled into the small mountain town of Beregstagi in Northern Sumatra to climb a well known volcano and to enjoy the end of Ramadan which would be marked by plenty of excellent food, good humor, and some local rock and roll. We spent that day wandering around the town, talking and laughing with the locals as they enjoyed eating in daylight once again. A favorite picnic spot was up on a hill overlooking the town, where one could rent a mat so as not to spoil his/her clothing, and enjoy fresh corn on the cob and home-made ice cream. It was

here that we noticed several people selling these little flutes. What first drew me to them was not the sounds they emitted but more the inventiveness of the materials. My college and graduate training was in art and photography, and I have always been interested in post-photographic uses for all those little plastic canisters. Here, thousands of miles away from home, was a quite inventive approach to the evil legacy of plastics.

The entire flute is or can be mode from salvageable plastics: ½-inch pipe makes the flute, with a few holes drilled for finger holes. An empty film canister makes the air chamber, and a piece of tubing is fitted into the side for a mouthpiece. A piece of balloon is then stretched over the top of the canister and the lid – which has been carved out – is snapped down to hold it taut. A few adjustments are then made to find the right distance from pipe to balloon for proper resonance and you have a lovely, deep bassy sound. Perfect for serenading a loved one after eating (Figure 18.7).

Figure 18.7 The maker of membrane reeds with Saul Robbins in Berstagi, Sumatra. Saul writes, "You can see the flutes in the foreground as he stores them before 'final assembly'. Also notice the woman and child [in her lap] with flute in hand, beginning early." (Photo © 1989 by Erin Dawn).

To learn more about the instrument and cultural context, I contacted Jack Body at the University of Wellington, New Zealand, who maintains an interest in Indonesian folk and children's instruments. He wrote back:

> I do have a similar instrument myself, which, if I remember correctly, I bought during a Sekaten festival in Yogyakarta in 1976 or '77. In showing it to Indonesian friends they simply say 'Oh yes, it's just one of those types of children's noisemakers.' And indeed in Indonesia there are many such ingenious inventions for children to make noise with. It is a remarkable design, though I would say it is simply a relatively recent invention, not derived from any traditional model or adult instrument that I know of. And like so many similar instruments, mine is made of scrap materials. The main chamber is on old film canister with its cap on, though with its center section cut out. The membrane is actually plastic wrapper which after all this time is still intact and totally functional. The mouthpiece is of transparent plastic and the protruding pipe of grey pvc. There are six holes which comfortably suit three for each hand. It seems to me the holes are distanced arbitrarily, the end one being rather close to the end of the pipe, thus effecting little change in pitch. I suppose I should have been more curious at the time I purchased it, but on other occasions I've found it difficult to get factual information – the sellers are suspicious of one's curiosity.

To pull some of this information together, we appear to have here an informal or children's instrument – albeit, a remarkably clever and inventive one. It's been around in fairly stable form for at least the 14 years or so since Jack Body purchased his model, and given that Saul Robbins picked up his just last year, it appears still to be going strong as a popular instrument type. Its geographic distribution is not all that narrow, since Yogyakarta, where Jack saw the instrument, is in central Java, roughly a thousand miles as the seabird flies from Saul's mountain town in northern Sumatra. No one that I have spoken to has been able to point to any traditional instruments employing the same sound production principle, so we may here have come across an instance of a relatively rare occurrence in organology: the spontaneous appearance of a new species – but this remains speculative. I might add that at this point I still haven't even learned the Indonesian name for the thing, or indeed whether it has a formal name (Figure 18.8).

And now, for a fuller description:

As shown in Figure 18.9, the instrument uses a main air column tube, with the modified plastic film canister described by our correspondents slipped over one end. To make this possible, the bottom of the canister has a hole cut in it, allowing it to fit snugly and air-tight over the main tube. The diameter of the pipe and canister are such that an air space of perhaps an eighth of an inch remains between the two

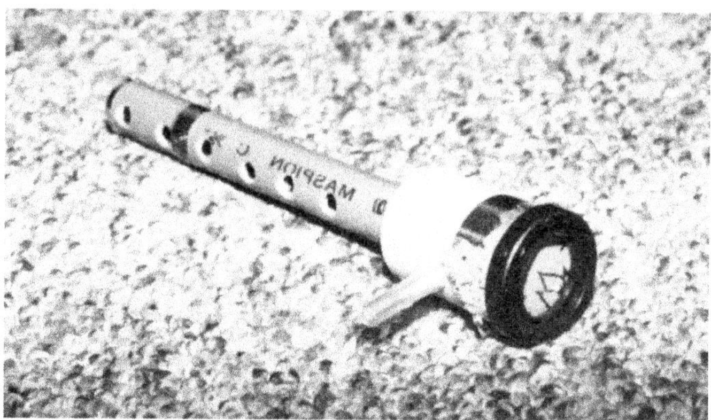

Figure 18.8 The instrument purchased by Jack Body in Yogyakarta, Java, in 1976 or 1977, showing the white film canister and the cellophane membrane held fast under the cut-out canister lid. (Note also the exotic notations on the main tube – a result of the original photographic print having been unintentionally reversed).

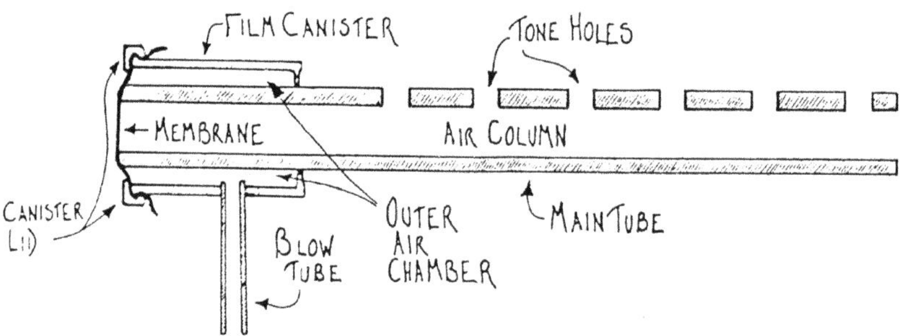

Figure 18.9 Cross section diagram of the Indonesian membrane reed. When the player blows the blow tube, air enters the outer chamber, and from there is forced to squeeze under the membrane and into the main tube air column.

all around. The snap-on top of the canister also has a hole cut in it, but of larger diameter than the main pipe. The membrane – balloon or cellophane – goes over the open end of the canister and is held in place when the canister lid is snapped on over it. The whole canister can then be slid down along the main pipe to a point where the end of the pipe would, were it not for the membrane, begin to poke out through the larger hole cut in the lid of the canister. Instead, in this position, it exerts pressure on the membrane and stretches it slightly as it pushes it outward. A small blow tube set snugly into a hole drilled in the side of the canister allows the player to direct air into the enclosed space between the canister and the main tube.

The operative differences between this membrane reed system and the arrangements described in the first part of this article lie in the air path by which air is forced under the membrane. When the player blows through the short mouthpiece tube, pressure builds up in the air chamber formed by the canister around the main pipe. One boundary of this chamber, as can be seen in the diagram, is in effect a ring-shaped section of the membrane. The pressurized air, having nowhere else to go, forces the membrane up and scoots underneath. Thus, it escapes in rapid pulses into the main pipe, exactly as it does with the other membrane reed arrangements described earlier.

The system is somewhat reminiscent of the early European reed cap instruments, or the reed pipes used in bagpipes, in that an air chamber intervenes between the player and the reed, and the only control the player has over the reed is through air pressure variation. The main adjustment to be made in getting the system to work well involves sliding the canister forward or backward along the main tube by small amounts. This has the effect of tuning the tension on the membrane, since it varies the amount that the main tube extends beyond the film canister end, and therefore how much it pushes outward on the membrane. The system as a whole is remarkably effective, in that a properly adjusted instrument speaks without hesitation, with a loud, full-bodied tone and well-defined pitch.

Acknowledgments

Thanks for assistance with this article go to Robin Goodfellow for first showing me the Indonesian membrane reed instrument; Saul Robbins for providing background information on the Sumatran version; Erin Dawn for the associated photograph; and Jack Body for photograph of and background information on the Javanese instrument and general insight into Indonesian children's instruments.

19

Circuit-Bending and Living Instruments

Qubais Reed Ghazala

This article originally appeared in *Experimental Musical Instruments* Volume 8 #1, September 1992.

The element of discovery has a special allure. The course of its pathways and revelations entice and mystify, tangling together art and science in a wild history of consequences. And so we find that to explore the origin of our subject – experimental electronics – even the dark forests of pre-history played a famous role. It's the sap of these ancient trees that over time hardened to form the precious golden material known as amber… the same amber that Thales was so intrigued by in Greece, back in 600 BC.

Figure 19.1 Qubais Reed Ghazala, bending circuits.

Figure 19.2 The Odor Box.

In the noted experiment, Thales discovered that nodules of amber rubbed with wool could, like a non-metallic magnet, attract small particles by means of some mysterious and unseen force. The energy at work within the amber was, of course, the same electricity that is at work in all our electronic instruments. And just as it was with Thales, the trail of discovery for the experimental electronic instrument designer is similarly based upon the control and manipulation of electrons; the name 'electron' in fact stemming from the Greek word for amber. But it's not only the customary patterns of these electrons, a scintillating fabric of energy traveling near the speed of light and named after tree sap, that I'm concerned with... the truth is, the most alluring realm of electronic discovery for me has been based upon sending electrons where they were never meant to be sent at all... I call this *Circuit-Bending*.

Maybe I should digress for a moment.

For better or worse (the verdict is not yet in), I started listening to electronic music when I was one or two years old. Do you picture me toddling down to the record shop and thumbing through the avant-garde bin? No, what actually happened was a little less picturesque, though I will remember it forever. My mother brought home a few of those kiddy records – colored vinyl 45s. One of them was entitled something like "Ohmichron and Nutmichron," the tuneful adventure of two lost travelers from outer space, temporarily stranded on planet Earth. What I really liked about it were the streams of early synthesizer sound effects that ran through the whole thing... abstract and gripping electronic music for kids! For some reason, these strange blooms of noise amazed me. I must have listened to Ohmichron and

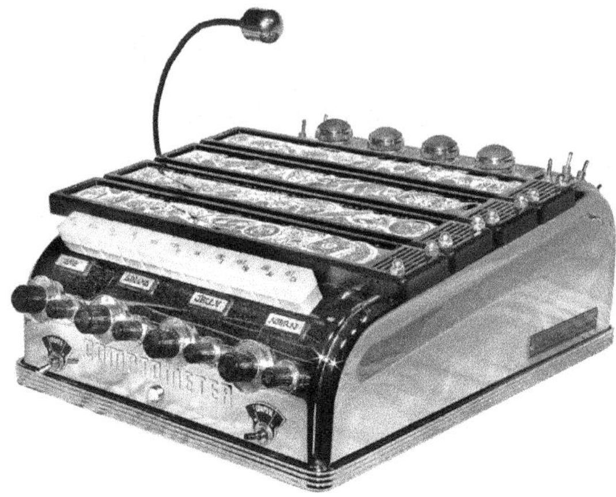

Figure 19.3 Dworkian register.

Nutmichron hundreds of times. From that point on, I did all kinds of things to hear or create odd sounds. But to try to make a long story short, my fascination with unusual music has, over the years, expanded into the collecting and creating of strange musical instruments, as well as recording instrumental music under the title of *Sound Theater*, with work now circulating on CDs and cassettes and, like Ohmichron and Nutmichron, a 7-inch single.

Like many modern composers, I consider the world to be my orchestra. Similar to the painter's palette, upon which all possible colors are accepted, is my palate, containing all possible sounds. And why not? Color and sound, both waveform phenomena, each rich in emotional and descriptive power, standing upon equal ground. Each to be manipulated as freely as the other in pursuit of aesthetic creation. Within the rich terrain of this conclusion, there lies a Never-Never Land…

I am certainly not the first person to have discovered that weird things happen to audio electronics when you short-out their circuit boards in just the right places. Such creative short-circuiting is what I refer to as circuit bending. This process is at the heart of many of my instruments. For right now, I'll just touch on the general practice of Circuit-Bending, to simply lay the groundwork.

In place of the amber, we now have an audio circuit board of some type to experiment with, which is fairly significant in more than one way.

In Thales' time, the arena of discovery was invitingly open to the layman. Today, however, it seems as though corporate monopolization and private R&D labs have successfully distanced the "threshold of invention" from the common person. Circuit-Bending changes all this. Additionally, to have a working complex circuit board to serve as the starting point for our explorations presents us with what in some

Figure 19.4 Eyed Inverter.

cases is an audio wonderland waiting to happen… the best boards amounting to intricate mazes or catacombs of fantastic sound-forms to uncover.

It's best to start with a circuit board that has two basic qualities. Number 1: It should be cheap enough that destroying it through the experiments, which is always a possibility, won't bother you too much.

Number 2: Choose a device that is easy to open (screws) and has lots of accessible internal electronics (which can sometimes be seen through the battery doors or other openings in the case). Just about any battery-powered low-voltage audio device is worth experimenting with. But *don't* try the following procedures with any devices that run on house current unless you're familiar with working in 110-Volt systems. As a general rule I use only devices operating on nine volts or less for Circuit-Bending. Also, I rarely spend more than five dollars on a device, many being bought for a buck or less. As you might have guessed, I don't buy these devices new. Places to search include second-hand shops, garage sales, rummage sales, flea markets, and charity retail outlets such as Goodwill, Salvation Army, St.Vincent De Paul, and Amvets. Also check warehouse auctions and going-out-of-business sales. Ask electronics store managers what they do with their damaged stock. You might be able to buy a box of usable parts or possibly even get their discards for free. Even department store dumpsters are worth a look!

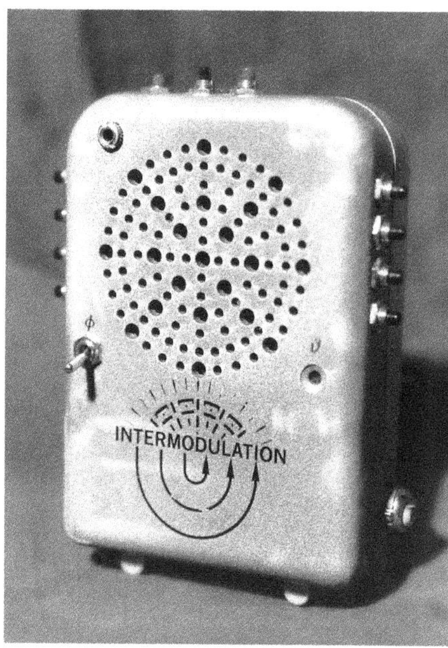

Figure 19.5 Intermodulation.

I seem to be able to find an endless supply of usable circuits through these local sources. Devices to look for include battery-powered amps for telephone, instrument, computer or related use (which usually run on a single 9-volt battery, and are quite small) as well as radios, tape players, and the occasional worn-out fuzz-tone, wah-wah, phase shifter, or other dated effects box. But I'm really impressed by the circuit-bending possibilities within children's audio toys. A couple of my most outrageous and complex instruments are based upon the intricate sampling electronics hidden within recent teddy bears! Space-guns, games, toy vehicles, small musical keyboards and other instruments, walkie-talkies, bike horns and other noisemakers, talking dolls, speaking educational toys, etc, can all be circuit-bent. Before I get into the actual process of circuit-bending, here's a very interesting concept to be considered. The concept is that of creating a "living instrument," an instrument very different from the sound-producing electronics that we're all familiar with. Because, as I stated before, the procedure of circuit-bending is one of an exploration through creative short-circuits, there is the possibility that the resultant modified device will contain new wiring that causes voltage and temperature thresholds of the original circuit to be surpassed. In effect, this thermal overload produces a force that can change the various electronic values of the circuit over a period of time ... minutes, years, decades. Most of the thermal changes are too small to be able to actually feel. In fact, if the bending of a circuit causes any

Figure 19.6 Incantor.

component on the board to warm up significantly I always remove the offending connection no matter how great the sound produced might have been. Not all circuit-bending produces such a device, but those modifications that result in the slow burn-out of an instrument, like the process of our own aging, leave us with an instrument that has a definite lifespan over which it and its voice will change, just as it is with you and me.

It could be said that all instruments, electronic and acoustic alike, have lifespans that will evolve in both a sonic and physical sense over time. Of course, this is true. But whereas there are examples of fine string and wind instruments whose tonal qualities are said to mellow and strengthen overtime, the greater concern of traditional fine instrument builders is always that of creating an instrument that will, first, produce a musical voice of superior quality, and second, be built in such a manner as to protect that voice for as long as possible from the destiny that time must inevitably demand. The circuit bent "living instrument" has made peace with time. There is no battle perceived. The finite lifecycle of a living instrument is something I accept as being natural and intrinsic to its being, function, and its purpose.

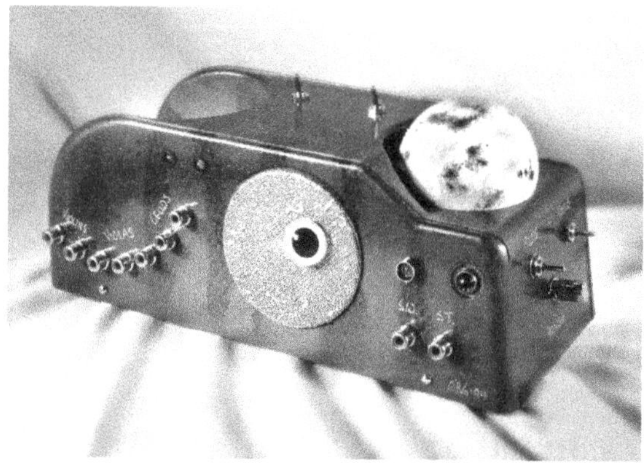

Figure 19.7 Video Octavox.

Figure 19.8 Cat Box Touch Pads.

We are not used to thinking of our instruments the way we think about our friends, pets or for that matter our house plants… that they're living entities and a little bit older and different every day. We expect our keyboard or reverb unit to do tomorrow exactly what it does right now. But if we're asked to consider an instrument as a thing that changes, ages, and someday dies, how then will our behavior toward it be altered?

I suppose a person could be tempted to try to "save the life" of a living instrument by not using it. I've felt this way at times and I can't tell you if my occasional decision to use a standard instrument instead of a living one for a piece of music has kept me from a further discovery or not. I can tell you this however… a living instrument seems to automatically create in the player, if not an unusual *degree*, then at least

Figure 19.9 Sound Dungeon.

an unusual *angle* of consideration. Anytime a musician builds an instrument there is automatically created a closer relationship between the person and the instrument than if the device were mass-produced and had simply been purchased. Add then to this the effect upon the user of operating an instrument in possession of a voice-evolution including a distinct infancy, childhood, adulthood, and an eventual collapse into silence. There have even been times when living instruments cried out upon initial power-up as their new circuit stabilized reminding me of a baby's first cries of birth.

So yes, while it's true that circuit-bent living instruments do inspire an uncommon respect in their creators, ultimately it is realized that for them to live at all they must be used, they must age, and they must eventually die. The breath of their existence would be lost if their "lives were saved" for fear of turning them on.

I am reminded of a similar situation and tough decision I had to make a few years ago. While hiking the forests along Lake Superior on a cold autumn day, I discovered an old deer skeleton off to the side of a sheltered and overgrown trail. All the bones were there except the skull, wish I could find nowhere around. I soon realized that the bones were probably the remains of a "trophy" kill, the head having been removed for mounting. Through my own experiences of recording and photography, I've developed a considerable degree of respect for the abilities of the deep woods tracker able to read the many stories written in the subtle disturbances of the terrain, but for the culmination of these skills to result in the killing of an animal for the sake of souvenir is an act I see as cruel and unconscionable. I felt a strong desire to try to salvage the dignity and intrinsic beauty of this animal that had been so grossly violated by the sportsman.

I sat down among the bones and struggled with the thought of collecting the time-bleached remnants of this organic living instrument, disturbing their placement to reassemble them into an art-piece of some musical or kinetic nature. But

I wasn't really sure if I should remove them at all. I sat for a long time tapping the bones, listening to them, pondering their place in the world… then asked myself: what if these were my bones? Would I rather they be lost, trampled underfoot, a silent testimony to a heartless and wasteful act, or would I rather their story be told and their voices be heard in the form of an instrument of some type to be played and pondered?

I collected the bones along with some driftwood and returned to camp to begin the construction of two great-sounding xylophones which today are on display in my home and have been used for recordings that are now heard far and wide. I can only hope that the deer would approve of my decision which, as it is in several ways with circuit-bent living instruments, favors active usage and exploration over the opposite – a slumber of wondrous potential.

The actual process of circuit-bending is really very simple. You don't have to possess any difficult skills or know even a bit of electronic theory. However, if you have never built an electronic circuit, I strongly recommend that you buy a couple basic kits to assemble before you begin building. These will teach you how to use the basic tools – low wattage soldering iron, wire, wire strippers, switches, components, and so on. While the techniques involved in electronics are within the grasp of everyone and take only a little practice to learn, their generalities are still outside the scope of this article and will be much better taught by writers devoted to the subject. Don't be put off by the mystique of electronics. You're not going to be electrocuted and nothing's going to blow up as long as you follow the simple rules. But, as I said, if you are an electronics novice, *do* familiarize yourself with the basics before you begin to bend.

So now let's assume that you're comfortable with the low-wattage soldering iron, you know how to wire a simple switch, and you have a drill for mounting switches onto a panel.

The first step is to open the audio device of your choosing to expose the circuitry. Once this is done, try to locate where the battery voltage enters the circuit-board and, as a general rule, avoid this area during your jumper tests, which are the first steps of exploration. To make the jumper wire, just strip about 1/2 inch of the insulation from each end of a wire that is itself about twice the length of the board. If your jumper is made up of many thin wires twisted together inside the insulation (stranded wire), it will help to solder or "tin" the exposed ends to firm them up and keep them together. Turn the device on and have it produce its normal sound (hold down keys, buttons, etc., with pieces of tape or paper wedge). In the case of an amp or other passive circuit just turn it up about one third of the way, or send a signal from another expendable device through it. You're now ready for the first audible circuit bending tests.

Choose a point on the circuit-board, away from the battery leads, while you touch one end of the jumper to bare metal – the path of a conductive trace on the board, or the metal lead of a component itself. Keep that end there. With the other end of the jumper, cautiously begin to touch various additional points on the circuit while observing any changes in sound. At first touch the traveling end of the jumper to the decided point only very briefly, just for a moment. If you hear a loud "pop" from the speaker, or if the speaker goes dead or just loudly hums, don't push your luck… go to the next point and try again.

Good circuit-bending points produce unusual tones without great changes in volume. Make a chart of successful bending points or code them right on the board with a marker. Once you've touched a number of points with the traveling end of the jumper, move the stationary end to a different point and start all over.

Once you're done with this process, if you're lucky, you will have discovered a number of jumper paths that result in interesting effects. The next and simplest thing to do is cut a series of jumpers to facilitate the creative short circuits you found, the idea being to solder these in place to gain access to the various new sounds produced. However, you will need to place a switch in the middle of each wire to enable you to choose which effect, or combination of affects, you want to turn on. I should note here that although individual jumpers alone may have no adverse effect upon the circuit, turning several, or even the wrong two on at once, can cause threshold problems as mentioned earlier. That having been said, the next logical step is to mount the switches on a case that will now protect the modified device. If there's room on the original case for these switches, fine. If not, put the whole device into a larger and closure of some type and mount the switches (volume control, speaker, power switch, etc.) on that. These are the beginnings, the very basics, of circuit-bending.

The elaborations possible with circuit-bending are endless. For example, you can interrupt or replace the jumpers with resistors or capacitors as well as other components for added effects. With a multi-position rotary switch, it's possible to turn one circuit-bending point into a series of effects by connecting a different component (resistor, capacitor, etc.) to each of the switch's terminals. You'll find circuit-points that alter functions other than audio. You'll discover points that light LEDs for status indicators, even contacts that will conduct current through your body for the ultimate living instrument bio-modulation possibilities.

Admittedly, if you do approach circuit-bending with only minimal prior experience, the initial attempts will probably be very haphazard. Still, circuit-bending can inspire experiment after experiment, teach you a lot about electronics, and help you create your own orchestra of unique original experimental musical instruments.

I'm a mushroom hunter. There are a few mushrooms at the side of the trail that can kill you. The dangerous ones are easy to avoid, leaving all the rest, with their various rewards, open to the curious explorer. It's the same with electronics. All of the following cautions should be respected and heeded.

Safety Rules of Circuit-Bending

ONLY USE BATTERY-POWERED DEVICES OF NINE VOLTS OR LESS. Higher voltages can be dangerous. House-current (110V line-supply) is definitely too much to fool around with if you're not an experienced technician.

ALWAYS WEAR PROTECTIVE GOGGLES. Only once in my 25 years of continual circuit-bending have I ever had a component explode, and it was due to a foolish mistake on my part ... I blew-up a transistor on a "Speak 'n' Spell" by attaching a power supply of twice the required voltage. It was a nice flash though.

BE AWARE OF THERMAL CHANGES. Feel, with caution, the working circuit before you begin to bend it. Some parts may naturally heat-up. If anything gets a lot warmer, it's probably wise to remove the cause.

FOLLOW THE BASIC CAUTIONS OF PROTOTYPING. Read up on circuit building. Learn which electrolytic capacitors are dangerous and avoid them. Such larger capacitors will be absent from most small circuit-bending targets, but it's good practice to treat any electrolytic larger than a cigarette filter with respect. The stored charge of a large capacitor, such as those that supply the zap for photographic strobes, has the power to melt the tip of your pliers and knock you across the room!

After all this, you might still ask – but why circuit-bend? How could a short-circuited toy compare to the superb electronics of pro audio equipment? To this, as an avid instrument builder and collector, I must answer that the question is not one of comparisons. The bottom line is that every instrument has its own voice, its own emotional sound-form to present. I should mention that I have no dislike for traditional circuit-building. I start many projects by drawing a schematic upon which to base the entire assembly, as is the standard procedure. This approach is usually rewarding and I end up with what I bargained for. But with circuit-bending you simply never know what you're going to get, and I find the discovery fascinating.

Most importantly though, in addition to the fun of the circuit-bending process are the bizarre sounds and special voicings often achieved, the likes of which certainly break new ground. If the modern synthesizer is to be accepted for what it is, a logically organized theory-true machine, then the circuit-bent instrument must be accepted as being based upon the chaos of wild chance and outlaw experiment ... elements dramatically reflected in its unorthodox phrasings. While many of my devices would certainly constitute an electrical engineer's nightmare, to me they are crystallized wonderlands of sonic eccentricity. And I must try to convey that in this society of mass-production and conformity there is something truly satisfying about sitting down with a uniquely-voiced living instrument that not only defies the design standards of its field, but also exists nowhere else in the universe.

Related book and articles

Qubais Reed Ghazala's full-length book devoted to Circuit-Bending is *Circuit-Bending: Build Your Own Alien Instruments* (Wiley, 2005). His articles on Circuit-Bending and other topics appeared in *Experimental Musical Instruments* on a regular basis and can be found in the following issues: Vol 8 #1-4; Vol 9 #1-4; Vol 10 #2, 3, and 4; Vol 11 # 1, 2, and 4; Vol 12 #2 and 4; Vol 13 #2 and 4, Vol 14 #2 and 4. These articles can be accessed through the Internet Archive at https://archive.org/details/emi_archive/.

20

The Giant Lamellophones

A Transatlantic Perspective

Richard Graham

This article originally appeared in *Experimental Musical Instruments* Volume 9 #1, September 1993. It has been revised and updated for this current edition.

In the study of organology, African lamellophones are a unique class of musical instruments. Conceptually these instruments are a handheld version of the African xylophone or marimba, consisting of a tuned series of metal, raffia, bamboo, or other vegetable lamella. These lamellae pass over a bridge which rests on a sound board and are tensed between a restraining bar, the bridge, and a back rest. Lamellophones are usually equipped with a box, calabash, coconut, or tin can resonator, at times various materials such as bottle tops or sea shells may be attached to the instrument as rattling devices, lending the instrument greater sustain and amplification as well as providing a subsidiary buzzing timbre. Lamellophones likely originated thousands of years ago in the raffia intensity zone of central West-Africa, eventually spreading across a huge swathe of the continent. But metallurgy revolutionized the manufacture of these instruments, and models with metal keys may have emerged as long ago as 1,400 years ago. Lamellophones enjoy a wide distribution area in Africa, especially in the Bantu-speaking areas which stretch across the entire center of the continent. To sound the instrument, the tensed lamellae are depressed and then released by the thumbs or up-picked by the index fingers, allowing the vibrating lamella to resound. Erroneously called "thumb piano", lamellophones appear in a variety of sizes and shapes, being constructed from locally available materials in both traditional and experimental forms, and have many different names in various local African languages, such as *likembe* in Angola, *mbira* in Zimbabwe, *malimba*, or *kadongo* in Uganda, *timbrh* in Cameroon, and *lulimba* or *ulimba* in Tanzania. They may possess any number of lamellae, and both the layout and tuning of the instrument's "keyboard" vary from place to place, although Andrew Tracey has suggested that an ancient "kalimba core" tuning scheme may have provided a template from which all subsequent tuning systems were derived (Tracey 1972:89).

Larger box-resonated lamellophones with dimensions exceeding 18 inches in length are rare in Africa, appearing mainly along a limited distribution area of the

Gulf of Guinea. Strong evidence suggests that these lamellophones are of a more recent introduction, and that their construction, performance techniques, and their musical contexts represent a remarkable departure from the core area of the instrument's central African distribution. In West Africa, there are two separate groups of large model lamellophones, some which are still small enough to be handheld, these models being derived from instruments which were exported from Eastern Nigeria and Cameroon via the maritime trade and the resettlement of liberated slaves in Sierra Leone, Liberia, and the Gambia. One such example is the large *kondi* lamellophone from Sierra Leone. Back in 1979, the late Collin Walcott, of the ECM recording groups Oregon and CODONA, kindly allowed me to examine his tin can resonated kondi (personal communication (8/27/79). The nine lamellae on Walcott's instrument were made from umbrella spokes. To play his instrument, Walcott placed it against his waist and depressed the lamellae with the index and middle fingers of both hands, employing a unique "pianistic" technique rather than the thumbs-only kondi playing technique traditionally used by the Loko people. On top of Walcott's lamellophone was a buzzing accoutrement made from a flattened bug spray can pierced with metal rings that gave it a pleasant buzzing timbre.

Yet another larger handheld instrument called *molo* is played by the Yoruba, also performed with the instrument's lamella facing outward and employing a pianistic technique which likely inspired Walcott's own approach. According to the musicologist Darius Theime,

> The performer will begin a two-hand ostinato played with the index fingers of both hands plus the middle finger of his right hand. After the ostinato pattern has been established, he will add a Yoruba proverb to it. A good player can do this so cleverly that it is difficult to tell which notes of the ostinato are omitted in order to release a finger to play the proverb. The proverb is 'played' by duplicating the variations in pitch from syllable to syllable of spoken tonal Yoruba by use of different pitches on the instrument.
>
> (Thieme 1967:43)

Still larger West African lamellophones, such as the Yoruba and Fon *agidigbo*, the Vai *kongoma*, and the Akan *prempensua* are so large they must be held in the lap or sat upon to be played, and lacking a better term, I dubbed those models "giant" lamellophones back in the 1980s. In these instances, the giant lamellophone's musical function is to provide a bass voice in creolized music, rather than the traditional hand-held instrument's usual function as a higher pitched lead voice in African music. Although the historic record of the Atlantic World is filled with literary and iconographic references to the presence of a plethora of hand-held lamellophones recreated and played there by enslaved Africans, most of these models have long since disappeared. Conversely, there is now a large international complex of giant lamellophones in the Atlantic world, with examples including the Cuban marimbula, the Jamaican rumba box, and with other local variants occurring in Aruba, Curacao, Colombia, the Dominican Republic, Haiti, Puerto Rico, and Venezuela. In most of these areas, the large box resonated lamellaphones are known as marimbula, an appellation deriving from the Bantu term “marimba”, meaning several notes (Kubik 1979:37), with the addition of the Spanish suffix “ula”, meaning “little”. Other instruments in this trans-Atlantic complex include the giant banja lamellophone of Trinidad, crafted from an old kerosene tin with bamboo tongues attached, which Andrew Pearse reports was used in the context of a wake by old Africans who played it while “singing about the dead.” (Pearse 1955: 34).

Complex issues regarding the ultimate origins of the giant lamellophones have yet to be completely resolved. In a 1991 conversation with the Austrian social scientist Gerhard Kubik, he suggested that the giant lamellophones were likely a Caribbean innovation which were subsequently reintroduced into West Africa (personal conversation 8/14/91). Kubik has since published his findings in a recent book chapter

Figure 20.1 Sun Facey playing his rumba box in Negril, Jamaica.

on the history of African and Atlantic lamellophones, confirming and expanding upon his earlier hypothesis (Kubik 1999: 20–57). Adding weight to Kubik's theory are a number of recent studies illuminating the repatriation of Brazilian and Cuban ex-slaves to Nigeria and expelled Jamaican Maroons to Sierra Leone (Bilby 2011, Cohen 2002, Matory 1999, Strickrodt 2008). Using the original handheld African models as a cultural template, Caribbean and Brazilian makers constructed these innovative giant Atlantic instruments with fewer lamellae, usually with only three to ten lamellae. These experimental instruments were made by enslaved or free craftsmen using spring steel tongues, often from clock or Victrola mechanisms, which were tensed between two wooden blocks screwed together, over larger resonators created from packing crates and other detritus of the industrialized New World (Figure 20.1). Working in Cuba in 1941, the folklorist Harold Courlander discovered extant examples of both the traditional handheld and the experimental giant lamellophone models during his field studies there (Courlander 1942: 239). So well into the 20th century we find that some Afro-Cubans were still making lamellophones that replicated traditional African handheld models alongside the new giant models.

It is to the transatlantic influence of Afro-Cuban music that we initially turn in order to reconstruct the creation and diffusion of the giant lamellophones. These instruments were associated with Abakuá lodges, the Western Cuban outpost of fraternal associations which originated in southeastern Nigeria and southwestern Cameroon (Ortiz V5 1955: 333). More recent scholarship by the cultural historian Ivor Miller suggests that Abakuá acolytes still revere and prominently display their ancestral African lamellophones;

> The altar spaces in Abakuá lodges display objects that evoke Cross River cultural history. They often contain a seven-keyed lamellophone to represent the original music of Ùsàghàdè; in the Calabar region where lamellophones were once common instruments.
>
> (Miller 2000: 288)

One of Miller's Cuban informants also reported a tuned set of three marimbulas once used as drum substitutes in the Uriabón Efí lodge of Havana in the 1940s. These included marimbulas which were graduated in size and named from smallest to largest, "puntera" with two lamellae, "bajo" with three lamellae, and "llamdor" with four lamellae, all named for their voices and musical functions in the familiar African Caribbean paradigm of "talking instruments."

Another likely source for the "invention" of the marimbula is found in eastern Cuba. Here agricultural workers, many of Bantu origins, first developed changüí music around 1900 (personal communication with Benjamin Lapidus 1/22/2020), a form which was instrumental in spreading the popularity of the giant lamellophones.

Figure 20.2 Marimbulero Jesús Chuchú Arístola with the Terceto Yoyo trio.

By the 1920s, the marimbula had diffused to the emerging urban son form in Havana. Played by a seated musician who at times also struck the instrument's box resonator for percussive effects, the marimbula's role in son was as an Africanizing "melo/rhythm" instrument, and the "anticipated bass" lines of the instrument's "tumbao" patterns soon revolutionized Cuban music. The oldest known recordings of the marímbula in Cuba were made in 1925 for the Victor recording company by the marimbulero Jesús "Chuchú" Arístola, along with his trio, the "Terceto Yoyo" (Figure 20.2).

With the advent of the Cuban son in the early 20th century, the music and instrumentation of the son conjuntos were heard and replicated through the medium of records, films, and touring groups in Jamaica, Puerto Rico, Colombia, and Argentina (List 1966:55, Pollak-Eltz 1978:29). Inter-island migrant workers also helped to spread Cuban musical instruments and performance practices throughout the Caribbean. As an associated phenomenon, the giant lamellophones spread with the son's international popularity, perhaps reinforced by existing lamellophone traditions in some areas, and achieving something of a cultural critical mass by 1930. In 1965, the musicologist George List recorded a Colombian marimbula player, Jose Isabel Castillo Martinez, who employed a "showy" technique wherein he passed his left hand underneath his bent knee to play the instrument, an attitude with clear connections to the Angolan likembe and other handheld lamellophones (List 1966: Figs 1 & 2). Remarkable also is the retention in English and French speaking areas of the Bantu/Spanish term, "marimbula", suggesting a central point of diffusion – Cuba.

During my own fieldwork in Jamaica, I recorded and interviewed Sun Facey, a delightful 6'6" musician who performs on his rumba box and kazoo with various mento groups in Negril. Sun played his giant lamellophone in a manner identical to the marimbula players of the 1920s Cuban son, although he has also added a few small tin jingles to the side of his box for additional percussive effects when he strikes them. Here in Jamaica the term "rumba box" is used to denote the instrument, rumba being another Bantu-derived word once used as an international marketing term for the son outside of Cuba. Today, one can purchase affordable "airport art" variants of the giant lamellophones on many Caribbean islands, often gaily painted as some of the local musician's models are.

Yet another scenario should also be considered regarding the evolution and diffusion of the giant lamellophones, this one focusing on the relationship between 19th century Brazilian lamellophones and the repatriation of Afro-Brazilians to Nigeria. Between 1501 and 1866, Brazil received and enslaved an estimated 4.9 million African people. As a consequence, no other Atlantic region had a greater distribution or diversity of African-derived musical instruments than Brazil did. Nowhere else in the Atlantic world would one encounter such a diverse array of lamellophones as in 18th and 19th century Brazil. Most of these were derived from Kongo/Angolan models, and a few Brazilian models were virtually identical to those instruments. Yet there were other African-derived models recreated by enslaved Brazilians, and these were a radical departure from the more common hand-held models.

Visiting Pernambuco State in 1817 on business, the French trader and author Louis-François de Tollenare took more than a casual interest in the enslaved population there, leaving us this detailed notice of a simple lamellophone;

> a wooden box – which looked like those in which German haberdashery is sent – with four small strips, six or seven inches long, each fastened to it by one end. These strips rested on a small cross-piece that served as a bridge. When the musician lifted one of these strips and let it spring back, he drew from it a muffled sound which made the belly of the box ring. These four strips of different lengths were no doubt tuned, but I could not guess which notes of the scale they yielded. The musician, squatting near his box, seemed to be paying it great attention and ran through his four notes very smoothly.
>
> (Tollenare 1971–73 Part II pp. 470–471)

That the instrument had only four lamellae, perhaps of vegetable matter rather than of metal, and that the wooden box resonator was sufficiently large to hold "haberdashery", here meaning either clothing or sewing implements, and that the musician squatted "near" his instrument, suggests something quite different from the usual hand-held lamellophones of 18th and 19th century Brazil. Again, what we find here in Tollenare's account is an instrument that is a bit of a departure from the other Brazilian lamellophones depicted and described by the various 19th century artists and writers (Theirmann 1971: pp. 90–94, Kubik 1999: pp. 55–57). One might speculate that this was an experimental model and perhaps just an outlier. But further evidence from 19th century Brazil suggests that other models, perhaps even closer to the giant lamellophones, were just on the horizon.

Kubik and other scholars point to the raffia intensity zone of West-Central Africa as not only the possible birthplace of African lamellophones, but also for models that may have inspired the development of the marimbula and other giant lamellophones. These models, with vegetable lamella, are rare in the Atlantic world, yet they do appear in Brazil. One such example is a watercolor by an anonymous artist dated from 1825. Entitled "an African vendor of music", this picture shows a walking musician playing his instrument alongside Guanabara Bay in Rio de Janeiro. The enslaved man is wearing tattered clothing and playing an unusual lamellophone. The instrument appears to have only four vegetable matter lamellae, possibly bamboo, set on a rectangular sound board, and the instrument's resonator, which appears to be wooden rather than a gourd, is curved and cradle-shaped with a protruding headstock. There are two rope handles with wrist loops that hang unused from the end of the instrument. The musician uses his thumbs to sound the instrument. This unusual instrument is nearly identical to a unique African model, in this instance a Fang model from Gabon, a few examples of which are found in European collections and documented by both Ankermann (1901: 32) and Kubik (1999: 42). These Fang models usually have seven raffia lamellae, and some have a head stock like the vendor's instrument in Rio. Still other Fang models are decorated with incisions and have a triangular sound hole cut into the side of the resonator, two features not apparent in the vendor's instrument.

In 1840, the German-Danish writer, revolutionary, and painter Paul Harro Harring spent several months in Rio de Janeiro, where he chronicled the lives of the enslaved. His complete Brazilian drawings weren't published for almost 100 years, but his picture titled, "Danse de Nègres Musiciens" depicts what actually may be two men engaged in a wrestling game, accompanied by two musicians. The standing musician is playing a monochord violin, perhaps inspired by the West African *goje*, but held and played more like a Fulani *nyanyeru* or a European violin. The seated musician has a box resonated lamellophone in his lap and is depressing the six lamellae of the instrument with the fingertips of his index, ring, and middle fingers. This "pianistic" approach, similar to Collin Walcott's technique, is a radical departure from the traditional thumb and index finger approach used to sound the smaller handheld models found throughout central Africa. But did Harro Haring witness

Figure 20.3 Yoruba agidigbo.

a Brazilian "dance", or something more like Brazilian capoeira wrestling, or perhaps the Yoruba *gidigbo* wrestling game, which was also accompanied in Nigeria by the agidigbo lamellophone, as Kubik first suggested in 1979 (Kubik 1979:31). Years of speculation have yet to solve that particular chicken or egg riddle. In an interesting sidebar, researcher Hal Rammel recently alerted me to a Brazilian marimbula player from the Pernambuco-based group, Turunas da Mauricéia. This, as yet unnamed musician performed a nice solo on the instrument in the Ginger Rogers/Fred Astaire film epic, "Flying Down to Rio" in 1933. Although the instrument may ultimately prove to be a Cuban import, it is nonetheless remarkable that a Brazilian musician chose to play it in an otherwise Brazilian music context (Figures 20.3 and 20.4).

Further facilitating transatlantic communication were the estimated 8,000 Afro-Brazilians who returned to West Africa from 1820 to 1899, and the historian Silke Strickrodt informs us that their considerable African resettlements;

> extended from the River Volta, in the west, to the Lagos channel, in the east, and comprised the coastal areas of today's Togo and Benin and parts of Ghana and Nigeria.
>
> (Strickrodt 2008: 39)

Figure 20.4 Paul Harro Harring's "Tanzende Neger" showing agidigbo lamellophone and goge fiddle. (From *Tropical Sketches of Brazil*, circa 1840s).

Some of these transatlantic migrations were undertaken by freemen voluntary, but others were clearly not. Following the Malê Revolt of 1835 in Bahia, a number of the revolt's organizers were also expelled to Ghana, Togo, Benin, and Nigeria. This, along with the large and lucrative maritime trade between Brazil and Nigeria in the 19th century (Turner 1942:55) may have facilitated the diffusion of the agidigbo and other giant lamellophones.

In conclusion, my research in the Caribbean, Brazil, and the US strongly suggests that the mono-directional approach of the "roots and retentions" paradigm fails to adequately address important transatlantic developments, especially the inter-ethnic African cultural exchanges and technology sharing in the Atlantic world and their subsequent impact on African cultures. At this juncture, further research is certainly required, but we do know that a variety of African-derived lamellophones were extremely popular in 19th century Cuba and Brazil (Ortiz 1952–55: 327–343), (Theirmann 1971: 90–94), and that these served as models for subsequent cultural developments in West Africa (Kubik 1999: 54–57). Regardless if the importation of the giant lamellophones into West Africa originated in Cuba or Brazil, and it is possible that both of these scenarios might prove, I propose that the dynamic flow of cross-cultural communication indeed moved back and forth across the Atlantic, enriching all of those cultures immeasurably. This includes innovative giant lamellophone designs with over five octaves of lamellae in Cuba, Haiti, and Puerto Rico, a revival of lamellophones in Brazil, the "Congotronic" electric likembes of Konono No.1 in Kinshasa, the African-inspired array mbira of the US with up to 150 lamellae, and the electric agidigbos of modern Nigerian Apala groups. The ever-changing nature of transatlantic African organology promises a bright future of such continuous experimentation, and I'm delighted to listen and learn as each new development unfolds.

Works Cited and Consulted

Ankermann, Bernhard. *Die afrikanischen musikinstrumente*. Berlin: Druck von A. Haack, Ethnologisches Notizblatt, 3, Heft 1, 1901.

Anonymous 1825 painter "An African slave vendor of music." Brazilian Biblioteca Nacional - SlaveTrade – BNDigital, 2020. http://bndigital.bn.br/projetos/escravos/galeriadesenho.html.

Bilby, Kenneth. "Africa's Creole Drum: the gumbe as vector and signifier of trans-African creolization." In *Creolization as cultural creativity*. Robert A. Baron & Ana C. Cara, editors. Jackson: University Press of Mississippi, 2011.

Cohen, Peter F. "Orisha journeys: the role of travel in the birth of Yorùbá-Atlantic Religions." *Archives de Sciences Sociales des Religions* 117, janvier–mars 2002.

Courlander, Harold. "Musical instruments of Cuba." *The Musical Quarterly*, Vol. XVIII, no. 2, April 1942, pp. 227–240.

de Tollenare, L. F. Notas dominicaes tomadas durante uma residencia em Portugal e no Brasil nos mannos de 1816, 1817, 1818. Parte relative a Pernambuco (Recife: Empreza do Jornal do Recife, 1905).

Goodman, Walter. *The Pearl of the Antilles, or an Artist in Cuba*. London: Henry S. King 1873.

Harro-Harring, Paul. *Tropical Sketches from Brazil*. 1840 Rio de Janeiro: Instituto historico e geografico Brasileiro 1965. Reprint.

Kubik, Gerhard. *Angolan Traits In Black music, Games and Dances of Brazil, a study of African cultural extensions overseas. Estudos de Antropologia Cultural 10*. Lisbon: Junta de investigacoes cientificas do ultramar 1979.

Kubik, Gerhard. "African and African American Lamellophones: history, typology, nomenclature, performers and intracultural concepts." In *Turn up the volume!: a celebration of African music*. Jacqueline Cogdell DjeDje, editor. University of California, Los Angeles. Fowler Museum of Cultural History. Los Angeles: UCLA Fowler Museum of Cultural History, 1999, pp. 20–57.

Lapidus, Benjamin. "The Changüí Genre of Guantánamo, Cuba." *Ethnomusicology*, Vol. 49, no. 1, 2005, pp. 49–74.

List, George. "The Mbira in Cartagena." *Journal of the International Folk Music Council*, Vol. 20, 1968, pp. 54–59.

Matory, Lorand J. "The English professor of Brazil: on the diasporic roots of the Yoruba Nation." *Comparative Studies in Society and History*, Vol. 41, no. 1, 1999, pp. 72–103.

Miller, Ivor. "A secret society goes public: the relationship between Abakuá and cuban popular culture." *African Studies Review*, Vol. 43, no. 1, Special Issue on the Diaspora, April, 2000, pp. 161–188.

Miller, Ivor. "Abakuá society: African sources and structure"; "Language: Efik-Carabali (Abakuá)." In *Cuba: people, culture, history. Encyclopedia. Charles Scribner's Sons*, Ralph Faulkingham, & Mitzi Goheen ed., 2010, pp. 287–292; 535–536.

Miller, Ivor. "Cuban Abakuá music." Médiathèque Caraïbe, Conseil Départemental de la Guadeloupe, June 2016. http://www.lameca.org/dossiers/abakua_music/eng/.

Ortiz, Fernando. *Los Instrumentos de la Muslca Afrocubana.* Vol. V. Havana: Dirección de cultura del Ministerio de educación, 1952–55.

Pearse, Andrew. "Aspects of change in caribbean folk music." *Journal of the International Folk Music Council*, Vol. 7, 1955, pp. 29–36.

Pollak-Eltz, Angelina. "The Marimbula: an Afro-American instrument." *The Review of Ethnology*, Vol. 5/4, 1978, pp. 28–30.

Ramon C. F. Caballero. "La juega de gallos, o el negro bozal", Recuerdos de Puerto Rico, Ponce, Puerto Rico. 1852.

Strickrodt, Silke. "The Brazilian Diaspora to West Africa in the nineteenth century." In *AfrikaAmerika: Atlantische Konstruktionen.* Ineke Phaf-Reinberger & Tiago de Oliveira Pinto editors. Frankfurt a. M.: Vervuert, 2008, pp. 36–68.

Thieme, Darius L. "Three Yoruba Members of the Mbira-Sanza Family." *Journal of the International Folk Music Council*, Vol. 19 (1967), pp. 42–48.

Theirmann, David. "The Mbira in Brazil." *African Music*, Vol. 5, no. 1, 1971, pp. 90–94.

Thompson, Donald. "The Marimbula, an Afro-Caribbean Sanza." In *Yearbook for Inter-American Musical Research*, Gilbert Chase ed., vii 1971, pp. 103—116.

Tracey, Andrew. "The Original African Mbira?" *Journal of the Society for African Music*, Vol. 5, no. 2, 1972, pp. 85–104.

Turner, Lorenzo Dow. "Some contacts of Brazilian ex-slaves with Nigeria-West Africa." *Journal of Negro History*, XXVII, January 1942, pp. 65–66.

Related articles

Following are other articles pertaining to instruments of the African diaspora that appeared in *Experimental Musical Instruments*: These articles can be accessed through the Internet Archive at https://archive.org/details/emi_archive/.

"Wind, Breath and strings Round and Flat" by Charles R. Adams in Vol. 1 #5

"Udu Drum: Voice of the Ancestors" by Frank Giorgini in Vol. 5 #2

"The Diddley Bow in a Global Context" by Richard Graham in Vol. 6 #5

"Trans-atlantic African Organology: The Tradition of Renewal" by Richard Graham in Vol. 8 #1

"Tumbas, Rumba Boxes, and Bamboo Flutes: Caribbean Instruments by Rupert Lewis" by Bart Hopkin in Vol. 9 #1

"'Sugar Belly' Walker and the Bamboo Saxophone" by Bart Hopkin in Vol. 9 #2 and included in the current collection

"The Benta: An African-Derived Glissed Idiochord Zither of Eastern Jamaica": Richard Graham in Vol. 9 #3

"The Bamboolin: A Jamaican Idiochord Zither" by Bart Hopkin in Vol. 9 #3

"Instruments of the Cuban National Folkloric Dance Ensemble" by Steve Smith in Vol. 12 #2

21

Tumbas, Rumba Boxes, and Bamboo Flutes

Caribbean Instruments by Rupert Lewis

Bart Hopkin

This article originally appeared in *Experimental Musical Instruments* Volume 9 #1, September 1993.

Popular music in Jamaica for many years has been dominated by the instruments of North American popular music – electric guitar and bass, the drums of the trap set, piano or electric organ, and more recently synthesizers and electronic drum machines. With that instrumentation, the ska, reggae, and dance hall styles emerging from the island have proven immensely fertile and popularly successful. Rare now are the unamplified calypso, mento, and quadrille bands of an earlier era.

Yet mento and calypso musicians can still be found here and there, in rural districts and sometimes in tourist places. In addition to guitars and the occasional saxophone, there was a place in these older Jamaican popular music styles for several instrument variants not seen up north. There were various and sundry types of shakers or scrapers, and hand-played, single-headed long drums both large and small. There were banjos – indigenous to some Caribbean islands just as to the US – made, as often as not, by local craftsmen. There were sometimes fifes and pennywhistles, locally made of bamboo or other available tubular materials. And, down in the bass, there was the rumba box – the big kalimba that appears in several of the islands.

As the older musical styles have given way to the contemporary, the number of people making the traditional instruments has dwindled. One of the few makers still active is Rupert Lewis. I visited Mr. Lewis in Kingston, Jamaica, on several occasions early in 1992 (Figure 21.1).

Rupert Lewis was born in Kingston. As a young man he learned woodworking and cabinetry. At the same time, he cultivated an interest in music and instrument making. It began, when he was a child, with home-made plucked string instruments made with herring tins and fishing line. Later he made himself a ukulele, teaching himself to play and eventually doing a lot of performing around Kingston. He took up the guitar sometime later, when someone who had brought a guitar to him for repair, deciding that Mr. Lewis would make better use of it than he, left it with him to keep. He studied music informally with teachers around Kingston, and later

Figure 21.1 Rupert Lewis.

through an overseas correspondence course from an institution called The United States School of Music. His growing reputation as a musician led to work on a tourist boat. For several years he worked at sea and in Miami, both in music and cabinetry. It was when he returned to Kingston that his instrument making came to the fore, as he supplied instruments to individual customers and to the island's leading music store.

Mr. Lewis still lives in Kingston today. He builds instruments on order, continues to work in cabinetry, and plays music whenever the opportunity arises.

Drums

So what has Rupert Lewis made since that early ukulele? His primary instruments have been drums, rumba boxes, and bamboo flutes. The bulk of his business in recent years has been drums – primarily *tumbas,* a type similar in design to what most people think of as conga drums. These he makes in a range of sizes, along with smaller bongo drums.

Mr. Lewis' father, in fact, had been a maker and player of drums. "When I started the drums," Mr. Lewis says, "it was an antiquated type of drums, with rope and things like that, like the African style of drums." These traditional laced drums had their heads tuned by *twitching* – that is, inserting short pegs of wood between the lacings and giving them several turns so as to twist two segments of the rope lacing together. The more turns, the greater the pull on the head. Later he graduated to metal tuning hardware in the style of contemporary conga drums.

His preferred wood for drum bodies is an aromatic hardwood known in Jamaica as cedar. (It is substantially harder and heavier than the wood called by the same name in the US and appears to be a different species entirely.) To create the outward curvature in the sides of the tumbas, Mr. Lewis must cut the staves on a band saw and then bevel the edges so that they fit together in the circular barrel shape. The Jamaican cedar has become harder to get in recent years, and so Mr. Lewis has turned increasingly to oak salvaged from barrels used in shipping. The barrels already have the slightly convex silhouette, eliminating the need for the band saw. But the staves must be narrowed and the angle of the bevel reset to allow them to form the narrower drum. To help in putting the pieces together, he uses a pair of circular metal templates, one sized for the top of the drum body, and one for the narrower bottom. Mr. Lewis takes pride in the perfection of the wood joints in his drums, as well as the levelness of the rim (Figure 21.2).

On most of Mr. Lewis' drums, the opening at the lower end is considerably narrower than what one might see on a conventional conga drum. In theory this should lower and focus the resonance frequencies of the enclosed air, making for a lower tone, with a strong but narrowly defined air resonance. Subjectively speaking, I found this to be the case with Mr. Lewis' drums: there seemed to be an extraordinarily deep, rich air resonance to the tone. He explains that he noticed the narrow opening on some of the Cuban drums whose tone he admired, and so incorporated the idea in his own drums to good effect.

For small drums Mr. Lewis uses goat skin heads. For larger drums, whenever possible, he uses muleskin. Muleskin, he argues, is less stretchy than cowhide. It holds a tuning better, produces a clearer tone, and resists tearing. He prepares his own skins, with a series of soaking and cleaning processes.

Figure 21.2 Drums by Rupert Lewis.

Rumba Boxes

African lamellaphones are known in the west by the nick-name 'thumb pianos' as well as by the names of African variants like kalimba, mbira, and sansa. Such instruments did not take hold in a lasting way in the US, except in later years as African imports. But in parts of the Caribbean and Latin America they did – particularly, for some reason, in larger bass forms called *marimbula* in Columbia and Puerto Rico, *marimba* in Haiti, and, in Jamaica, rumba box. A big lamellaphone can have a wonderfully full, oomphy sound, and a well-made rumba box carries the bottom of an acoustic dance band quite satisfactorily. Couple the rumba box with a guitar or banjo to play chords and some sort of percussion to keep the beat, and you have a very appealing little rhythm section. The rumba box's tone is not as loud as a string bass, and the pitch generally is not as clear, but the effect is more fat, bottomy and rhythmic. Most early rumba boxes had five or seven keys, and they were not always very deliberately tuned. To the extent that they were, the idea was to allow for triadic bass lines.

I was unable to get a photo of any of Rupert Lewis' rumba boxes for this article, but you can get some sense of what they look like from the photos in Richard Graham's article in this collection. Rupert Lewis makes rumba boxes in a range of sizes, including smaller kalimba-like instruments for children. A typical size for the bass instruments might be 22" wide by 18" high by 9" deep. The local cedar is the preferred wood, especially for the soundboard, at about 3/16" or ¼" thick. Plywood may be used for the back and sides if sufficient cedar isn't available. A sound hole near the center of the face is usually about 3" across.

Most of Mr. Lewis' rumba boxes have seven keys. The keys are made from the drive springs of old gramophones or large old pendulum clocks, typically about ¾" wide and less than 1/16" thick. Needless to say, it's not easy to locate sources of old gramophone springs these days.

Mr. Lewis keeps a cherished supply and is always on the lookout for more. The bridge that supports the keys is made up of two strips of wood or metal. The two strips are placed one on top of the other, with the keys sandwiched in between. Bolts pass through the bridge in the spaces between the keys. The bolts can be loosened to allow for sliding the keys further in or out, effectively lengthening or shortening the vibrating portion for tuning purposes.

Flutes

Diatonic sideblown flutes have figured in many of the older Jamaican music styles. Not only did they have a role in a lot of the mento, calypso, and quadrille music mentioned earlier, but they were central too in the jonkanoo dancing that traditionally happened in both city and country around Christmas time, as well as other

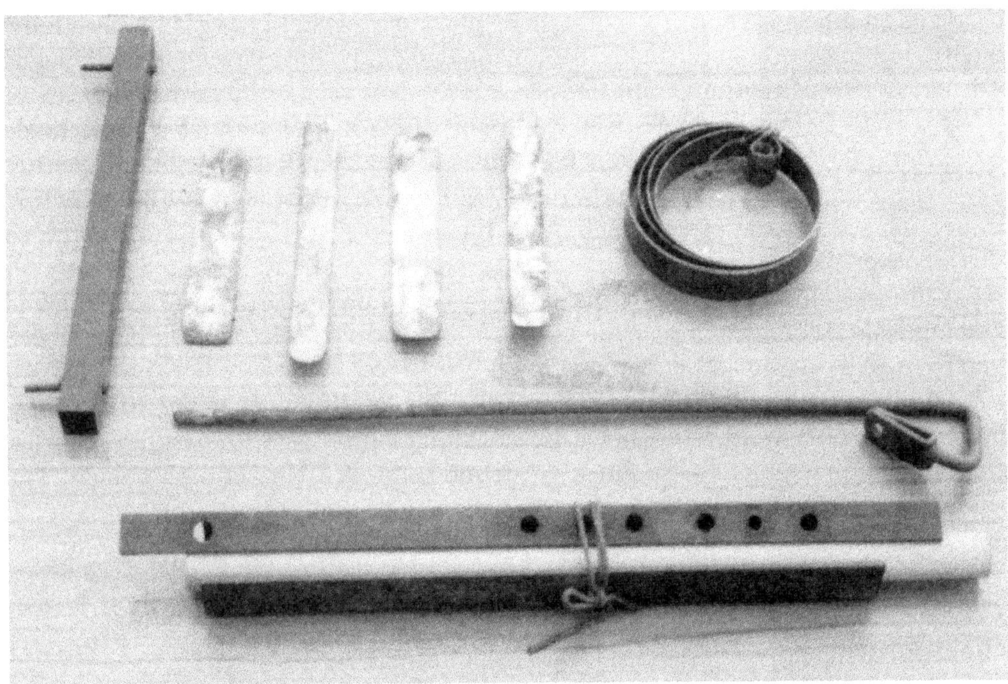

Figure 21.3 Rumba box hardware and flute-making hardware. In the top row, left to right, are an assembled rumba box bridge (two abutting wooden pieces with bolts), four spring steel keys, and a coiled spring from which keys can be made. Below are a poker for burning flute holes, and the assembled hole-placement template with a flute body waiting inside.

Figure 21.4 Components of the flute-making hardware disassembled.

localized traditions. An irresistible style of playing evolved in the island, with long, high-flying lines of rapid, highly ornamented melody full of a characteristically Caribbean sort of rhythmic phrasing.

Mr. Lewis' flutes are made from bamboo, with toneholes and blowhole made by burning through with a poker. For accurate location of the toneholes, he has made the iron template shown in Figures 21.3 and 21.4. The template consists of a trough-shaped bottom piece, which holds the flute in place, and piece of bar stock that fits over the top. This top piece has holes in it, sized and located correctly for the six finger holes and the blowhole. Long ago Mr. Lewis copied the hole sizing and spacing from a commercially made flute onto this iron template, and it has allowed him to make accurate and well-tuned flutes ever since.

The older Caribbean instruments are still with us, though they are not as common as they once were. Mr. Rupert Lewis, along with a very few contemporaries, continues to make them, experiment with and improve them, and play them. The sound is as sweet as ever.

Special thanks for assistance with this article to Marjorie Whylie and Carmen Verity.

Related articles

Other articles pertaining to Caribbean instruments which appeared in *Experimental Musical Instruments* are listed below. These articles can be accessed through the Internet Archive at https://archive.org/details/emi_archive/.

Rumba Boxes:

"The Giant Lamellaphones: A Global Perspective" by Richard Graham, in this collection

Congas and related:

"Congas According to Carraway" by Bart Hopkin, in *Experimental Musical Instruments* Vol. 4#4

"Polymorphous Percussion Construction; Making Drums out of Everything in Sight" by Zeno Okeanos, in *Experimental Musical Instruments* Vol. 14 #3

Caribbean Instruments:

"The Benta: An African-Derived Glissed Idiochord Zither of Eastern Jamaica" by Richard Graham, in *Experimental Musical Instruments* Vol. 9 #3

"The Bamboolin: A Jamaican Idiochord Zither": Bart Hopkin, in *Experimental Musical Instruments* Vol. 9 #3

"'Sugar Belly' Walker and the Bamboo Saxophone" by Bart Hopkin, in *Experimental Musical Instruments* Vol. 9 #2

22

"Sugar Belly" Walker and the Bamboo Saxophone

Bart Hopkin and Brian Wittman

This article originally appeared in *Experimental Musical Instruments* Volume 9 #2, December 1993.

A few years ago, one of the beloved figures of early Jamaican popular music passed away, and as so often seems to happen with once-popular musicians, he died in poverty. William Walker, known to all as Sugar Belly, developed on his own the instrument he called the bamboo saxophone, and played it with facility, style, passion, and joy. At the height of his popularity in the late 1950s, Sugar Belly was one of the important figures in the Jamaican music scene, turning his homemade saxophone into a natural vehicle for a distinctively Caribbean musical style.

Figure 22.1 Sugar Belly Walker (Drawing by Robin Goodfellow).

The Man

Sugar Belly was raised in Kingston. His economic background was lower class, and he was not well educated. In music he was entirely self-taught. It doesn't appear that he ever had any tutoring on a conventional saxophone. Just where he got the idea to create a bamboo saxophone is a bit of a mystery, since there is no traditional bamboo reed instrument in Jamaica, and no one that I have spoken to can recall seeing any other locally made saxophone-like instrument in the island. Sugar Belly's instrument seems to have been entirely his own in conception and design. In its construction, the instrument might seem simple and crude, but you know the tree by its fruit: from it Sugar Belly managed to bring the most fluid, warm, agile, and unquestionably sax-like music you could wish for.

In the early days, Sugar Belly played in talent exhibitions at Victoria Park in downtown Kingston, where years before Marcus Garvey had addressed the crowds. With increasing recognition he moved on to night clubs, such as the popular Glass Bucket located uptown at Halfway Tree. The leading popular music style in Jamaica at that time was *mento.* Mento evolved originally as a local extension of traditional quadrille dance music. In its mood it is closer to the easy good-times feel of the old Trinidadian calypso than it is to the more angular ska and reggae styles that were to develop later in Jamaica. Mento music is good natured, humorous, lively, and danceable. Sugar Belly's band originally used a typical mento instrumentation of banjo, guitar, and shakers, with the big bass kalimba known in Jamaica as a rumba box providing the bottom. Later he incorporated electric guitar and bass.

Through the 1960s, mento gradually faded in popularity. Sugar Belly brought popular songs from a broader range of local and international styles into his repertoire; still, as time passed, he and his band were heard from less and less. He later moved to the parish of St. Ann on the island's north coast, and it was there that he died circa 1990 following a long illness.

The Instrument

Although he never approached instrument making in a commercial way, Sugar Belly did make a fair number of bamboo saxophones over the years, keeping some to play himself and selling others.

The main segment of his bamboo saxophone (see the photos) is a straight section of bamboo, an inch or so in diameter and something over a foot long. Into this at one end is inserted a mouthpiece of a few inches long, made from a smaller piece of bamboo sized so as to fit snugly into the main segment. Where a commercial sax has cork to ensure a leakless fit between the mouthpiece and the main tube, Sugar Belly put several rounds of masking tape to provide an adequate gasket.

The reed end of the mouthpiece terminates in a simple angle cut. Although it is hard to see it in the photographs, and I neglected to verify this when I had the instrument in hand, the angle cut probably is not straight, but has a slight convex curvature. This provides what reed players call the "lay" of the mouthpiece, allowing for the required gap at the tip where the straight reed rises slightly from the gently curved mouthpiece surface. In lieu of a metal ligature, Sugar Belly took the time-honored approach of tying his reed to the mouthpiece with a piece of cord.

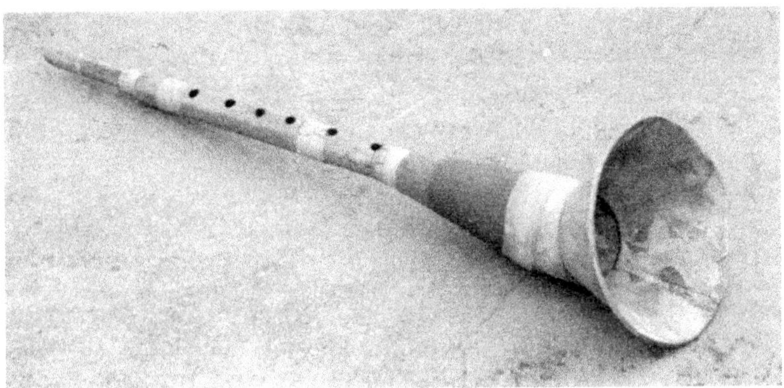

Figure 22.2 Bamboo saxophone by Sugar Belly Walker (from the collection of the Jamaica School of Music).

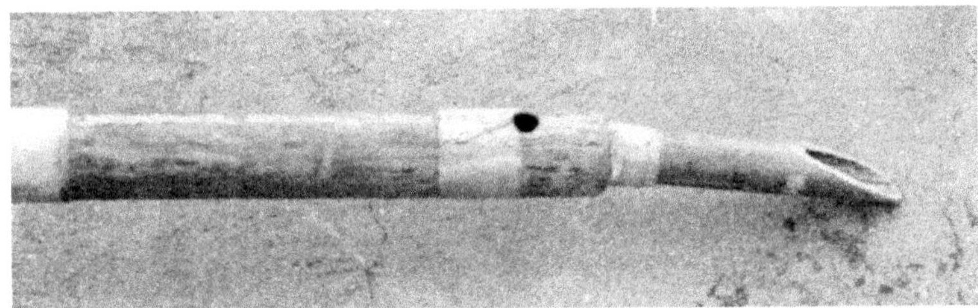

Figure 22.3 Detail showing the cut-away of the mouthpiece, and the single tonehole on the back of the pipe.

Originally, and for many years, Sugar Belly carved his reeds from bamboo. He turned to store-bought reeds only when his bamboo supply (which he found in the Hope River valley, not far from Kingston) gave way to coffee plantations and other forms of development. In Sugar Belly's hands, the instrument doesn't seem to have sounded very different with one sort of reed or the other.

At the other end of the main bamboo tube, Sugar Belly placed a conical commercial thread spool made of heavy cardboard. Once again, masking tape ensures a snug fit and a leakless joint. The conical spool in turn leads to a wider-angled funnel of tin. Sugar Belly had this one metal part fabricated for him by a tinsmith.

There are six finger holes on the front of the instrument and one on the back, made by burning. Sugar Belly's own playing was primarily diatonic and in major keys, but through cross fingerings and half-holing, he was able to include a bit of chromaticism here and there. The instrument lends itself to notes bent broadly through embouchure control, and Sugar Belly was a master of this as well. And he was also able to make excursions into a clear and dependable upper register.

One of Sugar Belly's saxophones can be seen in the instruments collection of the Jamaica School of Music at the Cultural Training Centre, 1 Arthur Wint Drive, Kingston 5, Jamaica. But no one seems to have followed him in making or playing a Jamaican bamboo saxophone. We are left with a few scratchy recordings, and the knowledge that it can be done.

Acknowledgments

This article would not have been possible were it not for assistance from Marjorie Whylie, who provided most of the information. Ms. Whylie knew Sugar Belly and his music when he was still around and playing. Beyond that, Marjorie's encyclopedic knowledge of Jamaican traditional and early popular music, always tempered with generous doses of insight, humanity, and humor, constitutes one of Jamaica's national treasures. And beyond that, she is a natural and inspired teacher. Thanks, Marjorie.

Thanks also to Derrick Johnson of the Jamaica School of Music, for opening the collection to me, providing additional information and assisting with photography.

On Sugar Belly

Notes from Brian Wittman, bamboo saxophone maker in Hawaii

I still remember the day Bart Hopkin sent me the tape of Sugar Belly, which went directly from the P.O. box to the car stereo as I rushed to pick up my kids from school. We all loved him immediately like a lost uncle. And all at once I felt sane and justified, for here was another man from earth who had built his life on the bamboo sax.

I have made over 15,000 such instruments in the past 20 years, all because of a single instrument I made on the whim of a child. The young lad lived with his mother in a tent in the woods and heard me playing the sax (the expensive metal variety). He approached respectfully and then boldly asked if perhaps I had a little one he could play. Why not? I fiddled around and whittled a small end-blown block flute out of bamboo. Its tone was wheezy and small and satisfied neither of us. I had a small grinding wheel I was using to shape some wooden boat cleats, and in sudden inspiration, I applied the flute to the wheel and ground off the whole corner of the mouthpiece at an angle, re-shaping it to take a sax reed. With a bit of string holding the reed, I blew a test note ... it screamed!

The child was delighted and couldn't wait to have it, so I passed it on, but immediately made myself another, this time a bit longer, and I made the mouthpiece first so I could hear the pitch as I located the finger holes. Somehow by chance I ended up with a serviceable scale in "E," and I couldn't put it down. I even played it one-handed as I drove into town, not noticing the speedometer was reading 80 until I heard the sirens.

Finally I arrived at the rehearsal studio where I was due, only to find a major hero, Mr. Airto Morierra (the Brazilian percussionist) just happened to be there jamming with my delighted band members. I jumped in on my new axe and found that its strong warm tone could be as full as a sax and amplified very well in an electric band setting. Airto was fascinated. So I offered this #2 instrument as a token of my respect for his music.

So I made a 3rd and played it on gigs. People would come up and ask about it ... Where did you get it? ... You made it? Can you make me one? ... What do you mean you don't have time – here's my money! So I ended up in business. A name developed from "bamboozaphone" to "bamboozafoon" to "bamboo zafoon" to just "zafoon," also spelled "xaphoon." I eventually moved closer to the bamboo forests and even took out a patent in several countries. And as I answered my mail and filled the orders, the years went by. My children were born into a house built of bamboo saxophones and heard them from the womb onward.

"Is that you on the tape, Dad?" No, that is the famous Sugar Belly, that I only now heard of just last week.

It is sad that we will never meet in this world. Yet I feel very thankful to know of Sugar Belly and to hear his music. We have touched the same whimsical source of magic from different worlds. Accidentally walked in the same moccasins. (Figure 22.4).

The Xaphoon

The instrument I have made commercially all these years is not much different than the first experimental models. I did construct several larger instruments, some with conical extensions (usually cow horn), but rather than complicate the design with a number of pieces, I have elected to maintain the "one stick" concept with the mouthpiece carved directly on the end of the instrument body. Fortunately the bamboo naturally lends itself to this type of construction if it is carefully chosen in the forest for the correct length and diameter.

After some experimentation, I eventually found a hole-placement and fingering system that will allow two complete chromatic octaves, though the instrument remains primarily diatonic. For example, it would be simple enough to play a C# note on a "C" instrument, or sketch through a riff in that key while following the chord changes, but it would not make much sense to transpose the entire tune to C#. There would be just too many crass-fingerings and lip adjustmentsI have generally restricted my output to "C" instruments, mostly to avoid confusing beginners with too many choices. I will gladly make instruments of any key, but only if the customer is still interested after having attained some skill on the "C." The "C" plays best in the keys of D, F, G, Gm, Dm, Am, and so on.

The tone of my instruments is remarkably similar to what I hear on the tapes of Sugar Belly, and my personal style of playing has some resemblance as well,

Figure 22.4 Brian Wittman makes the Maui Xaphoon®.

perhaps because of the inherent qualities and limitations of the instrument. However, some of my customers have surprised me by adopting radically different styles, from Baroque to Peruvian to Irish to African. I greatly appreciate the occasional tapes I receive from my customers. One can well imagine that the actual construction of 15,000 of anything can become tedious, so it has become the satisfaction of customers that drives me (as well as the opportunity to feed my family). It is truly rewarding to receive orders from distant places and it does get easier to make them now than I know how.

I can only wonder though, if perhaps my punishment in the next world will be to hear them all played at once.

Related articles

Other articles pertaining to saxophone-like instruments of bamboo or other natural materials which appeared in *Experimental Musical Instruments* are listed below. These articles can be accessed through the Internet Archive at https://archive.org/details/emi_archive/.

"Horn from the Sea: Bull Kelp, Part 2" by Bart Hopkin in EMI Vol. 6 #1

"The Development of Bamboo Saxes from Argentina" by Ángel Sampedro del Rio in EMI Vol. 12 #2

"Three More from Ángel" by Ángel Sampedro del Río in EMI Vol. 14 #3

23

Relating Timbre and Tuning

Bill Sethares

This article originally appeared in *Experimental Musical Instruments* Volume 9 #2, December 1993.

> Clearly the timbre of an instrument strongly affects what tuning and scale sound best on that instrument
>
> – W. Carlos

Introduction

If you've ever attempted to play music in weird tunings (where "weird" means anything other than 12-tone equal temperament), you've probably noticed that certain timbres (or tones) sound good in some scales and not in others. 17 and 19-tone equal temperament are easy to play in, for instance, because many of the standard timbres in synthesizers sound fine in these tunings. I remember when I first played in 16 tone. I had to audition hundreds of sounds before I found a few good timbres. When I tried to play in 10 tone, though, none of the timbres in my synthesizers sounded good. This article explains why and shows how to design matching timbres and scales. This suggests several possible new species of musical instruments.

The principle of local consonance describes a relationship between the timbre of a sound and a tuning (or scale) in which the timbre will sound most consonant. The principle answers two complementary questions. Given a timbre, what scale should it be played in? Given a scale, how can appropriate timbres be chosen? The ability to answer such questions will likely impact the way we design new musical instruments.

The presentation begins in the next section with an overview of the work of several acousticians, who have shown that people reliably judge the consonance of intervals composed of pure sine waves. These judgments are averaged into a "consonance curve" which is used to calculate the consonance of complex timbres. The results of such calculations agree well with the normal (musical) notion of consonance when applied to harmonic timbres. Thus, unisons, octaves, fifths, and fourths are highly consonant while seconds and sevenths are relatively dissonant.

Of course, this measure of consonance can also be applied to other (nonharmonic) timbres, and the succeeding sections show how to design timbres and scales. Several concrete examples follow, including finding scales for nonharmonic timbres (the natural resonances of a uniform beam, "stretched" and "compressed" timbres, FM timbres with noninteger carrier to modulation ratios), and finding timbres for equal tempered scales. This article is a less technical presentation of my paper "Local Consonance and the Relationship Between Timbre and Scale," which contains more mathematical details.

What Exactly is Consonance?

The standard musicological definition (see your favorite dictionary) is that a musical interval is consonant if it sounds pleasant or restful; a consonant interval has little or no musical tension or tendency to change. Dissonance, on the other hand, is the degree to which an interval sounds unpleasant or rough; dissonant intervals generally feel tense and unresolved.

In *On the Sensations of Tone*, Helmholtz offers a physiological explanation for consonance that is based on the phenomenon of beats. If two tones are sounded at almost the same frequency, then a distinct beating occurs that is due to interference between the two tones (piano tuners use this effect regularly). The beating becomes slower as the two tones move closer together and completely disappears when the frequencies are identical. Typically, slow beats are perceived as a pleasant vibrato while fast beats tend to be rough and annoying. Recalling that any timbre can be decomposed into sine wave components, Helmholtz theorized that dissonance between two tones is caused by the rapid beating of various sine wave components. Consonance, according to Helmholtz, is the absence of such dissonant beats.

More recently, Plomp and Levelt examined consonance experimentally, by generating pairs of sine waves and asking volunteers to rate them in terms of their relative consonance. Though there was considerable variability among the responses, there was a simple and clear trend. At unison, the consonance was maximum. As the interval increased, it was judged less and less consonant until at some point a minimum was reached. After this, the consonance increased up toward, but never quite reached the consonance of the unison. Plomp and Levelt called this tonal consonance, to distinguish it from musical consonance and from Helmholtz' beat theory.

Figure 23.1 shows an averaged version of the dissonance curve (which is simply the consonance curve flipped upside-down) in which dissonance begins at zero (at an "interval" of a unison) increases rapidly to a maximum and then falls back toward zero. The most surprising feature of this curve is that the musically consonant intervals are undistinguished – there is no dip in the curve at the fourth, fifth, or even the octave.

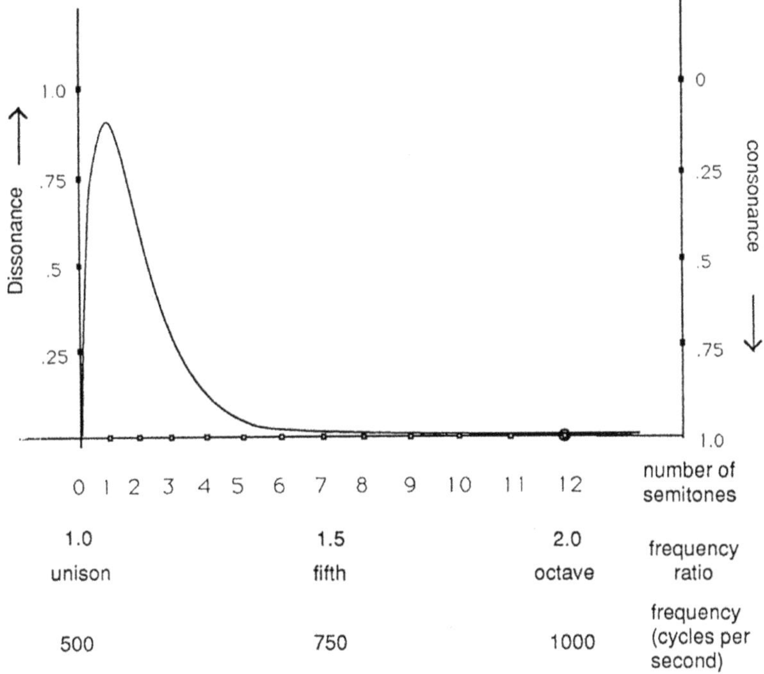

Figure 23.1 Dissonance curve for pure sine waves as a function of frequency difference. The consonance and dissonance scales are arbitrary.

To explain perceptions of musical intervals, Plomp and Levelt note that most traditional musical tones have a spectrum consisting of a root or fundamental frequency, and a series of sine wave partials that occur at integer multiples of the fundamental. Figure 23.2 depicts one such timbre. If this timbre is sounded at various intervals, the dissonance of the intervals can be calculated by adding up all of the dissonances between all pairs of partials. Carrying out this calculation for a range of intervals leads to the dissonance curve. For example, the dissonance curve formed by the timbre of Figure 23.2 is shown in Figure 23.3.

Observe that Figure 23.3 contains major dips at nearly all intervals of the 12-tone equal tempered scale. The most consonant interval is the unison, followed closely by the octave. Next is the fifth, followed by the fourth, the major third, the major sixth, and the minor third. These agree with standard musical usage and experience. Looking at the data more closely (see Table 23.1) shows that the minima do not occur at exactly the scale steps of the 12-tone equal tempered scale. Rather, they occur at the "nearby" simple ratios 1:1, 2:1, 3:2, 4:3, 5:4, and 5:3 respectively, which are exactly the locations of notes in the "justly intoned" scales (see Wilkinson). Thus, an argument based on tonal consonance is consistent with the use of just intonation (scales based on intervals with simple integer ratios), at least for harmonic timbres.

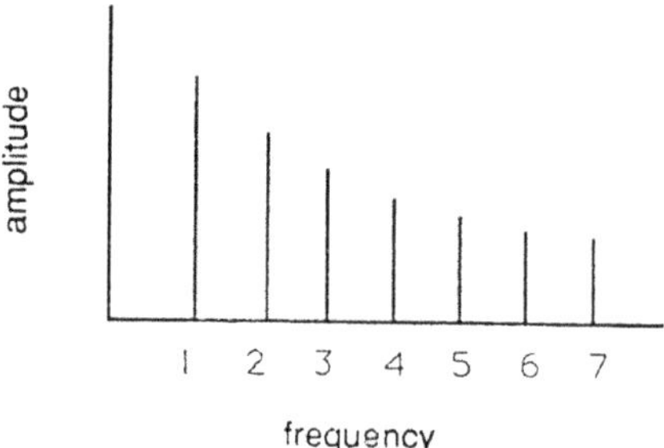

Figure 23.2 The standard harmonic timbre used to generate the dissonance curve of Figure 23.3. Amplitudes fall at a rate of 0.88. The frequency axis is normalized so that the root frequency is unity.

Perhaps the most striking aspect of Figure 23.3 is that virtually all the scale steps are coincident with local minima of the dissonance curve. Thus, the ear perceives intervals which occur at points of local minima in the dissonance curve as relatively consonant. This observation forms the basis of the principle of local consonance:

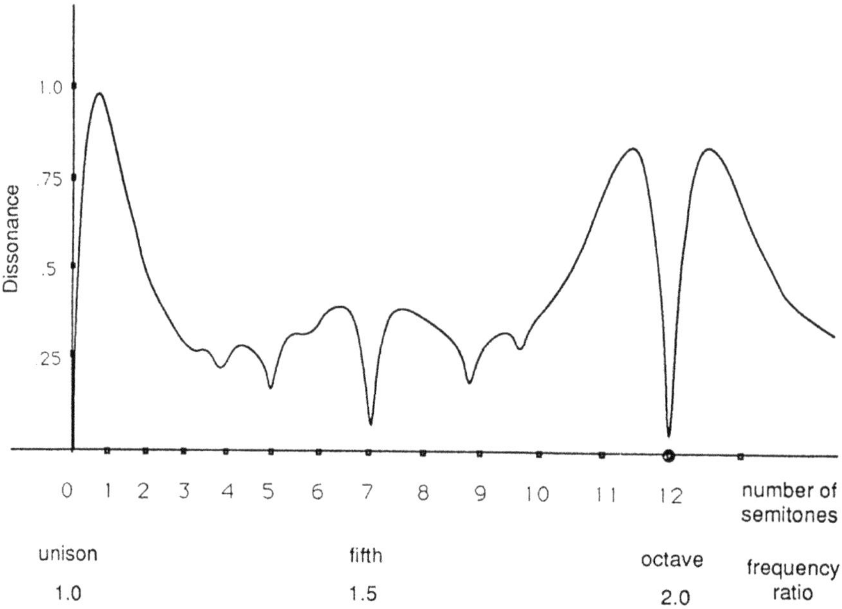

Figure 23.3 Dissonance curve for the timbre of Figure 23.2. The horizontal axis represents frequency difference. Dots mark the location of notes in the standard 12-tone equal-tempered scale. The vertical axis is arbitrary, and all dissonance curves are normalized so that the largest value occurs at unity.

Table 23.1 Location of minima of Figure 23.3, $\beta = \sqrt[12]{2}$

Location of minima	*Nearest 12-tone scale step*	*Nearest ratio*	*Interval name*
1.20	$\beta^3 = 1.189$	6:5	Minor 3rd
1.25	$\beta^4 = 1.259$	5:4	Major 3rd
1.33	$\beta^5 = 1.335$	4:3	Perfect 4th
1.40	$\beta^6 = 1.414$	7:5	Augmented 4th
1.50	$\beta^7 = 1.498$	3:2	Perfect 5th
1.67	$\beta^9 = 1.682$	5:3	Major 6th
1.75	$\beta^{10} = 1.782$	7:4	Minor 7th
2.0	$\beta^{12} = 2.000$	2:1	Octave

> A timbre and a scale are said to be related if the timbre generates a dissonance curve whose local minima occur at scale positions.

This notion of relatedness of scales and timbres suggests two interesting avenues of investigation. Given an arbitrary timbre T (perhaps one whose spectrum does not consist of a standard harmonic series), it is straightforward to draw the dissonance curve generated by T. The local minima of this curve occur at values which are good candidates for notes of a scale, since they are local points of minimum dissonance (i.e., maximum consonance). This might be useful to the experimental musician. Imagine being in the process of creating a new instrument with an unusual (i.e., nonharmonic) tonal quality. How should the instrument be tuned? To what scale should the finger holes (or frets, or whatever) be tuned? The principle of local consonance answers this question in a concrete way.

Alternatively, given a desired scale (perhaps one which divides the octave into m equal pieces, or one which is not based on the octave), there are timbres which will generate a dissonance curve with local minima at precisely the scale degrees. This is useful to musicians and composers who wish to play in nonstandard scales such as 10-tone equal temperament.

As the opening quote indicates, this is not the first time that the relationship between timbre and scale has been explored. Pierce's brief note reported synthesizing a timbre designed specifically to be played in an eight-tone equal tempered scale. Pierce concludes, "... by providing music with tones having accurately specified but nonharmonic partials, the digital computer can release music from the tyranny of 12 tones without throwing consonance overboard." Slaymaker investigated timbres with stretched (and compressed) partials, and Mathews and Pierce explored their potential musical uses. Recently, Mathews and Pierce examined a scale with steps based on $\sqrt[13]{3}$, rather than the standard $\sqrt[12]{2}$), which is designed to be played with timbres containing only odd partials. Carlos investigated scales for nonharmonic timbres by overlaying their spectra and searching for intervals in which partials coincide, thus minimizing the beats (or roughness) of the sound. This is similar to the present approach, but we provide a systematic technique that can be used to find scales for a given timbre, or to find timbres for a given scale.

It would be naive to suggest that truly musical properties can be measured as a simple tonal consonance. Even in the realm of harmony (and ignoring musically essential aspects such as melody and rhythm), consonance is not the whole story. Indeed, a harmonic progression that was uniformly consonant would likely be boring. Harmonic interest arises from a complex interplay of dissonance (restlessness) and consonance (rest). Perhaps the most important use of the principle of local consonance is to provide guidelines for exploring new tonalities and tunings.

How to Calculate Dissonance Curves

If you're thinking that there must be a lot of calculations necessary to draw dissonance curves, you're right. It's an ideal job for a computer.

Those familiar with BASIC or related computer languages may wish to look at the program appearing at the end of this article. (It is written in Microsoft's version of BASIC.) The program works by encapsulating the Plomp-Levelt consonance curve into a mathematical function that consists of a sum of exponentials. The *i* and *j* loops calculate the dissonance of the timbre at a particular interval *alpha,* and the *alpha* loop runs through all the intervals of interest. The first few lines set up the frequencies and amplitudes of the timbre. The variable *numf* must be equal to the number of frequencies in the timbre. Running the program as is generates the dissonance data of Figure 23.3 for the timbre of Figure 23.2. To change the start and end points of the intervals, use *startint* and *endint.* To make the intervals further apart, increase *inc.* All the dissonance values are stored in the vector *diss.* Don't change *dstar* or any of the variables with numbers.

Fortunately, there are some general patterns in the ways that dissonance curves can look. Let's examine a simple timbre with just two partials. As shown in Figure 23.4, the dissonance curve can have three different contours: if the partials are very close together then there are no points of local consonance, if the partials are widely separated then there are two local minima, if they are in between then there is just one. Using the program, you can reproduce these curves (or, of course, generate your own). Set numf=2 and freq(1)=500, freq(2)=505, amp(1)=10, amp(2)=10. This gives Figure 23.4(a), where the partials are too close to allow a point of local consonance. Setting freq(2)=1.15*500 shows that the point of local consonance occurs at an interval of 1.15, as in 23.4(b). Finally, setting freq(2)=1.86*500 gives

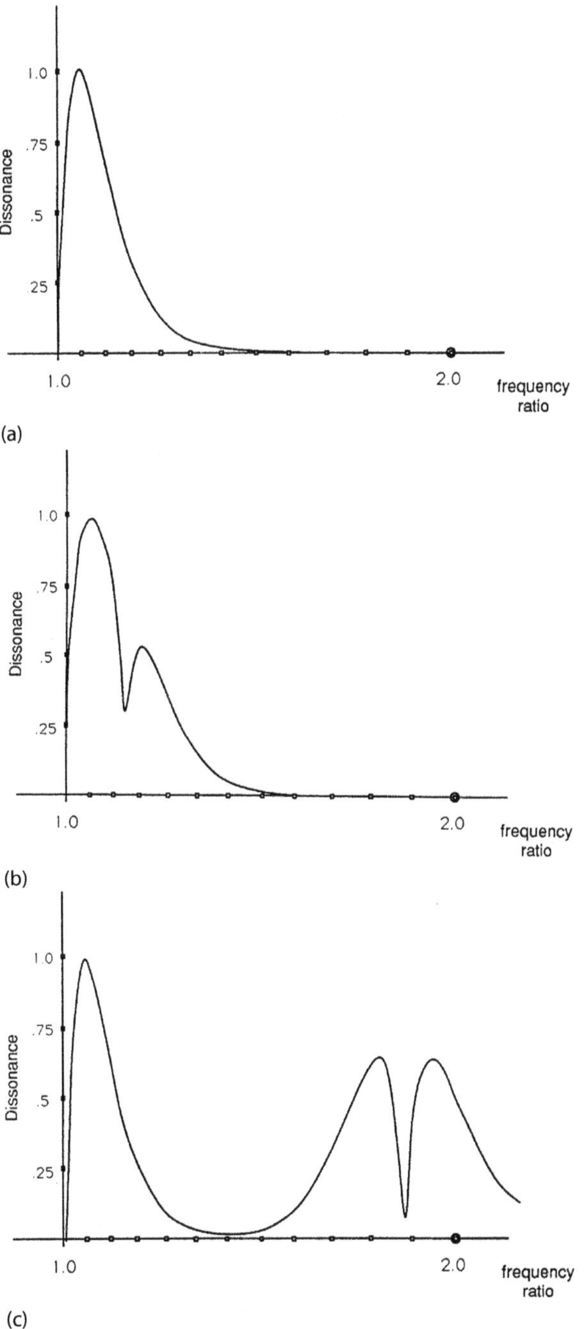

Figure 23.4 The three possible dissonance curves for timbres with two partials. a. For the timbre (f, t .01f), the partials are too close to allow a point of local consonance. b. Point of local consonance for timbre (f, 1.15f) occurs at $\alpha = 1.15$. c. Points of local consonance for timbre (f, 1.86f) occur at $\alpha = 1.86$ and a second "broad" minimum due to sparcity of partials

23.4(c), with two points of local consonance. The steep minimum occurs at an interval of 1.86. Notice that the second minimum is shallow, and is a result of the large distance between the partials of the timbre.

You can listen to Figure 23.4 with a synthesizer or tone generator. First, find a tone that is as close to a sine wave as possible. (If using a sample-based machine without such a humble waveform, try an organ or flute sample). Assign two tones to each keypress, one at frequency f, and one at a major seventh above f. (A major 7th is an interval of 1.86, just as in 23.4(c)). Listen to the consonance of the various intervals in this timbre. The first few are very rough. The next few are somewhat aharmonic, but not unpleasant. Then the dissonance rises and plummets quickly, at the interval of 1.86. The octave, at an interval of 2, sounds very dissonant and unoctavelike. For this timbre, the major 7th plays the role normally occupied by the octave, at least in terms of consonance. This is something you can hear for yourself.

Properties of Dissonance Curves

Here are some general properties of dissonance curves. Suppose that the timbre F has n partials located at frequencies $(f_1, f_2, \ldots, f_n)$.

Property (1): Dissonance curves may have up to $2n(n-1)$ minima.

Property (2): The unison is the global minimum (the lowest possible value of the dissonance curve).

Property (3): As the interval grows very large (as α goes to infinity), the dissonance approaches a value that is no more than the intrinsic dissonance of the timbre itself.

Property (4): Up to half of the local minima occur at intervals α for which $\alpha=f_i/f_j$ where f_i and f_j are arbitrary partials of T. Up to half of the local minima are the broad type of Figure 23.4(c).

The fourth property is particularly interesting because it says that points of local consonance tend to occur at intervals which are simply defined by the partials of the timbre. In Figures 23.4(b) and 23.4(c), for instance, local minima are found at $\alpha = 1.15$ and $\alpha = 1.86$, respectively, which is the ratio between the two partials. The broad minima tend to vanish for timbres with more than a few partials. Figure 23.3, for instance, consists exclusively of local minima caused by coinciding partials. Thus, dissonance curves usually have fewer than $2n(n\text{-}1)$ local minima. In Figure 23.3, for instance, there are only nine local minima within the octave of interest, considerably fewer than the theoretical maximum of 84. It is possible to achieve the bound. For instance, the timbre $(f, 2f, 3f)$ over the range $0 < \alpha < 6$ exhibits all 12 possible minima.

From Timbre to Scale

This section constructs examples of scales appropriate for a variety of timbres and explains various consonance related phenomena in terms of the principle of local consonance.

Harmonic Timbres

The points of local consonance for the harmonic timbre with partials at $(f, 2f, \ldots, 7f)$ are located at simple integer ratios. The results of the previous section explain this elegantly. Candidate points of local consonance are at intervals α for which $f_i = \alpha f_j$. Since the partials are at integer multiples of f, $\alpha = x/y$ for integers x and y between 1 and 7. The principle of local consonance says that the most appropriate scale tones for harmonic timbres are located at such α, and indeed, all the points of local consonance of Figure 23.3 occur at such values, as tabulated in Table 23.1. In a sense, this provides a physical basis for justly intoned scales. In terms of tonal consonance, the ear is fairly insensitive to small deviations in frequency, and the 12-tone equal tuning can be viewed as an acceptable compromise between the consonance-based desire to play in justly intoned scales and the practicalities of instrument standardization.

Stretched and Compressed Timbres

Slaymaker (1968) and Mathews and Pierce (1980) have investigated timbres with partials at $f_j = fA^{\log 2(j)}$. When $A = 2$, this is simply a harmonic timbre, since $f_j = f2^{\log 2(j)} = jf$. When $A < 2$, the frequencies of the timbre are compressed, while when $A > 2$, the partials are stretched. The most striking aspect of compressed and stretched timbres is the lack of a real octave. This can be seen clearly from the dissonance curves, which are plotted in Figures 23.5(a), (b), (c), and (d) for $A = 1.87$, 2.0, 2.1, and 2.2 respectively. In each case, the frequency ratio A plays the role of the octave, which Mathews and Pierce call the *pseudo octave.* Real octaves sound dissonant and unresolved when A is different from 2 while the pseudo octaves are highly consonant. More importantly, each curve has a similar contour. Points of local consonance occur at (or near) the 12 equal steps of the pseudo octaves. "Pseudo-fifths," "pseudo-fourths," and "pseudo-thirds" are readily discernible. This suggests that much of music theory and practice can be transferred to compressed and stretched timbres, when played in compressed and stretched scales.

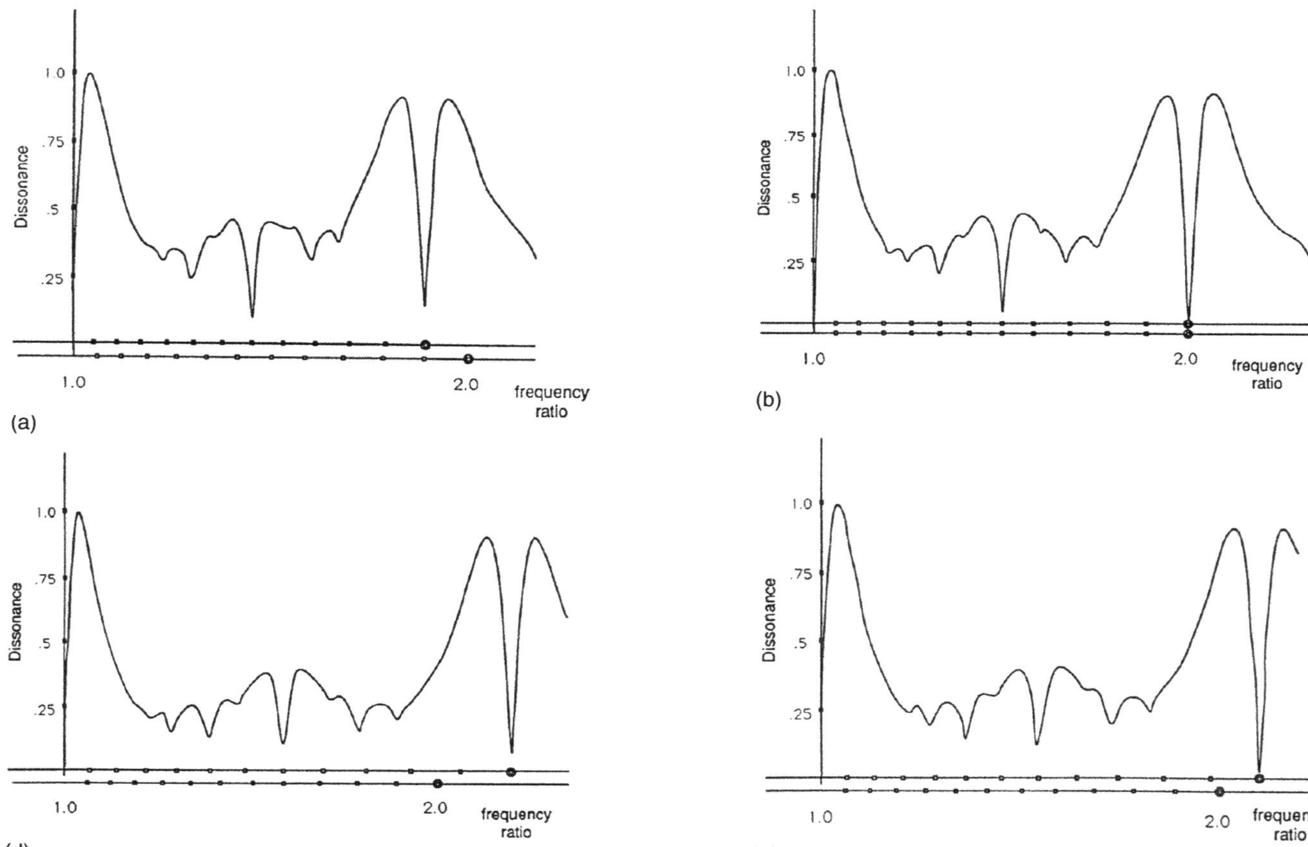

Figure 23.5 Dissonance curves generated by stretched and compressed timbres. Each Figure has two horizontal axes. The top axis shows the standard 12 tone equal-tempered divisions of the octave (frequency ratio 2:1). The bottom axis shows twelve equal divisions of the pseudo-octave with a frequency ratio of *A*:1.

a. A = 1.87 (pseudo-octave = major 7th). b. A = 2 (true octave). c. A = 2.1 (stretched octave). d. A = 2.2 (stretched further).

A Tuning for Uniform Beams

It is well known that glockenspiels, marimbas, and other instruments which consist of beams with free ends have partials which are not harmonically related. The principle of local consonance suggests that there is a natural scale, defined by the timbre of the xylophone, in which it will sound most consonant. The first seven frequencies of an ideal beam which is free to vibrate at both ends are given by Fletcher and Rossing as

$$(f, 2.758f, 5.406f, 8.936f, 13.35f, 18.645f, 24.82f)$$

Two octaves of the dissonance curve for this timbre are shown in Figure 23.6. The curve has numerous minima which are spaced unevenly at

$$1, 1.27, 1.33, 1.4, 1.49, 1.65, 1.76, 1.86, 1.96, 2.09, 2.25, 2.47, 2.76, 3.05, 3.24, 3.45, \text{and } 3.98,$$

which suggests that this would be the most natural sounding tuning for a xylophone, at least in terms of consonance.

Tuning for FM Timbres

One common method of sound synthesis is frequency modulation (FM) (see Chowning). Noninteger ratios of the carrier and modulating frequencies give nonharmonic

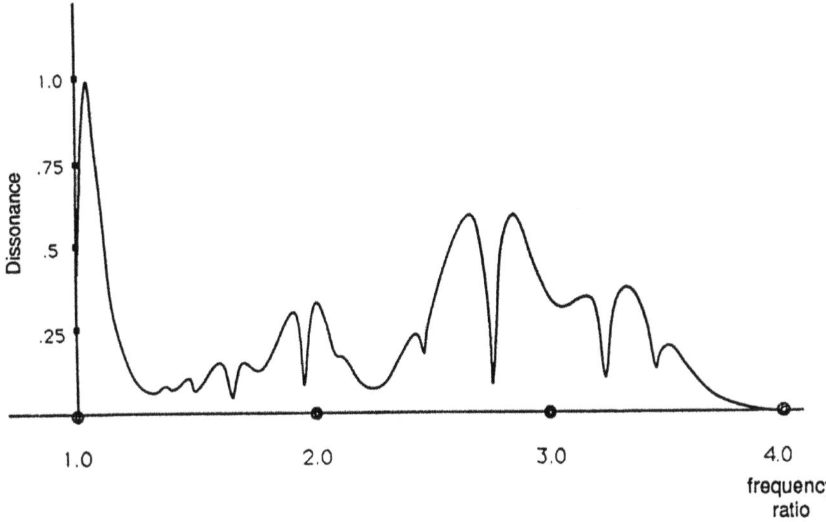

Figure 23.6 Dissonance curve for a uniform beam.

Table 23.2 Partials for the FM timbre with c:m of 1:1.14 and modulating index I=2

	Frequency	*Amplitude*
c-m	0.4	.57
c	1.0	.22
c-2m	1.8	.35
c+m	2.4	.57
c-3m	3.2	.13
c+2m	3.8	.35
c-4m	4.6	.03
c+3m	5.2	.13
c+4m	6.6	.03

timbres that are typically relegated to percussive or bell patches because they sound dissonant when played in traditional 12-tone harmonies. The principle of local consonance suggests that such sounds can be played more harmoniously in scales which are determined by the timbres themselves. For example, consider a simple FM tone with carrier to modulator ratio *c*:*m* of 1:1.4 and modulating index *I*=2. The frequencies and amplitudes of the resulting timbre are tabulated in Table 23.2, and three octaves of the dissonance curve are plotted in Figure 23.7. The appropriate scale

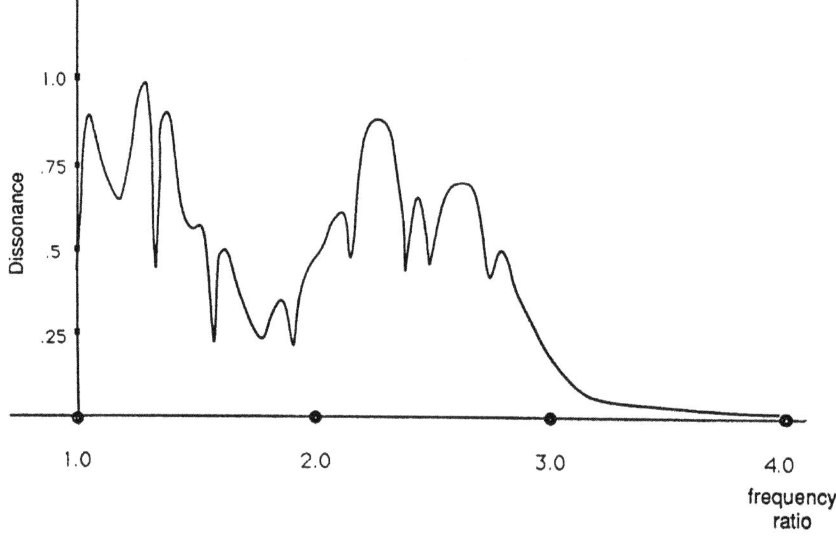

Figure 23.7 Dissonance curve for the FM timbre of Table 23.2.

notes for this timbre occur at the local minima of the dissonance curve, which can be read directly from the figure.

From Scale to Timbre

The optimal scale for a given timbre is found simply by locating the local minima of the dissonance curve. The complementary problem of finding an optimal timbre for a given scale is not as simple. There is no single "best" timbre for a given scale. But it is often possible to find "locally best" timbres which can be specified as the solution to a certain optimization problem. For certain classes of scales (such as the m-tone equal tempered scales), the properties of the dissonance curve can be exploited to solve the problem efficiently.

Timbre Selection as an Optimization Problem

Suppose that a set of m scale tones is specified. A naive approach to the problem of timbre selection is to choose a set of n partials $(f_1, f_2, ..., f_n)$ and amplitudes $(a_1, a_2, ..., a_n)$ to minimize the sum of the dissonances over all the intervals in the scale. Unfortunately, this can lead to "trivial" timbres in two ways. Zero dissonance can be achieved by setting all the amplitudes to zero, or by allowing all the partials to have arbitrarily high frequencies (recall property 2). To avoid such trivial solutions, some constraints are necessary:

Constraint 1: Don't allow the amplitudes to change; that is, choose a fixed set of amplitudes before carrying out the operation.

Constraint 2: Force all frequencies to lie in a predetermined region.

The revised (constrained) optimization is then: With the amplitudes fixed, select a set of n frequencies $(f_1, f_2, ..., f_n)$ lying in the range of interest so as to minimize the cost

$$J = w_1\left(\text{sum of dissonances over all intervals}\right) + w_2\begin{pmatrix}\text{number of points of}\\ \text{local minima}\end{pmatrix}$$

where the w_1 and w_2 are weighting factors. Minimizing this cost J tends to place the scale steps at local minima as well as to minimize the value of the dissonance curve. Experimentally, we have found weightings of $w_1/w_2 = 1000/1$ to be reasonable.

Minimizing the cost J is a n-dimensional optimization problem with a highly complex error surface. Fortunately, such problems can be solved adequately (though not necessarily optimally) using a variety of "random search" methods such as "simulated annealing" (see Kirkpatrick) or the "genetic algorithm" (see Goldberg).

The genetic algorithm (GA) seems to work well. The GA requires that the problem be coded in a finite string called the "gene" and that a "fitness" function be defined. Genes for the timbre selection problem are formed by concatenating binary representations of the f_i. The fitness function of the gene $(f_1, f_2, \ldots, f_n)$ is measured as the value of the cost J above, and timbres are judged "more fit" if the cost J is lower. The GA searches n-dimensional space measuring the fitness of timbres. The most fit are combined (via a "mating" procedure) into "child timbres" for the next generation. As generations pass, the algorithm tends to converge, and the most fit timbre is a good candidate for the minimizer of J. Indeed, the GA tends to return timbres which are well matched to the desired scale in the sense that scale steps tend to occur at points of local consonance and the total dissonance at scale steps is low. For example, when the 12-tone equal tempered scale is specified, the GA converges near harmonic timbres quite often. This is a good indication that the algorithm is functioning and that the free parameters have been chosen sensibly.

Timbres for an Arbitrary Scale

As an example of the application of the genetic algorithm to the timbre selection problem, a desired scale was chosen with scale steps at 1, 1.1875, 1.3125, 1.5, 1.8125, and 2. A set of amplitudes were chosen as 10, 8.8, 7.7, 6.8, 5.9, 5.2, 4.6, 4.0, and the GA was allowed to search for the most fit timbre. The frequencies were coded as 8-bit binary numbers with four bits for the integer part and four bits for the fractional part. The best three timbres out of ten trial runs of the algorithm were

$$(f, 1.8f, 4.9f, 14f, 9.87f, 14.81f, 6.4f, 12.9f),$$
$$(f, 1.5f, 3.3f, 10.3f, 7.8f, 7.09f, 3.52f, 3.87f),$$
$$(f, 2.39f, 9.9275f, 7.56f, 11.4f, 4.99f, 6.37f, 10.6f)$$

The dissonance curve of the best timbre is shown in Figure 23.8. Clearly, these timbres are related to the specified scale, since points of local consonance lie near the specified scale steps.

Timbres for Equal Temperaments

For certain scales, such as the m-tone equal tempered scales, properties of the dissonance curve can be exploited to quickly and easily design timbres, thus bypassing the need to run an optimization program.

Recall that the ratio between successive scale steps in 12-tone equal temperament is the twelfth root of 2 (about 1.0595), in symbols $\sqrt[12]{2}$. Similarly, m-tone equal temperament has a ratio of $\beta = \sqrt[m]{2}$ between successive scale steps. Consider

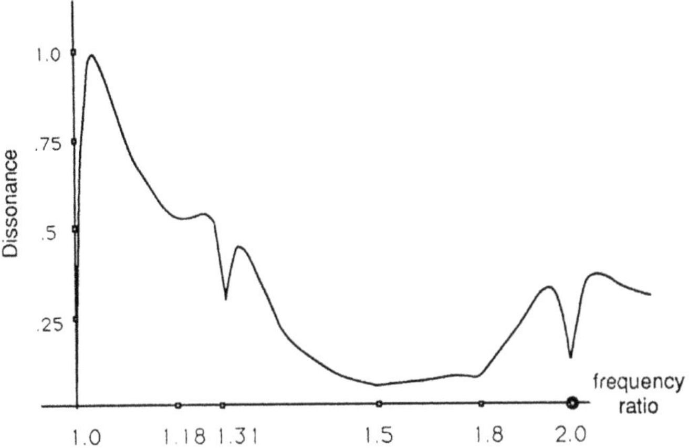

Figure 23.8 Dissonance curve for the scale with steps at 1, 1.18, 1.31, 1.5, 1.81, and 2.

timbres for which successive partials are ratios of powers of β. Each partial of such a timbre, when transposed into the same octave as the fundamental, lies on a note of the scale. Such a timbre is said to be *induced* by the m-tone equal tempered scale. For example, harmonic timbres are induced timbres for the justly intoned scale.

Induced timbres are good candidate solutions to the optimization problem. Recall from property 4 that points of local consonance tend to be located at intervals α for which $\alpha = f_i/f_j$ where f_i and f_j are partials of the timbre T. Since the ratio between any pair of partials in an induced timbre is β^k for some integer k, the dissonance curve will tend to have points of local consonance at such ratios: these ratios occur precisely at steps of the scale. Such timbres tend to minimize the cost J.

This insight can be exploited in two ways. First, it can be used to reduce the search space of the optimization routine. Instead of searching over all frequencies in a bounded region, the search need only be done over induced timbres. More straightforwardly, the timbre selection problem for equal tempered scales can be solved by careful choice of induced timbres.

As an example, consider the problem of designing timbres to be played in 10-tone equal temperament. 10-tone is often considered one of the worst temperaments for harmonic music, since the steps of the ten-tone scale are distinct from the (small) integer ratios, implying that harmonic timbres are very dissonant. The principle of local consonance asserts that these intervals will become more consonant if played with correctly designed timbres. Here are three timbres induced by the 10-tone equal tempered scale. Let $\beta = \sqrt[10]{2}$.

$$\left(f, \beta^{10} f, \beta^{17} f, \beta^{20} f, \beta^{25} f, \beta^{28} f, \beta^{30} f\right),$$
$$\left(f, \beta^{7} f, \beta^{16} f, \beta^{21} f, \beta^{24} f, \beta^{28} f, \beta^{37} f\right),$$
$$\left(f, \beta^{7} f, \beta^{13} f, \beta^{17} f, \beta^{23} f, \beta^{28} f, \beta^{30} f\right)$$

The dissonance curve of the first timbre is shown in Figure 23.9. All of these sound quite nice when played on a 10-tone equal tempered scale. Not surprisingly, the same tones sound quite ugly when played in a standard 12-tone scale.

Analogous arguments suggest that the consonance of 12-tone equal tempered tuning can be maximized by moving the partials away from the harmonic series to a series based on $\beta = \sqrt[12]{2}$.

New Instruments, Anyone?

Any arbitrary timbre (set of frequencies and amplitudes) can be realized with the aid of a computer. Is it always possible to design acoustic instruments that will have a given timbre? How about brasses? Fletcher and Rossing proclaim that "If the flaring part of the horn extends over a reasonable fraction of the total length, for example around one third, then there is still enough geometrical flexibility to allow the frequencies of all modes to be adjusted to essentially any value desired." With stringed instruments, the trick is to find a variable thickness string that will vibrate with partials at the desired frequencies. The partials of a drumhead can be tuned by

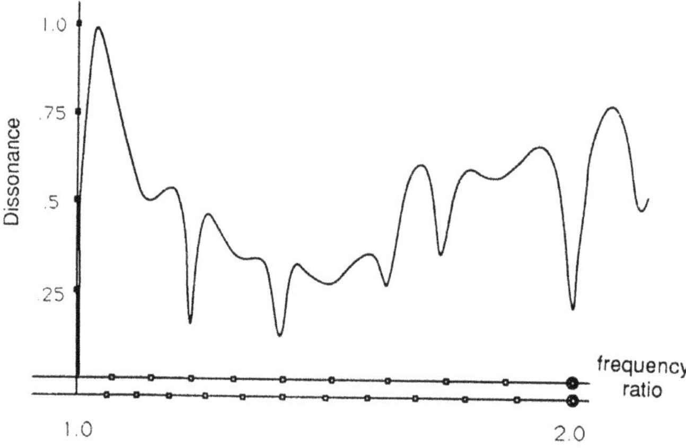

Figure 23.9 Timbre designed to be played in 10-tone equal-tempered scale. Note that points of local consonance coincide with the 10-tone scale (top axis) but not with the 12-tone equal-tempered scale steps (bottom axis).

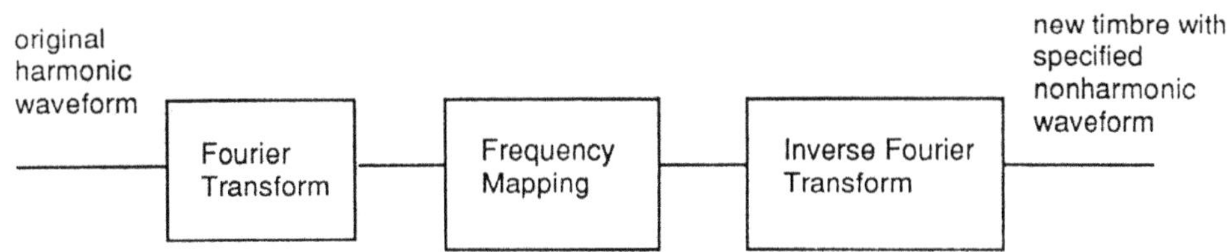

Figure 23.10 A resynthesis approach to the generation of nonharmonic timbres.

weighting or layering sections of the drumhead. The partials of reed instruments can be manipulated by the contour of the bore as well as the shape and size of the tone holes. Bells can be tuned by changing the shape and thickness of the walls. Exactly how to engineer acoustic instruments with specified timbres is an interesting issue.

Even if it is not always possible to make acoustic instruments with an arbitrarily specified timbre, it is possible to synthesize the timbres. One approach is diagrammed in Figure 23.10, where a harmonic waveform (which may be a sample of an acoustic instrument) is transformed into its constituent frequencies. The frequencies are changed in a systematic way that maps the partials into the specified timbre and then transformed back into a useable waveform. The result would be a nonharmonic timbre with much of the character of the original instrument.

The principle of local consonance shows how to imagine a number of differently tuned orchestras, digital or acoustic, each with instruments designed with a particular timbre and played in the related tuning. How about a band of instruments tuned to stretched or compressed tunings? An orchestra optimized for seven- or ten-tone equal temperaments? A wind instrument with the timbre of a drum? A trumpet with the harmonic structure of a steel beam? The consonance curve shows how to properly tune the instrument. Using a computer to generate the timbres gives the ability to audition the design before building, saving time in the design and specification of nontraditional instruments.

Conclusions

This paper developed the principle of local consonance which is based on Plomp and Levelt's notion of tonal consonance. Two complementary computational techniques were proposed: a way to find consonant scales given a specified timbre, and a way to find consonant timbres given a specified scale. One implication is that the musical notion of consonance of intervals such as the octave and fifth can be viewed as a result of the timbre of the tones we listen to. The justly intoned scales can similarly be viewed as a consequence of the harmonic timbres of musical instruments.

The advent of inexpensive musical synthesizers capable of realizing arbitrary sounds allows exploration of nonharmonic acoustic spaces, and the principle of local consonance provides guidelines on how to sensibly relate tuning and timbre. More ambitiously, it is easy to imagine new nonharmonic instruments capable of playing consonant music. The computational techniques of this paper allow specification of timbres and tunings for such instruments.

Appendix

Microsoft basic program for calculating dissonance curves

This version in Microsoft Basic written by Bill Sethares. This program first appeared in "Relating Tuning and Timbre," *Experimental Musical Instruments*, Vol IX #2, Dec. 1993. Related work appears in "Local Consonance and the Relationship Between Timbre and Tuning," Journal of the Acoustic Society of America 94(3) pp, 1218-1228, Sept. 1993.

```
DIM freq(10),amp(10),g(10),diss(1500)

' Variables you must set:
' n=number of frequencies in timbre f
' freq(i) = frequency value of ith partial
' amp(i) = amplitude of ith partial

' loop through all intervals from startint to endint

dstar=.24: s1=.0207: s2=18.96: c1=5.0: c2=-5.0
a1=-3.51: a2=-5.75: index=-1
PRINT " Interval Dissonance"
startint=1: endint=2.3: inc=.01
FOR alpha=startint TO endint STEP inc
index=index+1: d=0
FOR k=1 TO n
g(k)=alpha*freq(k)
NEXT k
' calculate dissonance between f and alpha*f
FOR i=1 TO n
FOR j=1 TO n
IF g(j)<freq(i) THEN fmin=g(j) ELSE fmin=freq(i)
s=dstar/(s1*fmin+s2): fdif=ABS(g(j)-freq(i))
arg1=a1*s*fdif: arg2=a2*s*fdif
IF arg1<-88 THEN exp1=0 ELSE exp1=EXP(arg1)
IF arg2<-88 THEN exp2=0 ELSE exp2=EXP(arg2)
```

```
dnew=MIN(amp(i),amp(j))*(c1*exp1+c2*exp2): d=d+dnew
NEXT j
NEXT i
diss(index)=d
PRINT alpha, d
NEXT alpha
```

References

W. Carlos, "Tuning at the Crossroads," *Comp. Music J.*, Spring, 29–43 (1987).

J. M. Chowning, "The Synthesis of Complex Audio Spectra by Means of Frequency Modulation," *J. Audio Eng. Soc.*, 21, 526–534 (1973).

N. H. Fletcher and T. D. Rossing, *The Physics of Musical Instruments*, Springer-Verlag, New York, NY (1991).

S. Goldberg, *Genetic Algorithms in Search, Optimization, and Machine Learning*, Addison-Wesley, New York, NY (1989)

H. Helmholtz, *On the Sensations of Tone*, Dover, New York (1954).

S. Kirkpatrick, C. D. Gelatt and M. P. Vecchi, "Optimization by Simulated Annealing," *Science*, 220 (4598), 671–680 May (1983).

M. V. Mathews and J. R. Pierce, "Harmony and Nonharmonic Partials," *J. Acoust. Soc. Am.* 68, 1252–1257 (1980).

M. V. Mathews, J. R. Pierce, A. Reeves and L. A. Roberts, "Theoretical and Experimental Explorations of the Bohlen-Pierce Scale," *J. Acoust. Soc. Am.* 84, 1214–1222 (1988).

J. R. Pierce, "Attaining Consonance in Arbitrary Scales," *J. Acoust. Soc. Am.* 40, 249 (1966).

R. Plomp and W. J. M. Levelt, "Tonal Consonance and Critical Bandwidth," *J. Acoust. Soc. Am.* 38, 548–560 (1965).

W. A. Sethares, "Local Consonance and the Relationship Between Timbre and Scale," *J. Acoust. Soc. Am.* 94(3) 1218–1228 (1993).

F. H. Slaymaker, "Chords from Tones Having Stretched Partials," *J. Acoust. Soc. Am.* 47, 1469–1571 (1968).

S. R. Wilkinson, *Tuning In*. Milwaukee, WI: Hal Leonard Books (1988).

Acknowledgments

The author would like to thank Tom Staley for numerous discussions on tuning, timbre, and tonality. Don Hall, Bart Hopkin, and Bill Strong were quite helpful in improving various versions of this work.

Related articles

The ideas set forth in this article are most fully set out in the author's book *Tuning, Timbre, Spectrum, Scale* (Springer; 2nd edition, 2005).

Other articles by Bill Sethares which appeared in *Experimental Musical Instruments* are listed below. These articles can be accessed through the Internet Archive at https://archive.org/details/emi_archive/.

"Tuning for 19 Tone Equal Tempered Guitar" by Bill Sethares in EMI Vol. 6 #6

"Scrapyard Percussion" by Bill Sethares and John Bell in EMI Vol. 8 #1

"The Sound of Crystals" by Bill Sethares and Tom Staley in EMI Vol. 8 #2

24

Fire music

Norman Andersen, Bart Hopkin, Michel Moglia, and Etiyé Poulsen

The following group of short articles on pyrophones originally appeared in *Experimental Musical Instruments* Volume 10 #1, September 1994.

Introductory Notes

by Bart Hopkin

Here is how to make flame sing: obtain a glass tube, one or two inches in diameter, and perhaps two or three feet long, open at both ends. Light a propane torch or similar burner and insert the nozzle about one fourth of the way into the lower end of the tube. If conditions are right, you will hear the tone. It will begin not abruptly, but with a growing volume. The pitch will correspond to the resonant frequency of the tube, but higher than what one would expect as the resonant frequency for a tube of the given length due to the heating of the air. Gather together a tuned set of such tubes, develop the mechanisms to shut the flames on and off in a controlled manner, and you will have created a flame organ.

How does the flame tone come about? Michael Meadows, in a letter to Experimental Musical Instruments several years ago, suggested the explanation that follows. Further notes appear in the footnote below. The flame rapidly heats the air in its immediate vicinity, causing it to expand. The rapid expansion creates a pressure wave which travels to the end of the tube and partially reflects back, just as in an open-ended organ pipe. The expansion of the air and resulting rarefaction in the vicinity of the flame simultaneously creates a reduction in oxygen, causing the flame to lose intensity. This cooling, coupled with the continued rise of the heated air, draws in more oxygen-rich air from below. The air reaches the flame just as the reflected pressure wave (region of denser air) does. The flame intensifies, and the heating and resulting expansion are repeated. This is very much analogous to the process occurring in other tubular wind instruments: the frequency of a driving force, such as the puffs of air passing through a trumpet player's lips, comes into agreement with the resonant frequency of a tubular air column; the two reinforce on another, and a strong, focused oscillation at the resonant frequency is established.

With the flame organ it is the oscillation in flame intensity and the resulting localized expansion and contraction of the air, rather than the air pulsing through the trumpeter's lips, that provides the initial impulse. In either case, the primary factor determining the resonant frequency is the length of the tube.

I sent this description of flame organ mechanics to EMI's acoustics referee, Professor Donald Hall of Sacramento State University. He responded with these additional notes:

> Yes, that seems a pretty good explanation of one model of what drives a flame tube. There is perhaps room for a couple of additional thoughts. The suspicion that it is a little more complicated than that might arise from suggesting that if changing air density to alter the combustion rate is the whole story, then the place where the flame should be most effective would be at the middle of a tube open at both ends. That pressure antinode is where the density fluctuations would be greatest.
>
> On the other hand, one might think of a competing picture in which it is not the density but the motion of the air that enhances the flame by bringing fresh air into the vicinity to replace that which has been depleted of oxygen. Consequences of that model are (1) that the flame should be most effective if placed right in the mouth of the tube where the velocity oscillation is greatest, and (2) that maximum motion occurs twice per cycle (not mattering whether right or left direction). The latter means an inconsistency in the idea: the hypothesized wave gives rise to an effect that should reinforce a different wave an octave higher in pitch rather than itself. That inconsistency can perhaps be removed if there is a steady stream of air through the tube, so that left/right motion superimposed upon that (steady plus some more vs. steady diminished by same amount) makes a total speed which is faster/slower at times half a cycle apart, thus now the right frequency to be self-reinforcing. Maybe that has something to do with such tubes working in a vertical position with air rising through them as in a chimney but not when horizontal (or at any rate maybe not as well).
>
> Insofar as such tubes actually are found to respond best to a flame about a quarter pipe length inside (one-eighth wavelength, halfway between node and antinode, a place where both density and velocity are fluctuating), it may indicate that each of the competing explanations is partially true, with a situation where they can help each other out being better than either can do by itself.

The sounds of such an arrangement according to people who have worked with flame tones are highly varied. The system can be refined so as to dependably produce clear, steady tones at the frequency of the tube's fundamental. Or the mechanism can be adjusted to bring out harmonics. On the other hand, you can take a less controlling approach, and let the system come forth with a menagerie of whoops, shrieks, and moans. One consistent characteristic: the attacks are not sharp; rather, each tone grows as the resonance establishes itself.

The earliest references to "burning harmonica" or "chemical harmonica" come to us from the late 1700s. A century later, the physicist Georges Fréderic Eugéne Kastner published *Les flammes chantantes* (Paris, 1875), a description of his fire organ, the *pyrophone.* A photograph of this instrument appeared in Kenneth Peacock's article on color organs [appearing following this one in this book]. It appears as a moderately large console containing a small keyboard, with ten glass pipes rising from it. Later references to fire music generally take Kastner's pyrophone as a starting point.

Of modern fire organs, there are not many. One has been created by engineers at the Tokyo Gas Company. It is fully functional and played regularly in public. In the following pages, you will read about three more, created by contemporary artists-in-fire.

Bibliography, sort of

Published information on flame organs is rather scarce. Most references are brief. Following are a few sources that touch on the topic. Please forgive the incompleteness in some of the references; interested readers may be able to fill in the missing information, or perhaps come up with further references.

Bragg, William: *World of Sound* (Salzwasser Verlag, Paderborn, 2013, originally published 1920)

Hauch: Article in *Kopenhagen (phys. chem. naturh. und math.), Abhandl. Aus der neuen Sammlung der Wissenschaften, ubersetz von D.P. Scheel und C.F. Degen*, Kopenhagen, 1798, Vol 1, 1st part, p. 55.

Hopkins, George M.: *Experimental Science: Elementary Practical and Experimental Physics* (Lindsay Publications Inc., Bradley, IL 1987; originally published in New York, 1906.

Kastner, Georges Fréderic Eugéne: *Les flammes chantantes* (Paris, 1875).

Strutt, John William (Lord Rayleigh): *Theory of Sound* (Macmillan and Co., London, 1894).

Sachs, Curt: *Reallexicon der Musikinstrumente* (originally published by J. Bard, Berlin 1913).

Acknowledgments

Several people at different times have provided me with information that I have drawn on here, among them: Norman Andersen, François Baschet, Donald Hall, Dennis James, Shig Kihara, Michael Meadows, Michel Moglia, ltiyé Poulsen, Leo Tadagawa, and Trimpin.

In the following report, Etiyé Dimma Poulsen describes the contemporary fire instrument, L'Orgue a Feu, created by the French sound artist Michel Moglia. Following that, Michel provides some insight into the philosophy behind his work.

Michel Moglia's Fire Organ

by Etiyé Dimmo Poulsen

The Fire Organ consists of a pyramid structure which carries between 250 and 300 tubes (in stainless steel) at different diameters and lengths. The instrument has a wide sonorous range consisting of six untempered octaves (if we refer to the classical musical scale).

The sound is produced by the flame of a burner which functions with propane gas, using the propane gas cylinders made for hot air balloons. The burner, which is held in the hand, heats the chosen tube which in return gives a sound corresponding to the length of the tube in question.

In fact the flame isn't always placed in the tube, as it is with the pyrophone. The flame heats a metallic mass which is placed in a given position in the tube. Once you remove the burner, an exchange of heat is produced in the tube which finally evokes the vibration of the air stream. The vibration lasts several seconds; this enables you to play other notes simultaneously so as to give a harmony. The harmonies change continually as if living due to the varying thermal changes. One can provoke microtonalities, evolutive thumpings, varying loudness, forces and pitches, screams, calls

Figure 24.1 Michel Moglia's Fire Organ. (Photo by Nicholas Sersiron).

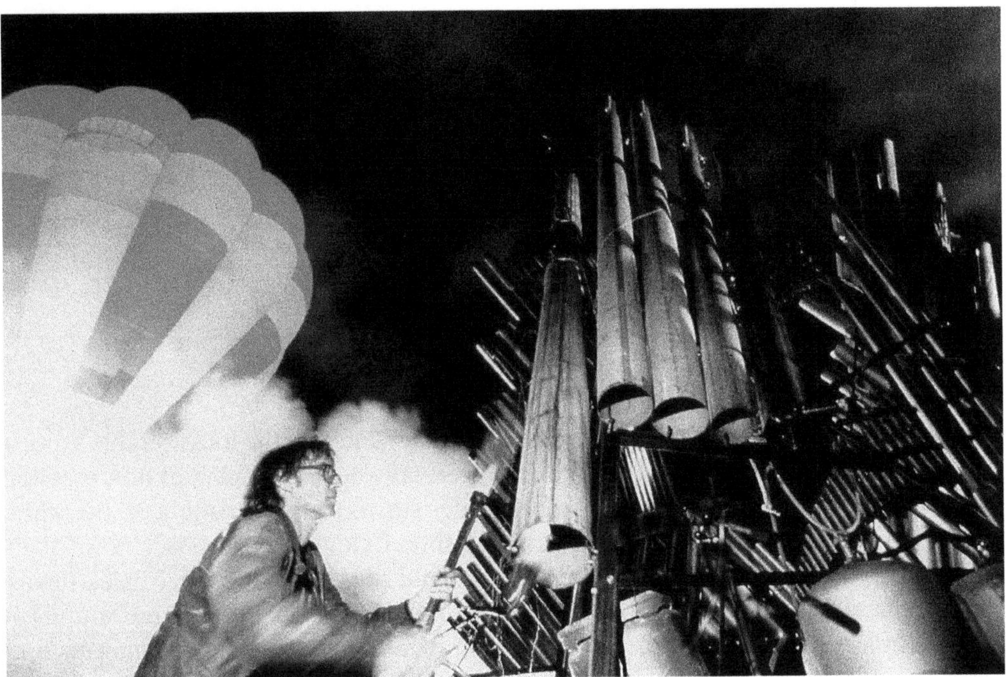

Figure 24.2 (Photo by Nicholas Sersiron).

of animals, etc ... Different types of burners allow the player to make effects similar to those of respiration of mammals.

Conceptually the Fire Organ has little to do with the pyrophone: It neither sounds nor looks like a classical instrument with a keyboard. The sound it emits is closer to nature than culture. It allows one to play with aleatoric effects caused by the flame.

The flame is partly free and can vary in length and intensity. The music of the Fire Organ includes violence and mildness, the force and the fragility of the flame, both on a visual and sonorous level. Certain flames are several meters long due to the employment of liquid gas. The Fire Organ also has a percussive system generated by mini-gas explosions which take place in titanium tubes (using solenoid valves).

Finally, the Fire Organ, accompanied by a surrounding of musicians specialized in new sonorous research, is used especially in ritual performances by night in spectacular sceneries and places; for example, a concert which took place in Cei, Ural, in front of an audience of 10,000 people. A huge Fire Organ of more than 10 meters high was placed in an electric power station, which permitted us to divert the gas of the power station for our "thermal chants."

The aim of the Fire Organ is to try to find a new sonorous atmosphere which differs voluntarily from what we commonly call "music," because the lastly mentioned is nothing but a vibratory coded system used in a precise cultural universe.

This creation is based on the following ideas:

Energy, different potentials
Control, chance
Order and disorder

Michel Moglia's Thermal Chants

by Michel Moglia
Translated from the French by Etiyé Poulsen

The infinite vibration of the flame, of time and life ...

As time goes by, all sonorous phenomena which are related to the innumerable aspects of combustion and transfers of energy have gradually become a kind of an obsession to me. Ranging from the roaring of the thunder storm to the whistling of the wind, passing by the heavy beatings of motors on a ship, followed by the staccato of thermal motors, going all the way to the breathings and calls of mammals, whales, fawns, of humans, right to the singing of birds at the break of dawn ... All these sounds at times formless to an inattentive ear, all this immense and mysterious sonorous universe, has become a kind of reference to me, a family which I have wanted to belong to as an active creator and participant, conscious of the forces which surround me.

That's how I decided one day to compose with my screams, my own thermal chants. A way to mix with the others in a huge thermal symphony.

This intuitive desire to be in harmony with the elements that have contributed to the beginning of life, which are life in themselves, seems remote to most western musical approaches in which one often struggles to control the tempo. They force themselves to regulate the modulation of the sound by means of sophisticated technology and skilled interpreters submitted to the cultural norms of society.

The thermal chants owe a lot to unwritten "primitive music" which takes into consideration exterior effects in their musical game. They equally owe a lot to John Cage who undoubtedly was one of the first composers to have consciously shared with his interpreters the idea of randomness in music ...

Opposed to all kinds of systems founded on ideas such as domination and control, my aim consists in playing with the flame, time, my life, and chance effects: elements that seem fluid and that one ought to accept as variables undecided beforehand.

Of course one is free to organize a kind of rule to this game depending on one's own emotional wishes. It's fascinating to compose a piece of music accepting equally to compose with the latent uncertainly. One can compare it with the attitude of sailors who at times put up their sails against the wind and succeed despite (or maybe thanks to) everything in reaching their aim, the aim of getting to the end of the initiatory voyage ...

To accept mystery and the role it plays in our lives is essential. The ritual, the sacrifice, the risk of being condemned to the stake still roam in our minds, for the better and for the worse. Therefore, both the violence and the danger are founding elements of the thermal chants. The most dangerous and unexplored territory on earth is life; it is us ...

Composing a thermal chant by using either classical, primitive or post-industrial instruments, organizing the breathing, voices, screams, releasing the roars of motors or concentrating on listening to the crystalline rustling of running water or on the vibration of a dancing flame, to "sculpt" the whole in a visual and sonorous universe, all this is undoubtedly a means by which I try to find my own roots, which stretches beyond our strong and omnipresent cultural system.

Playing with Pyrophones

By Norman Andersen

The fascination began as a child; it was difficult to resist the allure of matches. My childhood studies of what burned and what didn't may not have been scientific, but I was learning something in the process. Fortunately, I curbed my young appetite for fire after nearly burning down the large pine tree in front of my parent's house.

Fire studies continued but in somewhat more sophisticated fashion, as I proceeded through high school and college. I went into the field of fine art and on the

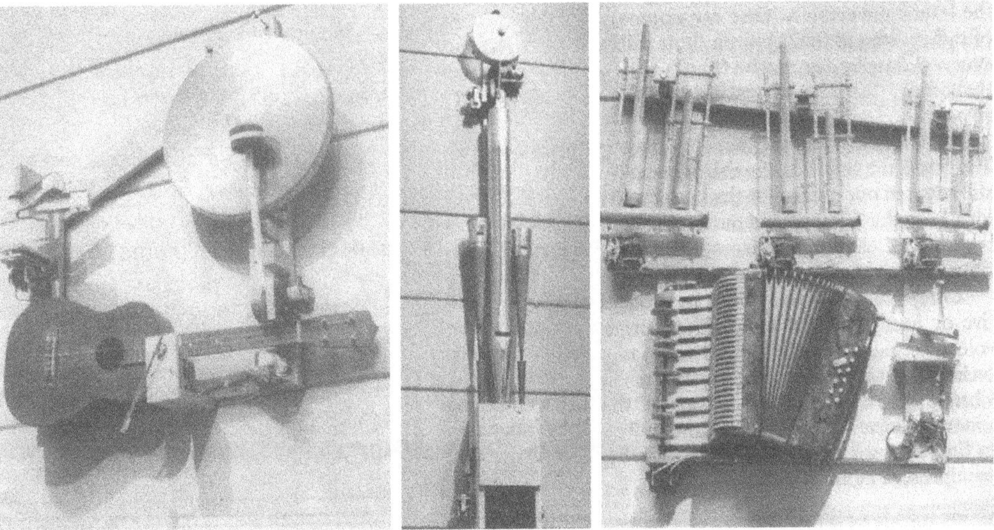

Figure 24.3 A, B and C. Details from installation "Clamorama" 1990 by Norman Andersen. Three of the four electrically actuated acoustic modules that made up the sculpture are shown here.

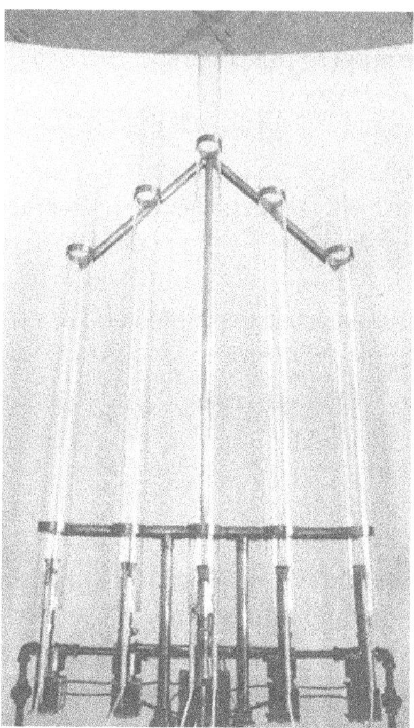

Figure 24.4 Detail of Norman Andersen's "Siren" sound sculpture 1990: Gas-fired pyrophone unit.

Figure 24.5 "Production of Sounding Flames" – an illustration from George M. Hopkins' *Experimental Science - Elementary Practical and Experimental Physics*, 1904 (republished by Lindsey Publications, Bradley IL, 1987).

side, had a very strong interest in music and sound. Ultimately, I found myself making kinetic sound-producing sculpture and satisfying the integration of a number of interests. My fascination for fire patiently waited.

Opportunity arose with a potential sculpture commission offered by the Duluth Water & Gas Company in Minnesota in 1988. Among the criteria for a potential work of art was the desire to have water and gas (presumably flame) used in the sculpture. Experience had taught me the difficulties of using water in sculptural projects (too numerous to mention here), but the challenge of gas and fire was too much technical punishment to pass up (besides, I knew few artists with the information necessary to attack this problem). The project was right up my alley, and after considerable thought and site research, I began to conjure up a softly spoken sound sculpture using resonating tubes with pitched gurgling water, a sheet of falling "rain," and, for real excitement, a glass pyrophone. All these visual and auditory elements would be controlled by an electromechanical random sequencing system like others I have typically employed in my sculptures.

My information on the pyrophone was limited.[1] A friend had provided me with photocopies from an old book entitled *Experimental Science*[2] (Chapter VIII, "Sound") which described a method by which sound could be produced using a thin flame within an open-ended tube. I called upon my personal pyrotechnical history and began experimenting. Not having a science lab and subsequent ready

Figure 24.6 Detail showing configuration of burner, valve and igniter on one of Anderson's pyrophone tubes.

supply of natural gas at my disposal, I began studies with the use of a basic propane torch. I could CAREFULLY position the torch in a vise with the flame pointing straight up, then lower a variety of heat-proof tubes over the flame until something happened (hopefully not an explosion). It wasn't long before I had a sound, and it was an interesting sound at that! I tested many variables: length of tube, diameter of tube, size of flame, different torch heads, and position of flame with respect to the open bottom of the tube. All these variables led to different sound characteristics (timbre) within a pretty stable pitch relative to the length of tube.

One of the earliest results from my tests was that the *thin flame* approach suggested through book research proved ineffective for me; I could get no satisfying sound response. It seems that a broader source of heat is required, such as that created by the standard tip on a propane torch. I surmised (perhaps a scientist among you can verify) that in order to set up the necessary standing wave needed to vibrate air and generate sound, heating the air across most of the width of the tube is required. This is contrary to the type of *convection current* that results from the narrow flame heat source.

Because my limited research wasn't helping to any great extent, I decided I was pretty much on my own with this technology, winging it in my customary empirical fashion. The next hurdle for my particular application was to figure out how to get natural gas (under moderately low pressure) to imitate the heat pattern of the torch used in my experiments. By this time, I had tapped into the pilot light fittings of my home's furnace for a ready supply of natural gas (it was summer, and the heat wasn't on).

I wanted to take advantage of the well-developed technologies already existing for gas appliances. I also thought that commercially available systems must somehow be safer and would be more to the liking of the Duluth Gas Company, than parts I might design and build from scratch myself. I knew that I had to add air to the gas in order to obtain a nice hot blue flame, so I began by investing in a couple of mixer valves like you would find inside a gas range or cook-top. These valves provide the essential Venturi (a narrow orifice) which accelerates the low-pressure gas between vents in order to draw in combustion air. Lots of great components can be purchased at appliance part stores, but it's probably best not to say what you're going to do with them. Parts store guys always want part numbers, so just getting something generic can be tricky. My usual approach is to ask for the cheapest one, since it is most often the simplest in configuration and most useful to the experimenter.

Once I had the mixer valve, I inserted the venturi end into a ½" diameter copper tube about 6" long, in which I had drilled two 3/8" holes to accept air alongside the Venturi. I could control the airflow with a small metal hose clamp, and the gas flow with the commercial valve (Figure 24.6).I felt like I was getting down to a controllable science but then had to concentrate on devising some kind of a *burner head*. I tried many options using different kinds of *caps* with patterns of holes in them (almost imitating a range top burner in miniature). A few of these designs began to yield some results but were subject to uneven lighting and burning.

The standing waves within the pyrophone tube can set up a pretty strong pulse of pressure on the flame. With many of my test burners, the sound waves would extinguish the flame, or at least enough parts of it, to prevent generation of more sound. The problem was getting frustrating and difficult. The flame needs to be able to vibrate in intensity as a response to the waves it creates. One is encouraging a kind of a feedback system because of the controlled airflow of the tube. I asked myself, How do you generate a broad plasma of flame heat, and also keep it ignited in an environment that is trying to crush it? Scanning my memory of hot, glowing, burning things, I suddenly remembered camping trips and the old Coleman lantern with its ingenious glowing mantle. This was the answer I was looking for. I had no intention of using actual lantern mantles; in addition to being too large and delicate, I didn't want bright white light in my sculpture. I reasoned that if a delicate ash mesh could contain and maintain a flame, other kinds of mesh might also work, as long as they could withstand the heat. I found a piece of old steel window screen, cut out a small circle, formed it into a kind of *dome* shape over a rounded dowel rod, and inserted it in the burner end of my copper

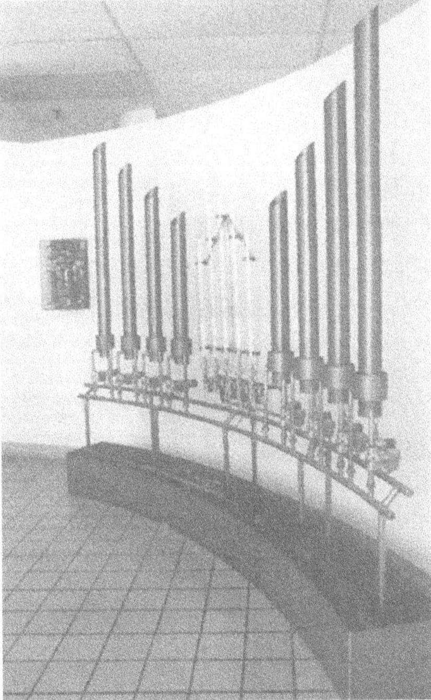

Figure 24.7 "Siren" 1990 sound sculpture by Norman Andersen. Pyrophone is surrounded by blue anodized aluminum tubes which emit changing pitches of gurgling water sounds.

tube. PRESTO! I was very pleased to discover the solution to my problem. The gas flame was small, hot, and very controlled. The wire screening quickly glowed a dim orange color, and provided plenty of re-ignition potential. All parts of the burner head stayed lit consistently, it used very little gas, and the best part was the interesting sounds that could be made.

Unfortunately, another tricky problem still had to be solved for my pyrophone to be fully functional; how could these gas burners be reliably and safely ignited. Originally I had considered using high-voltage arcs generated by neon sign transformers, but I discovered something better.

The *ignition system* for my sculpture again made use of appliance technology. I discovered the wonderful electronic high-voltage igniters which are also found in kitchen ranges in the place of pilot lights. These igniters can be bought as replacement parts, and consist of a ceramic bound electrode with special high-temperature wire, and an electronic module operating on 120vac (standard household current). The modules are very compact, and upon applying the input voltage, they respond with repeated "firings" of high voltage (low current) sparks from the electrode to any ground (which a burner is likely to be). The sparks come at intervals of one to two per second.

Having ignition is meaningless without a gas/air mixture to light. The amount of gas needed to keep a good sound going is very small, so careful adjustment must be maintained to ensure proper ignition. Timing is crucial, and anyone playing around with this stuff better consider it carefully. You have to get the gas supply switched "on" just in advance of the ignition sequence (or perhaps simultaneously since the igniters repeatedly fire). I used *solenoid valves* which independently feed each pyrophone mixer valve and its burner. I hope it is obvious to readers that you don't want a glass tube full of gas/air mixture to get ignited, unless you are creating a pyro-cannon or a pyro-fragmentation bomb. Remember, both ends of the pyrophone tube must remain open. Common sense should prevent accidents. Think about what is happening where and when, sniff for gas frequently, and when in doubt, ventilate and try again. I teach at the Minneapolis College of Art and Design, and one of my critical rules for my students is: "No one gets hurt!" This may be sound advice for instrument builders as well.

On my particular instrument/sculpture titled "Siren" (Figure 24.7), you may note that there are only three different lengths of pyrophone tube, two at 2', two at 2.5', and the tallest at 3'. As any instrument maker understands, this choice limits us to only three pitches. Wrong.... There are so many kooky variables with pyrophone technology, that quite a number of different timbres and pitches might be achieved even from one tube (frequencies can double or halve and dominant harmonics can fluctuate just like in organ pipes). All of "Siren's" tubes are 2" in outside diameter, and are made of Pyrex with flame polished ends. Pyrex tubing is available from chemistry supply companies, and it may be best to have them cut it. Clear tubes are not necessary, of course, but it's nice to see what's going on inside, especially since the flame is usually located at least a couple inches up from the bottom. Remember that combustion gasses have to pass through the length of the tube, so don't get any ideas about stopping them. On my sculpture, I have adjusted the pyrophones to yield some interesting sounds:

The tallest pipe sounds something like a low-pitched distant foghorn (it is breathy and vibrant).

The 2.5 footers operate simultaneously as a pair, sounding something like hooting ships' whistles at a distance (also with some white noise).

The shortest pair also operate together, making a kind of yelping sound which might be associated with coyotes or even wolves (apparently caused by a secondary or harmonic wave which oscillates the principal pitch at a slow rate).

It may be helpful to the reader unfamiliar with Minnesota to know that Duluth (the permanent home of the sculpture) is an important harbor on Lake Superior which boasts a wild and scenic shoreline. When these strange pyrophone-induced sounds are combined in irregular and unpredictable sequences with the hydraulic aspects of the sculpture, the result is a strange mechanical concert of natural sounds. The occasional clinking of 21 solenoid valves which control gases and fluids is an unavoidable aspect of the technology. I like to think that the mechanical sounds help to ground us in our place, reminding the viewer of our technological clunkiness relative to nature.

While I have outlined my experimentation and results with the pyrophone, I hope it is obvious that I have only scratched the surface of possibilities for this technology. I hope that some brave reader may consider additional development, and perhaps even come up with a new playable instrument (I'm sure someone somewhere has already). It should be noted that pyrophone sounds are "building" sounds; it takes a few seconds for the air in the column to be heated enough to begin to speak. I think it would be quite difficult to build an instrument with staccato capabilities using pyrophone science. I would imagine a pyrophone instrument to have similar musical characteristics to a glass harmonica or bowed glass bars, only breathier.

I hope that this article has been helpful to those who might venture into the realm of pyrophones. Perhaps others may find interest in recognizing that many of us share the same creative problem-solving process. I'm sure there are some readers who have already experimented with pyrophone science, and it would be fun to hear about your results. While I will make no promises, I would like to write about other sculptures, noisemaking systems, and instrument projects that I am involved in. Meanwhile, as always, I look forward to learning the latest from the rest of you in *EMI*. Keep up the good and important work.

Notes

1 I might add that I later found the image of an early pyrophone (ca. 1870) in an article, "Instruments to Perform Color Music," by Kenneth Peacock, in *Leonardo* Vol 21 #4 p. 396-406, 1988. A revised version of the article, with pyrophone photo included, also appeared in EMI Volume VII #2, Sept. 1991 [and is included in this collection].

2 George M. Hopkins, *Experimental Science – Elementary Practical and Experimental* Physics (Bradley, Il: Lindsay publications, 1987. Originally published in New York, 1907).

Related articles

One other article in this collection, while not focused primarily on pyrophones, includes information on the subject of pyrophones: "Famous Early Color Organs" by Kenneth Peacock.
Additional articles pertaining to pyrophones which appeared in the *Experimental Musical Instruments* journal are listed below. These articles can be accessed through the Internet Archive at https://archive.org/details/emi_archive/.

"The Pyrophone Explained" by Michael Meadows in EMI Vol. 3 #4, page 5

"More About Pyrophones", letters-to-the-editor from several writers in EMI Vol.3 #6 page 2

"More on Pyrophones", in "Notes form Recent Correspondence" in EMI Vol 6 #5, pages 3-4

Untitled notes on pyrophones from Trimpin in EMI Vol. 10 #1

"The Flame Componium and Reflections on the Pyrophone" by Reed Ghazala in EMI Vol 10 #3, pages 20-26

25

A Short Introduction to the Bambuso Sonoro

Hans van Koolwijk

This article originally appeared in *Experimental Musical Insruments* Volume 10 #1, September 1994. It was translated by John Lydon and edited by Elise Reynolds.

It is fascinating to see how sound is produced and learn how it can be altered; to hear how sound behaves in different spaces; and to feel how it effects its listeners.

Bambuso Sonoro was designed and built by me, Hans van Koolwijk. It still is being further developed, refined, and enlarged. The instrument arose from a need to enable a single player to play more than one bamboo flute at a time. First five flutes, then ten, and today, in 1994, more than one hundred.

It is an unpolished instrument in which the visual element is intimately linked to the auditory. You must be able to see the sound, so to speak. You have to be there, preferably close by or between the flutes, see the player's exertions, experience the "difficulty-of-the sound."

The main differences between this instrument and a pipe organ is that it works with variable wind pressure, thus maintaining many of the capabilities of mouth-blown flutes. For instance, it can play crescendos, glissandos, and harmonics. It produces many chance subsidiary sounds, and each flute has its own character. This is in part because the bore of the pipes is irregular due to the knots in the bamboo.

The air is generated by a ventilator that feeds various dividing chambers – together forming the console – which distribute it among the flutes. The wind supply can be diminished to various degrees with slides on the console. This makes it possible for the player to effect a broad range of nuance in the tone of each individual flute: ranging from a single, thin tone to one with an incredibly broad harmonic spectrum. When all of the slides are open, the Bambuso produces an enormous mass of sound: a stationary but nevertheless highly differentiated sound. The sound can be brought into motion using a system of valves with springs and membranes. The flutes produce rhythmic patterns, the speed of which is dependent upon the amount of air being supplied. The rhythmic patterns alter slowly and the keys influence each other at certain degrees of low air pressure, as though the flutes were fighting each other. Two flutes may become mutually entangled in a quickly pulsing rhythm as though they were wrestling, an effect that lasts until they release their grip and each passes into its own rhythm.

There are a number of ways of altering pitches.

Opening a slide more, so that the air pressure increases and the flute sounds a higher overtone. Some flutes are rich in overtones.

Moving a stop in one of the flutes, closing it off, so the column becomes shorter or longer.

Figure 25.1 Bambuso Sonoro, created by Hans van Koolwijk. (Photo: M. van der Hoeven).

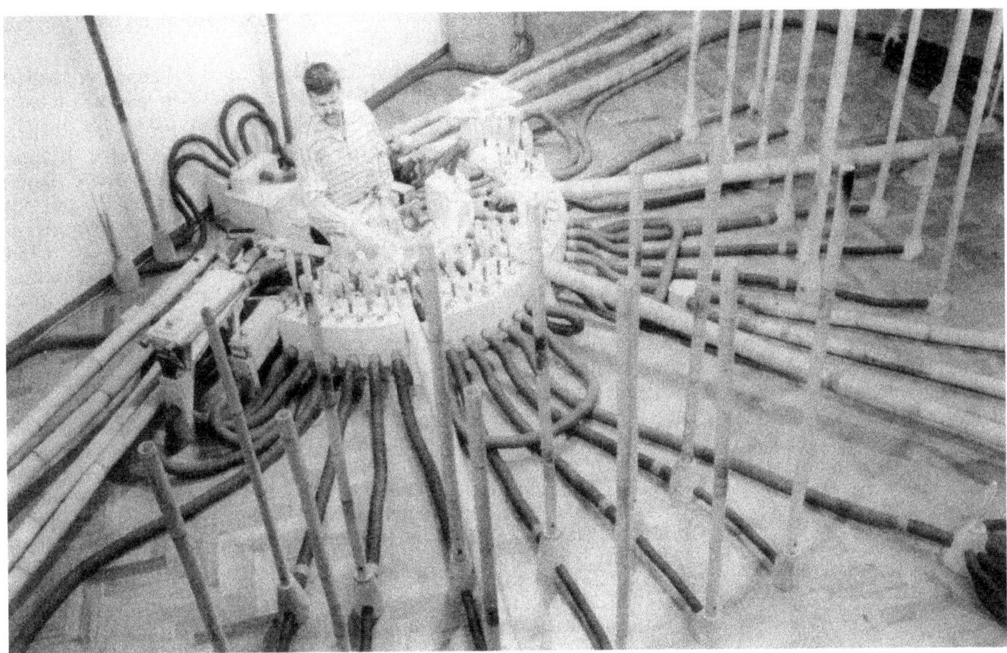

Figure 25.2 (Photo: Reyn van Koolwijk).

This is the same process used in "glissando flutes" and, in combination with a key, "bird flutes." They can really screech, and because of their slightly varied pitches, they also produce different tones.

With another system, the player passes a somewhat too small cork (a variable node) through the pipe, almost totally obstructing the wind passage so that the flute starts warbling. Different harmonics sound depending on the position of the cork and as it passes further, the sound jumps to the following harmonic. The many ways of playing this instrument also give the performance visual substance: image and sound reinforce each other.

The unruly and at times unpredictable nature of the Bambuso is a product of its chance subsidiary sounds, the various sound qualities, divergent harmonics, glissandos, interference of tones, rhythmic patterns, and so forth. The conglomeration of these qualities makes this instrument especially well suited for the playing of unconventionally notated scores where the expression of a particular sphere is more important than the reproduction of exact pitches.

Related articles

One other article on bamboo organs appeared in *Experimental Musical Instruments*: "A Bamboo Organ" (no author credited) in Vol. 9 #2. This very short article with one poorly reproduced photograph was a reprint from a 1926 issue of *The Etude*. It can be accessed through the Internet Archive at https://archive.org/details/emi_archive/.

26

Augustus Stroh and the Famous Stroh Violin

Or "The Inventors of Abnormalities in the Field of Violin-Building Have Not Yet Become Extinct"

Cary Clements

This article originally appeared in *Experimental Musical Instruments* Volume 10 #4, June 1995.

> It is the experience of all inventors of devices for the improvement of the violin that one has the greatest difficulty in opposing the traditions which surround the instrument.
>
> – Stroh catalog

The first time I laid eyes on a Stroh violin was at the Smithsonian Institution about ten years ago. As I was strolling through the musical instrument section, I came across one reposing in a glass case. What is this thing? This thing with a horn. Its awkward beauty impressed itself on my senses. Its whole look made me think of that vague era long ago before the electronic age, when record players had horns. And motorcars had horns that you could see. But violins?

The second time I saw a Stroh violin was only a few months ago. I had been researching this article for a few months and had seen quite a few pictures in books and magazines. But I needed to see one in person. Somehow I felt things would click in my head if I could examine up close the thing that I had spent many hours pursuing in libraries and archives, and with many letters to museums here and in Europe.

And when I did, I fell in love all over again. They really are things of such graceful beauty. Even in its old age, this instrument radiated a sense of charm and elegance that you may find hard to imagine on something with a foot long horn that would be more at home one of Henry Ford's Model Ts or Edison's phonograph.

This particular example of the Stroh violin did not arrive in the 1990s without suffering some indignities along the way. Among the many extra holes drilled into its body and horn over the years were a set designed to hold a tube that was used to

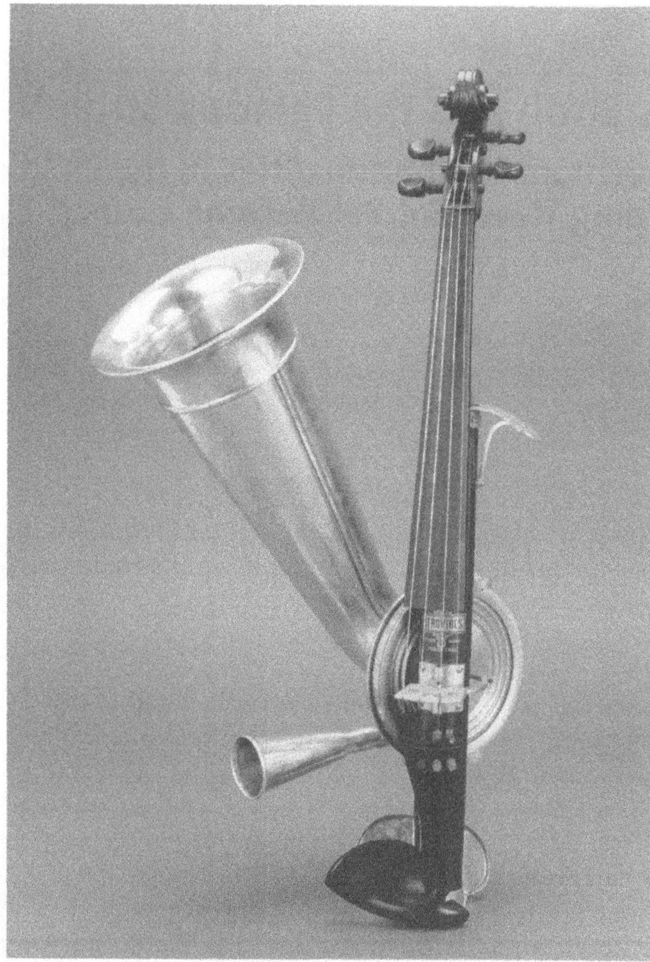

Figure 26.1 This pristine example of a Stroh violin is from the Shrine to Music Museum in Vermillion, South Dakota [now called the National Music Museum]. Notice the 'Stroviol' trademark decal on the body between the bridge and fingerboard and the small monitor horn protruding from the back of the diaphragm housing. This instrument has a bridge lever that is supported between two set screws. (Photo credit: National Music Museum, The University of South Dakota. Simon Spicer, photographer.).

blow up a balloon in the horn. Attached at the mouth of the horn was a sharp pin that was hinged so that when you pressed the 1929 Lincoln penny that was soldered to the back of it the balloon would pop.

So ... imagine a performer on stage, or maybe a musician out on the street that has gathered a small crowd. He's playing a tune, maybe not that well, when a bright rubber object starts to grow out of the horn of his fiddle. Howls of laughter. And then it suddenly pops at a certain point in the melody right on cue. People turn to each other smiling. How did he do that?

This is the impression that a lot of people have of the Stroh violin – that it's just a gimmick, some wise guy's idea of a novelty musical instrument designed to appeal to the quirkiness in us all.

Or that somehow it was meant to be an improvement on the traditional violin; that if it had appeared 20 years earlier, it would have been immortalized in Edward Heron Allen's classic book *Violin-Making: As it was, and Is* in the chapter called "The Violin ... its Vagaries and its Variegators" as a pretender to the throne of Stradivari.

Invented by a prolific and often overlooked inventor and scientist, Augustus Stroh, and introduced in 1901, the Stroh violin became cutting edge technology – the standard of an industry – for over 20 years. The place where the Stroh violin gained a foothold and stayed for such a long time? The recording studio.

The Early Days of the Recording Industry

When Edison invented the phonograph, some of the uses he envisioned for it were more along practical lines such as being able to record dictation for later retyping. Little did he realize that by the 1890s the sale of recorded discs and cylinders of music would be very popular.

In order for a good recording to be made though, the music had to be played very loudly. This would explain why there were a lot of banjo recordings made in the 1890s. For anyone who's never been to a bluegrass festival, the banjo is a very loud instrument. Therefore, it recorded well. I've been told that a lot of singers made at least one record back then and that the only ones to make more than one record were the singers that could sing loudly.

Obviously, the recording studio of the early 1900s was very different from the recording studio of today. One key thing to understand about the Stroh violin story is the fundamental difference between recording techniques before and after 1925. In 1925, a major development in recording technology occurred, and brought about a new era in that field. This innovation, electrical recording, fundamentally changed the way records were made and drastically improved the frequency response of recorded discs. It also made the Stroh violin obsolete as a recording tool. Prior to this innovation though, recordings were done in a much different way, and it was under these circumstances that the Stroh violin ruled.

The earlier method has become known as acoustic recording. It is called acoustic because there were no microphones and no electrical amplification of the sound before it was recorded.

Music was played into a horn. These sound waves vibrated the diaphragm at the end of the horn, and a stylus that was attached to the diaphragm cut a groove into the recording medium of soft wax. In a nutshell this was how it was done long before the days of the LP or hi-fi, let alone the CD or DAT machines.

Acoustic recording was not very sensitive and required musicians to crowd together as close as possible to the recording horn, with the louder instruments being just a little further back, in order for a decent recording to be made. One drawback to this early recording method was that a traditional violin did not record well under these circumstances.

All recordings done using the acoustic process – that is, all recordings prior to 1925 – could reproduce frequencies from 350 to 3,000 Hertz only. A performing band could easily generate frequencies from 30 to 12,000 Hertz. Recording engineers struggled unsuccessfully for many years to reproduce frequencies above 3,000 Hertz.

A large portion of the violin sound occurs in this higher range – that is, above 3000 Hertz. Therefore, most of the higher harmonics that make up the sound of a violin were impossible to record, and what could be recorded was not loud enough to be heard over the other instruments in the band.

What was needed was a louder violin. Augustus Stroh understood this and it was he who presented to the recording world a solution to the problem of capturing the violin sound on wax.

If you were to play a record or cylinder made prior to 1925 of a band with strings, chances are the violin you would hear was not the spruce and maple violin of old, but a mahogany and aluminum instrument that was invented with the intention of improving the recorded sound of the violin.

This was the Stroh violin. It replaced the sound box of the traditional violin with an aluminum diaphragm and a large trumpet horn. This mechanically amplified violin was used in the recording industry from its introduction circa 1901 until the mid-twenties when the electrical recording system was invented.

Not only was it a louder instrument, but the player could point the movable horn and direct the sound where it needed to go – that is, into the recording horn. Most pictures that you see of early recording sessions that included strings often show one or more Stroh violins.

The first known use of the Stroh violin on a recording done in the US was in April of 1904 by the violinist Charles D'Almaine. On that day he recorded four sides for the Victor Co. Sometimes the label of Victor 78's advertised the fact that a Stroh violin was used on the recording.

The art of making an acoustic recording was in the placement of the musicians in relation to the recording horn. The challenge was to get the right balance of sound, and having them close enough to the horn to be audible.

Joe Batten, an English conductor and musician whose recording career began in the days of acoustic recording, explains:

> The real perplexity of a recording session was to get singer and instrumentalists as close to the all-too-small horn as possible. The singer had the premier place, but his discomfort was always apparent, with the violins a foot away, the bassoon midway between his mouth and the recording horn the clarinets perched on high stools eight feet from the ground with the bells of each instrument six inches from his right ear, and the flute standing a foot behind him. Only the cornets and the trombones were kept at a respectable distance, the cornets standing ten feet away, and the trombones, perched on stools like the clarinets, twelve feet away.
>
> \- from *Joe Batten's Book*

Figure 26.2 Recording session at the Edison Recording Studio, 79 5th Avenue, New York City, circa 1907-1910. Singer Harry Anthony (John Young) sings while Eugene Jaudas conducts the orchestra that includes three Stroh violins. (Photo credit: U.S. Department of the Interior, National Park Service, Edison National Historical Site.).

Before Stroh violins came into use, the violin sound would just become lost on recordings in such a large group of instruments. Lest you think that Mr. Stroh's invention was the perfect solution to the problem, note what Fred Gaisberg – whose jaunts through Europe and Asia in the early part of this century recording musicians for Emile Berliner are legendary – had to say:

> There stood the recording machine on a high stand; from this projected a long thin trumpet into which the artist song. Close by, on a high movable platform, was an upright piano. If there was an orchestral accompaniment, then half-a dozen wind instrumentalists, also on high stands, would be crowded in close to the singer. Perhaps one Stroh violin, its trumpet bearing close on the singer's ear, would be the sole representative of the string section, and he would be left inaudible if he did not exaggerate heavily the pizzicato, glissando and vibrato characteristics of his instrument.
>
> – from *The Music Goes Round* by F.W. Gaisberg.

Alas, the Stroh violin was not accepted with open arms as the answer to all violinists' and sound recordists' dreams. In fact, there was disdain for the instrument then that still remains today whenever a violin builder makes anything that varies from the norm.

One newspaper report from 1904 announcing a new recording by the violinist Kubelik actually made the point that he was not playing a Stroh violin. Perhaps as a concession to the usefulness of the Stroh violin for recording, Kubelik did later use it to record.

How the Stroh Violin Works

When held in playing position the neck, fingerboard, bridge, and strings of the Stroh violin are in exactly the same position they would be on a regular violin.

The bridge sits on a rocking lever that on some models rides on a knife edge. On other models, the rocking lever is held between two adjustable set screws. As a bow is drawn across the strings, the lever and bridge are free to oscillate and these vibrations are transmitted from the end of the lever through a small connecting rod to the center of the aluminum diaphragm.

The conically shaped diaphragm is held in a cast aluminum housing that is screwed to the cylindrically shaped wooden body of the violin. The housing is open on the side on which the connecting rod attaches to the diaphragm and is closed on the other side except for an opening in the center that the big horn attaches to.

The horn, or "trumpet-shaped resonator or tube," as the UK patent for the Stroh violin states, is there "to augment or distribute the sounds emitted by the diaphragm." Without the horn, the Stroh violin would not be as loud as it is. To understand this, think of the difference between a normal-sounding voice and what a person talking through a megaphone would sound like.

On some models, there is a smaller monitor horn that attaches to the housing on the same side as the big horn. It points back at the player's ear and makes it a little easier for him to hear the instrument.

What Does a Stroh Violin Sound Like?

In a word, like a violin. Even veteran collectors of acoustic era discs and cylinders cannot tell by listening whether it is a Stroh violin or not being played on a particular recording. It is generally accepted that most acoustic recordings that included violins were done with Strohs and that it was the Stroh violin that made it possible for the violin to be audible.

To be honest, the only Stroh violin that I've heard in person had some problems with the diaphragm bottoming out on the housing and muting its tone, but it too sounded like a violin. It was just not as loud as it should have been.

I do have a tape of the late Irish fiddle player Julia Clifford playing her Stroh violin. Listening to this recording, done in the late 1980s, there would be no way of knowing that it was a Stroh she played, unless you were told so.

Two Augustus Strohs?

Who invented the Stroh violin? This question needs to be asked because there is some confusion between the roles played by the two Strohs, father and son, central to the story of the Stroh violin.

Augustus Stroh was the inventor of this instrument. His son Charles Stroh was the first manufacturer of the Stroh violin. Charles is often erroneously credited with its invention.

Augustus' full name was John Matthias Augustus Stroh. Sometimes he is referred to as J.M.A. Stroh, and sometimes as A. Stroh, but for the most part he is known as Augustus Stroh. Charles' full name was Augustus Charles Stroh. Perhaps since his father went by Augustus, he went by Charles. Sometimes you see him referred to as Charles Augustus Stroh. But Augustus Charles Stroh is the name on his birth certificate.

To avoid confusion, I will refer to John Matthias Augustus Stroh, the father, as Augustus and to his son as Charles.

Figure 26.3 "The Late Mr. Augustus Stroh" (Photo credit: Self-portrait from *Engineering: An Illustrated Weekly Journal*).

Figure 26.4 "Mr. Charles Stroh" (Photo credit: from *The Talking Machine Review* No. 36, August 1975).

J.M.A. Stroh

Augustus was 73 years old when the final patent for the Stroh violin was issued in 1901. He was born in Frankfurt-am-Main in 1828. He originally went to London on holiday in 1851 to visit The Great Exhibition.

Held at the newly built Crystal Palace, The Great Exhibition was intended to showcase the marvels of the then-emerging industrial world and was organized and presided over by the Prince Consort Albert, a fellow German. Impressed by England and its scientific institutions, the young Stroh decided to settle there, eventually becoming a British subject in 1869.

Apprenticed as a watchmaker in Germany, he set up shop in London at 2, Carlisle St., Soho Square, from 1857 to 1861. In order to pass the exam for his apprenticeship in Germany, the young Stroh was required to construct a special form of vertical clock escapement. So well did he do at this, that the examiners allowed him to keep the instrument when completed – an unprecedented concession. His interest in horology, the art of constructing instruments for indicating time, remained, for he was issued a patent in 1869 for an electric clock escapement.

Sir Charles Wheatstone

From 1858, he worked closely with Sir Charles Wheatstone. Wheatstone is best known as the inventor of the concertina, and stereoscopy, as well as for making many improvements to the telegraph.

Music being one of the things that these two men had in common, they were issued a joint patent in 1872 for an accordion-style instrument that could slide between notes. Years later, while eulogizing Augustus Stroh at the Institution of Electrical Engineers, Mr. W.M. Mordey remarked: "He was associated with Wheatstone in the invention of what is called the English concertina – an instrument that I believe musical people consider is really a musical instrument." In fact, Wheatstone had invented the concertina years before he met Stroh.

It is told that Stroh made a watch and presented it to Wheatstone that had the dimensions of an English half crown, both in diameter and thickness. A half-crown is slightly larger than a fifty-cent piece.

In 1860, Stroh and Wheatstone set up a shop together at 29, Tolmers Square, Hampstead Road, London, to manufacture the telegraphic equipment that they designed together. It is generally agreed that while Wheatstone got the credit for such inventions as the ABC telegraph, and the Wheatstone automatic high-speed telegraph – machines that were on the cutting edge of communications at that time – he could not have done it without Augustus Stroh's great mechanical skill.

In fact, Stroh was awarded a Gold Medal by the International Jury of the Paris Exhibition of 1878 for the high-speed telegraph. It is not unusual to read reports from that time that heap great praise on Stroh: "the prince of mechanicians," "the greatest mechanic of the day," "Clever Mr. Stroh," "one of our ablest mechanicians."

After Wheatstone's death in 1875, Stroh maintained the Tolmers Square factory for a while, then sold it to the General Post Office in 1880. Contrary to what has been reported elsewhere, Augustus Stroh was never an employee of the GPO. He was, however, associated for a time in the late 1870s with Sir W.H. Preece, Assistant Engineer and Electrician of the GPO.

The Phonograph

Late in 1877, Thomas Edison announced his latest invention to the world – the tinfoil phonograph. An Englishman named Henry Edmunds who was in America at the time visited Edison's workshop and witnessed a demonstration of this new device. Upon returning to England in the New Year, Edmunds told this story to *The Times*, and an article about the phonograph appeared on January 18, 1878. After reading this, Preece retained Edmunds to consult with Stroh to make a phonograph.

Augustus Stroh built the first phonograph to be made in England, and this machine was shown at the February 1, 1878 meeting of the Royal Institution. This was a replica of Edison's first phonograph, with the addition of a heavy hand crank that acted like a flywheel to even out the speed of rotation.

By February 27, Stroh had built an improved version of the phonograph, adding a fan governor and a clockwork motor. This machine was demonstrated on that date at the Society of Telegraph Engineers by Preece and Stroh and was noted to produce a quality of sound much better than the hand-cranked phonograph.

This very machine, later modified by Stroh to record on wax or tinfoil over wax – as opposed to tinfoil alone – is today the subject of research to determine if the recording on its mandrel is indeed the oldest existing sound recording in the world, as reported by the 1994 edition of the Guinness Book of World Records. It is owned by a Southern California collector.

The first company to manufacture the tinfoil phonograph in England was the London Stereoscopic Company. It is generally believed that the phonographs sold by the London Stereoscopic Company were designed by Augustus Stroh.

Figure 26.5 "The Stroh Violin Being Played". This photograph appeared with the Stroh Violin article in *The Strand Magazine* In 1902. I have a hunch that this photograph was taken by Augustus Stroh himself and that the model is one of his daughters. Perhaps it is Amelia Louisa Stroh who would have been in her early thirties when this picture was taken.

Synthetic Vowels

Around this same time, Stroh and Preece were also investigating the synthetic production of vowel sounds. Machines were built and demonstrated at the Royal Society in London.

An article that appeared in February of 1879 entitled "Studies in Acoustics. I. On the Synthetic Examination of Vowel Sounds" was later referred to in the 1885 English edition of the monumental work by Hermann Helmholtz *On the Sensations of Tone*. Alexander J. Ellis, English translator of Helmholtz, describes how he was given a personal demonstration of the apparatus used in this experiment by Augustus Stroh himself.

One of the machines used in these experiments consisted of a pen attached to a series of levers and wheels that mechanically produced a complex sine wave and drew it on a piece of paper. Somehow Stroh and Preece were able to transfer these synthetically produced waves into the groove on the edge of a small brass disc that could later be played through a needle and diaphragm producing vowel sounds.

Anyone who seeks out the above-mentioned article and examines the drawings of these machines cannot help but feel the mechanical genius of Augustus Stroh who designed and built the apparatus for these experiments

Figure 26.6 The 'Stroh' String Bass· (Credit: *Melody Maker*. Reproduced in *The Guinness Book of Jazz*.).

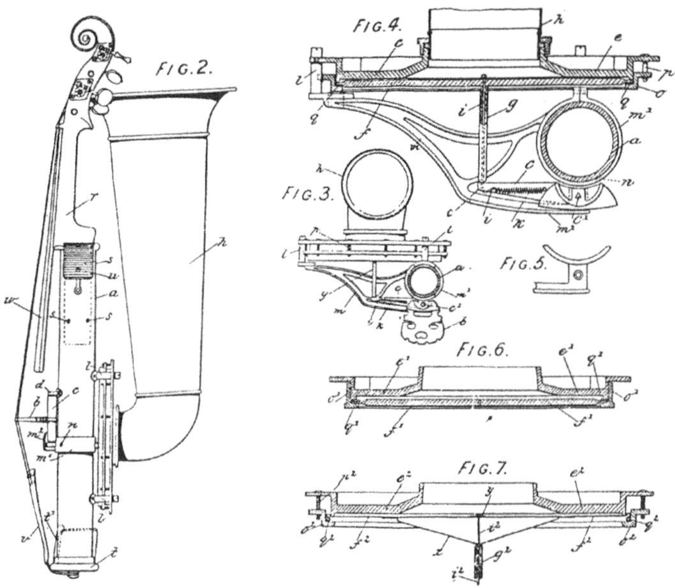

Figure 26.7 Detail from the British patent for the Stroh violin. A good source for copies of British patents going back hundreds of years is, believe it or not, the Los Angeles Public Library.

Figure 26.8 Augustus Stroh Tinfoil Phonograph. This is a drawing of the machine demonstrated at the Society of Telegraph Engineers by Preece and Stroh on February 27, 1878. It became the model for the tinfoil phonographs sold by the London Stereoscopic Company starting a few months later. It was later modified by Stroh to record on wax or tinfoil over wax; nobody is sure which or when it was done. It is currently owned by Mr. John Woodward of California who seeks any information at all about its history because the recording on the mandrel of this machine could be the oldest existing sound recording in the world. (Drawing from *Engineering: An Illustrated Weekly Journal*).

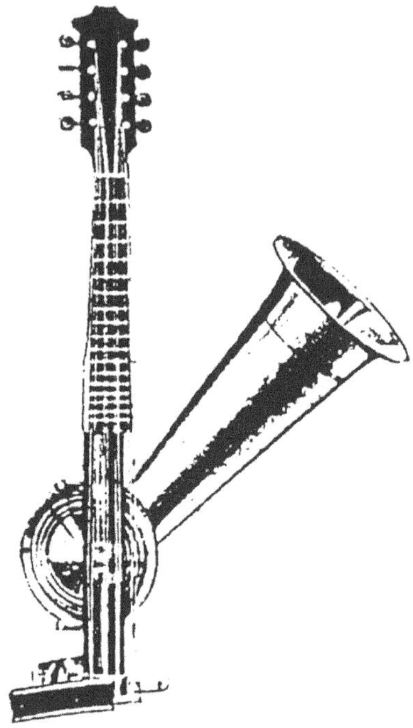

Figure 26.9 The "Stroh" mandoline. This fine instrument would have cost you £10 in 1929 including the case.

Retirement

After selling the premises at Tolmers Square, Stroh retired from business to pursue his many other interests. He then spent many hours in his home workshop at 98, Hayerstock Hill, London, N.W.

One of the many interests that he pursued was photography. He built his own cameras and was able to produce color photographs. The photograph entitled "The Late Mr. Augustus Stroh" (Figure 26.3) is in fact a self-portrait.

Stroh's interest in sound recording did not end at this point. It's known that he continued his development of the phonograph, building and improving many machines up until at least 1903. This work eventually led to the invention of the violin that bears his name.

Charles Stroh

John Matthias Augustus Stroh married Miss Emma King on June 16th, 1860 in the parish of St. James, Westminster in London. At the time, he was a 32-year-old watchmaker with a shop in Soho Square. Emma, the daughter of a bookseller, was

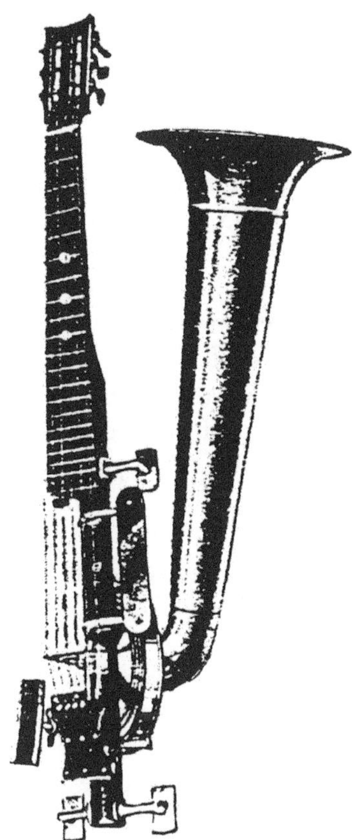

Figure 26.10 The "Stroh" guitar. (From the Stroh catalogue).

30 years old. Their first child, a girl, was born two years later in the summer of 1862. She was also named Emma. Three other children followed in the next five years. Julia was born in 1863. The only son, Augustus Charles, was born in 1864 and was named after his father and his maternal grandfather, Charles King. The youngest, Amelia Louisa, was born in 1867.

Son of the inventor of the Stroh Violin, Charles Stroh, was its first manufacturer and, as mentioned above, is often mistakenly credited with its invention. As a youth he assisted with his father's tinfoil phonograph experiments.

He was apprenticed as a telegraph engineer at the Government Telegraph Works in London at the age of 16 and then decided to seek his fortune in Australia. Perhaps he was motivated to leave England by a desire to get out from under the shadow of his by then well-known father. He lived in Australia for 15 years, returning to England around the turn of the century, just in time to become the first maker of the Stroh Violin.

Production Begins

Augustus Stroh's knowledge of the state of the art in sound recording at the time allowed him to see the need for the development of a recording violin. He received UK patent 9418 in 1899 for the basic violin and then two years later received patent 3393/1901 for the conical diaphragm that was used in the violin when it went into production. US patent no. 644,695, for the Stroh violin was issued to J.M.A. Stroh on March 6, 1900.

At the time Augustus was over 70 years old and probably had little interest in undertaking a new business venture. It seems natural that his son, just returned from a spell overseas and in need of work, would be the one to take this on.

Charles opened a shop at 94, Albany St. in London circa 1901 to make the violin. Production of the Stroh violin and, in later years, other Stroh-style instruments remained at this address until 1942. One popular Stroh instrument was the phonofiddle. This was a one-stringed cello, sometimes called a jap fiddle, that was used by musicians that performed in music halls or in vaudeville.

Albany St. runs north-south on the east side of Regents Park. Number 94 was across from Chester Gate. Unfortunately, the building has since been demolished and replaced by a block of flats.

By 1906 however, Charles had become a director of the Russell Hunting Record Co. Russell Hunting was a recording artist at that time who was famous for the Michael Casey monologue records. These were humorous routines in which Mr. Hunting did all the voices. Charles had become a fan of his while living in Australia, and little realized that he would one day work with him.

It was announced in 1906 by the *Phono Trader and Recorder* that Russell Hunting had "acquired and taken over the whole business in relation to the manufacture" of the Stroh violin. In December of 1908, however, the Russell Hunting Record Company went out of business. From this point on it is unclear what Charles Stroh's role in the manufacture of his father's invention was. In 1909 advertisements appeared in the London trade paper "Talking Machine News" announcing that the "sole maker" of the Stroh violin was now a Mr. George Evans, successor to Charles Stroh.

George Evans

George Evans and Company took over production of the Stroh violin in 1909. Under his leadership, the company expanded its line of instruments from the violin and phonofiddle to include the guitar, banjo, mandolin, ukulele, and upright bass – all of these with diaphragm and horn! I believe that it was after 1909 that the trademark "Stroviol" began to be used. It appears that instruments made prior to this date do not have the Stroviol decal that was later put on the slim body between the bridge and the end of the fingerboard.

It was noted above that in the mid-1920s recording technology fundamentally changed. The new method of recording used microphones and vacuum tube amplifiers and was much more sensitive than the previously used acoustic method.

The sole reason for the Stroh violin's prominence in the recording studio up to this point was its loudness. With the new recording technology, it was no longer necessary for musicians to crowd close to the recording horn. Now a microphone could be set up in the middle of the room to pick up the sound and the band could spread out a little and play more comfortably.

The Stroh violin after more than 20 years of use in the recording studio, and after being used on practically every record with strings made during that era, was now obsolete. But only obsolete for studio work. Recording studios now sold all of their Stroh violins to antique stores or consigned them to cold storage in the basement.

Post-Acoustic Era

The Stroh violin went out of use for recording in 1925 but continued to be made until 1942. All along it had been used for live performance, but from 1925 until 1942, this would have been the only demand for it. Certainly, there was an advantage to having a louder instrument for performing, especially in the days before good public address systems.

> THE GREATER volume of tone produced by the "Stroh" instruments makes them especially suitable for the small orchestras found in kinemas, dance rooms, restaurants, etc. Very often the exigencies of space make it imperative that the orchestra be kept small. The exchange of one, two or three instruments of the ordinary kind for others of the genuine "Stroh" pattern will give the effect of on orchestra more than trebled in size.
>
> – from the Stroh catalog circa 1927.
> (Spelling for "kinema" is from the original.)

By the early forties, though, the violin no longer held the same position in popular music that it had in earlier times. Julian Pilling touches on this topic in his brilliant article "Fiddles with Horns" published in the *Galpin Society Journal* in 1975. At the same time, the electric guitar was starting to be played more and more in orchestras, and so after 40 years of production, the Stroh violin, and its younger siblings, the Stroh guitar, mandolin, and so on, ceased to be made.

A Question to Ponder

Both Smith and Brozman, in their respective books on Rickenbacher guitars and National resonator instruments, mention that George Beauchamp had seen a violin with a horn used in vaudeville and wanted to have a guitar made with a resonator that was louder than the usual instrument. More than likely this was a Stroh violin that he saw. Apparently unaware that Stroviol also made a horned guitar, Beauchamp sought out John Dopyera in the mid to late 1920s to have him build a resonator guitar. Thus, began what was later to become National and Dobro. Could it be said that in an indirect way, Augustus Stroh and his violin was an influence on the development of the resonator guitars made by National and Dobro?

A Legend Dies

John Matthias Augustus Stroh died on November 2, 1914. He is buried in London's Hampstead Cemetery in a family grave with his wife Emma, their daughter Emma, his mother-in-law Margaret King, and his sister-in-law Louisa King.

His life coincided with what is sometimes called the Age of Progress, an era that began in all of its innocence with the Great Exhibition in 1851 and ended in 1914 with Europe plunging into the depths of war. Stroh was a product of this time of great advances in science and industry and his contributions, often unrecognized, were none the less great. Stroh reaped the benefits of his contributions though. At the time of his death, his estate was valued at more than £93,000 – quite a considerable sum for 1914. But for all of his work with the telegraph and the phonograph, his best known legacy is still the one that bears his name: The Stroh violin.

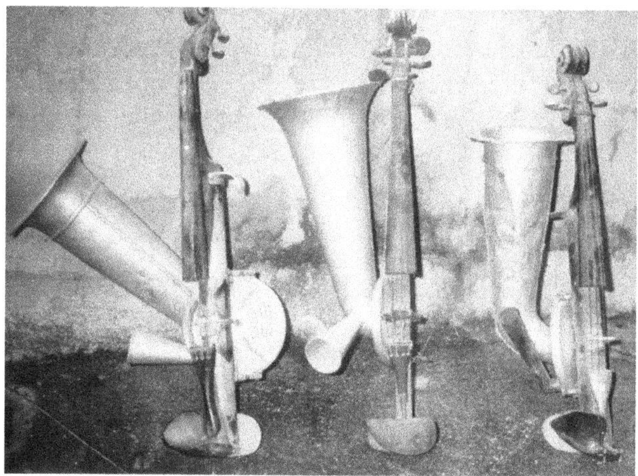

Figure 26.11 This trio of beauties was found on a trip through Burma by Dennis Griffin. The instrument on the left has an unusual cover plate on the diaphragm housing with the initials U.B.C. stamped into it. The same initials are also stamped on the diaphragm. It also has tuning gears on a plate. The violin on the right has a very thick neck scroll and metal friction pegs as well as an unusual monitor horn. Both have the distinctive seamed large horn that is very typical on Stroh instruments but don't have the usual 'Made in England' stamped on the inside of the shoulder rest. The violin in the middle has a spun horn and is either a cheaper model made at the Stroh factory or is a German copy of the Stroh. (Photo courtesy of Dennis Griffin).

Figure 26.12 Here's a view of the Stroh violin that you don't see too often. The small hole in the housing next to where the horn mounts is for the smaller monitor horn which is missing on this violin. This instrument belongs to Al Dodge. (Photo by Cary Clements).

Figure 26.13 Close-up view of the bridge, rocking lever, connecting rod and the aluminum diaphragm of the Stroh violin. Notice the knife edge on the bottom of the lever. It rides in the groove in the curved plate that is screwed to the circular body of the Instrument. (Photo by Cary Clements).

Figure 26.14 "Mr. Hats McKay on THE 'STROH' HAWAIIAN or STEEL GUITAR." In the late 1920s the Stroh guitar cost £15. An example of this instrument is in the collection of the Roy Acuff Museum at Opryland, USA In Nashville, Tennessee. Neither it nor the Stroh violin seem to have had much of an impact on country music. (Photo courtesy of the National Music Museum, University of South Dakota.).

Acknowledgments

For generous assistance in the making of this article, special thanks to Al Dodge, Ralph Novak, John Woodward, Craig Ventresco, Dave Radlauer, Bob Greenburg, Dennis Griffin, Dr. Margaret Downie Banks, and Ji Marsh.

Related articles

This article's author, Cary Clements, wrote three more articles on Stroh instruments appearing in *Experimental Musical Instruments.* These articles can be accessed through the Internet Archive at https://archive.org/details/emi_archive/.

"Historical Patents for Horned Violins" by Cary Clements, in EMI Vol. 13 #2

"The Homemade Clements Plywood Centennial Augustus Stroh Violin" by Cary Clements, in EMI Vol. 14 #3

"Extra, Extra – Stroh Violins Still Being Made!!!" by Cary Clements, in EMI Vol.14 #4

In addition, letters from various correspondents concerning Stroh violins, some including photographs, appeared in the letters sections of *Experimental Musical Instruments* Vol. 11 #1 and Vol. 11 #3

27

Drums for the 21st Century

Kris Lovelett and Christopher White

These three short articles on drums designed by Ken Lovelett originally appeared in *Experimental Musical Instruments* Volume 11 #2, December 1995.

Protocussion, a small and growing drum company in Upstate New York near Woodstock, has developed a new and exciting technique in the art of drum making. The drums that take the name of the man who created them, "Lovelett," were patented in 1991 and feature many of the unique and innovative ideas that have come to represent Protocussion.

Sound rings that protrude from the Lovelett Drum are the first noticeable feature. As can be seen in the photograph, these sound rings are like ledges extending out from the side of the drum all around its circumference. They accept the lug screws that hold the drum head in place. Through the process of translocation, vibrations from the drum head and rim are transferred back into the drum via the lug screws into the sound ring, which creates a more resonant drum shell. Protocussion's sound rings eliminate the need to attach metal lugs, many of whose internal springs vibrate, to the drum cylinder. They also reduce stress on the rest of the drum. As a result, a much more clean and pure sound is obtained.

Another innovation is the bearing edge of the drum where the drum skin passes over the rim of the shell. Protocussion's Ken Lovelett has found that with a variety of bearing edge shapes, a multitude of sound characteristics may be created. In addition to a standard bearing edge, when a rounded edge is applied, overtones are substantially reduced, while an edge that is peaked or pointed toward the center will give the tone a greater amount of ring.

Multi-annular ring construction is another patented feature which gives the Lovelett Drum its unique appearance as well as its strength, making it the strongest wood drum shell made. In multi-annular construction, the drum shell is built up by laminating rings of solid wood on top of one another, rather than making the shell in the traditional way from multiple wraps of thin, bent plywood.

Due to its innovative construction, the drum will never lose its cylindrical shape whether a drum head is or is not mounted, thus making the drum extremely durable and impervious to stress.

This construction technique has also led to a plethora of possibilities in this versatile Lovelett Drum. Drums can be made to any depth, from a pancake drum of just under two inches made of only a sound ring and a bearing edge, to a large drum measuring 48 inches or more. Aside from depth, shell thickness may also vary. Drums of larger or smaller diameters and drums of increased or decreased thickness may be crafted by creating thicker or thinner multi-annular rings. Moreover, when a cylinder made of material other than stacked multi-annular rings, such as metal, plexiglass, or wood, is used together with the Lovelett sound rings, even more variation of sound can be attained.

Aside from variations of shape, construction, size, and sound, different finishes can also be applied such as opaque shellacs, wood varnishes, and a special solid paint that will buff out scratches. These finishes will not only enhance the drum's look but also protect the wood.

With all the Lovelett Drum's innovations, a world of possibility has been opened for all percussion instruments. Utilizing obvious modifications, tambourines and banjos are just a few of the many drum-like percussion instruments that can be crafted with the techniques described in Lovelett's patent (U.S. 4,993,304, issued February 19, 1991) (Figures 27.1 and 27.2).

Protocussion's combination of revolutionary sound rings, variable bearing edges, durable multi-annular construction, and a selection of beautiful finishes all illustrate

Figure 27.1 The Lovelett Drum.

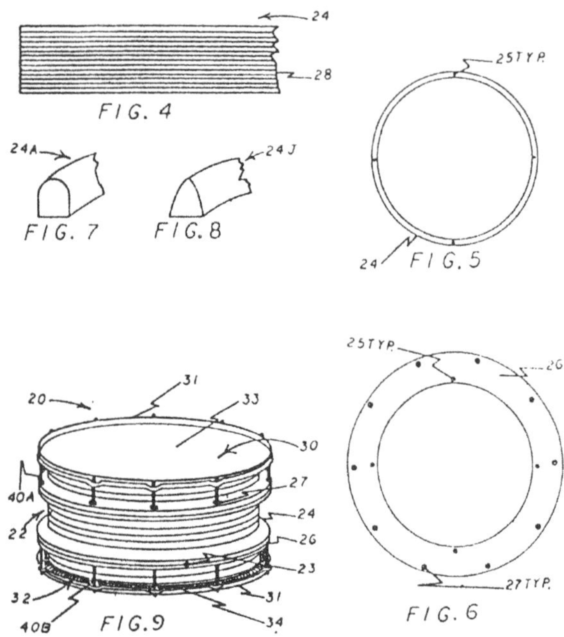

Figure 27.2 Drawings from the Lovelett Drum patent, showing various facets of the design. U.S. Pat. No. 4,993,304.

Protocussion's commitment to new and unique methods in crafting not only the Lovelett Drum but also an entire generation of percussion instruments.

The Busker

By Chrisopher White

A "Busker" is an English street musician, especially a "One-Man Band." It is also the very appropriate name given to a new sound sculpture by Ken Lovelett. Ken's Busker is a sonic sculptural assemblage that finely balances between the poles of art and function. The object/instrument dichotomy is rarely held in such satisfying equipoise (Figure 27.3).

Ken always has his ear attuned for musically usable sounds and his eye open for structurally interesting forms. The Busker owes its overall shape to a clothes rack purchased at Big Scott's going-out-of-business sale 12 years ago. Ken saw the chrome spiral stand and immediately envisioned it festooned with objects which

Figure 27.3 The Busker.

could be struck, plucked, scraped, and otherwise caused to make musically satisfying sounds.

During the intervening years, Ken has scavenged, purchased, adapted, and constructed innumerable sound makers and sound shapers. As individual items are added to the inventory, Ken continually experiments with their incorporation within larger structures like The Busker. Eventually, each of Ken's sonic sculptures begins to establish its own, unique, if related, character. Certain musical tonalities and contexts develop for each assemblage alongside their individual visual and spatial logic.

When one looks at the Busker, the first feature to catch the eye is likely to be the teardrop-shaped conical steel resonator on top. It both provides a witty visual "hat" for the man-sized Busker and colors many of the near-field sounds. It focuses and, therefore, seemingly amplifies those sounds as well as adding reverberance. The resonator tops a 20" bicycle wheel with tuned spokes. Ken has positioned cardboard attached to clothespins with handles below the wheel. It's the old "Boy on a Bike with Baseball cards" setup filtered through a musician's aesthetic. The handles attached to the clothes pins are used to change the angle and pressure of the cards hitting the tuned spokes, giving a greater dynamic range. The sound produced is simultaneously similar to a ratchet and steel drum.

Below this are four PVC tubes cut to different lengths and played with three-inch Styrofoam ball mallets. Next are three aluminum chimes and a large spring stretched to about three feet. Also there is a six-sided aluminum star that has many overtones. A sound source Ken discovered many years ago which has been useful in a number of sonic assemblages is the tin drum. As in the case of The Busker, Ken usually groups a few tin drums of different sizes together. They are cardboard canisters with tin tops and plastic bottoms. Ken paints each drum individually in brightly colored patterns. The tin drums are also attached to foot bellows that pump air into the drums creating pitch shifts. The sound is similar to a steel drum with a glissando. Continuing down the arc of The Busker's spiral from the tin drums there is an eleven-inch vibraphone bar. There is also a five-inch, five-toothed metal gear that has a very high expanding pitch.

Next is a 12-inch aluminum disc with a long-lasting mid-range pitch. Two pan-shaped discs have short but distinct pitches. The next item is an exhaust pipe from a Kero-sun Moniter 20™ heater – another example of Ken finding multiple purposes for an item, since he is a registered Kero-sun™ dealer. The pipe is musically versatile because it has an aluminum guiro sound when scraped and a nice bell note when struck. Next is a drum with a steel top head and a cardboard bottom head that is 14 inches in diameter and two and a half inches thick. Like many of Ken's other drums, it is also connected to a foot bellows for a low glissando effect. The Busker ends in a "skirt" made of 23 copper chimes in graduated lengths.

Ken is currently readying a new venue for displaying his sound sculptures, the *Sonart Gallery*. The gallery is intended as a space for both exhibitions of and performances on sonic sculpture. Ken has numerous works of his own which will be on display, and in addition, he invites anyone with a similar approach or concept to contact him. While certain aspects of the operation remain uncertain, Ken hopes to create an advisory board and expand the scope of Sonart Gallery's activities significantly over the next couple of years. It is hoped that this will become a premiere showcase for innovative instruments which embody visual/sculptural excellence as well as having musical use and interest.

The Nakers

By Kris Lovelett

Featured in the June 1994 issue of *Experimental Musical Instruments* (Volume IX #4) was an article by Ken Lovelett dealing with one of his sound sculptures called the "Naggara Drums." The sculpture consists of five clay drums connected to two foot bellows and nestled in a finely crafted wooden table with an arc rising above and spanning across. Ken has received many inquiries about this particular instrument.

In response to this interest and his on-going inventive spirit, Ken has created a new sound sculpture, "Nakers," that shares many characteristics similar to the Naggara Drums. Like its predecessor, the Nakers is fashioned in a similar style, but now with only two clay, dumbek-shaped drums of equal size that rest in another finely crafted wooden table. While one drum of the Nakers has a very thin head, the other's head is much thicker, allowing for a varied timbre attributed to varying head thicknesses. Each drum is also coated with a bright, brass glaze that not only beautifully reflects light but also enhances and compliments the natural grain and finish of the table. The glaze of the drums, however, is not the only thing that will catch the eye of anyone who sees the Nakers, for a hand-carved cherry wood block lies between the two drums whose purpose is to break-up the sound of the drums rhythmically and sonically while also adding character to appearance and sound (Figure 27.4).

Aside from visual appeal, special felt mallets and bamboo brushes created for the Nakers can add even more variety of sound. One can strike the shoulder of the mallet against the clay rim while simultaneously hitting the felt tip on the drum's head to create a unique sound. Variation can be furthered when combinations of four mallets or brushes are used in unison, or in independent rhythmic combinations. In addition, two wooden pedals, each with a styrofoam ball attached to the end of a dowel, extend up into the bottom of each drum,

Figure 27.4 The Nakers.

and when a pedal is depressed, the styrofoam ball is lowered. This increases the volume of air in the drum, thereby changing the drum's pitch, giving even more musical potential and variety to the Nakers. Unlike the Naggara Drums, however, the Nakers produces a more definitively pitched note, rather than a glissando effect. This is especially evident when a pedal of the Nakers is held in the same position.

When considering the numerous possibilities including the pitch-varying pumps, drums with different head thicknesses, wooden block, and special techniques such as hitting both the rim and head, using four mallets, or using bamboo or felt in combination, the Nakers are uniquely able to offer any array of pitch, timbral, or rhythmic effects.

Related articles

The following additional articles by Ken Lovelett and/or about his drums and musical sculpture appeared in *Experimental Musical Instruments.* These articles can be accessed through the Internet Archive at https://archive.org/details/emi_archive/.

"The Bellatope" by Ken Lovelett in Vol. 9 #4

"More Drums for the New Millenium" by Ken Lovelett in Vol. 12 #3

28

The Flutes and Sound Sculptures of Susan Rawcliffe

Susan Rawcliffe and Gene Ogami

In this article, Susan Rawcliffe provides descriptions to accompany Gene Ogami's photos of some of her ceramic wind instruments. The article first appeared in *Experimental Musical Instruments* Volume 11 #2.

Three Person Whistle

Light green engobe with colored splotches, clear glaze, and gold. (Engobes are clay slips in a wide range of colors. I often add splotches of additional colors with oxides. By using patches of glaze, my pieces have smooth, shiny areas mixed with the dull textured clay surfaces of the engobes.) The little balls on the bottom of this piece are the whistles: there are two for each of three players. I am fascinated by the concept of placing performers in intimate and unusual juxtapositions. Each whistle has one fingerhole, and all six are tuned to play together. By placing the players in close contact, the effect of the combination tones generated by the high-pitched whistles is intensified. After observing reactions to my space flutes and to my sets of such single whistles for play by small groups of people, I find that standard experiences of combination tones include: a sense that sound is physically moving the ear drum; an impression that the sounds seem to move through the head from ear to ear; a feeling that the sounds are generated inside the head, and, in small groups, the sensation among the players that they do not know who is playing which sounds (Figure 28.1).

Ball and Tube Flutes

Decorated with engobes and glazes; four fingerholes each. The ancient examples of these flutes as discussed in my article "Complex Acoustics in Pre-Columbian Flute Systems" (*Experimental Musical Instruments* Vol. VIII #2, pg. 10) generally consist of two balls joined by one open tube. I extended the design by making four balls with three tubes or, as shown here, five balls with four tubes. This makes a small compact flute that can play unexpectedly low tones. Because of its irregular shape, it plays irregular overtones. Thus, the scale cannot be precisely controlled and each flute seems to have its own built-in melody. They have a haunting, intimate, thoroughly lovely sound (Figure 28.2).

Figure 28.1 Three-Person Whistle. 1995, Ceramic, 12½" x 14" diameter.

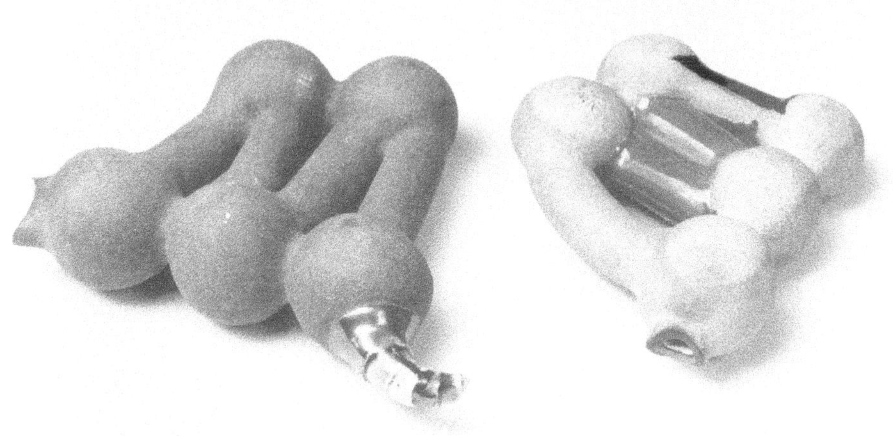

Figure 28.2 Ball & Tube Flutes. 1989 Ceramic, 1½" x 6" x 5¾".

Water Flutes

Decorated with various colors of engobes and glazes. I also make these flutes in a smaller version: approx. 5" high by 14" long. These flutes are another example of my process of evolving and adapting a pre-Columbian concept such as the ball and tube flute. Adding water was an inspired thought. In play, the performer tilts the flute from side to side. Water moving inside the flute effectively changes the length and volume of the instrument and thus the pitch. Because the resulting movement of the water is partly random, the sound does not always change in predictable ways. The performer sets the mood of the flute by controlling the air pressure to determine the partial that is played, and by controlling the amount of physical movement, which determines the activity level of the resultant sounds. In general, this flute has a mind of its own and is truly interactive. Because of the nature of the physical movement required by the player, a performance becomes almost a dance. It is great fun to play and to watch being played (Figure 28.3).

Flute Pile

Beige engobe with splotches of color and glaze. Seven ocarinas tuned to play together. Each is tuned to play a chromatic scale over a range of a ninth with perfect 5th & 4ths, using an ocarina fingering pattern that I developed. There are six finger-holes. The diatonic scale requires two cross-fingerings; with no cross-fingerings, a pentatonic scale is played. The more cross-fingerings required, the greater the scale as tuned is compromised. The ocarina is greatly affected by air pressure changes. I developed this fingering pattern for my pitched necklace ocarinas. See my article *op. cit.* pgs. 8-9 for a discussion of vessel flutes (Figure 28.4).

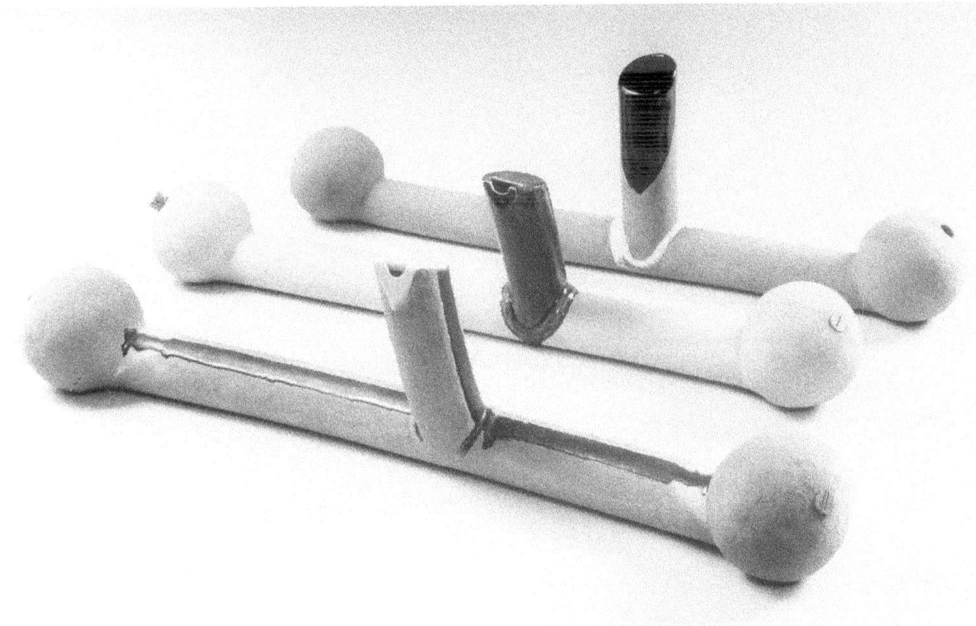

Figure 28.3 Water Flutes. 1993 Ceramic, 6-7" high x 22-23" long.

See Beastie

Dark gray engobe with colored splotches and tortoise shell glaze stripe. In this two-person instrument, the players cannot actually see each other while blowing. Each blowing end contains a chamberduct flute, with the exit tube adapted to play sideways. This type of flute produces sounds which range from soft and airy, to rich and reedy, to extremely raucous. Please see my article *op. cit.* pgs. 11-13 for a thorough discussion of the acoustical innards of the chamberduct flute as found exclusively in the pre-Colombian organology. The *Experimental Musical Instruments* cassette #VIII includes the sounds of two different such flutes (Figure 28.5 A, B).

Whiffles

Decorated with various colors of engobes and glazes. The lower chamber is an ocarina with five fingerholes tuned to play a chromatic scale over an octave using my standard ocarina fingering pattern. The upper chamber is an open tube placed around the mouthpiece. The instrument is held sideways; the player blows through the small tube protruding from the side. One hand fingers the holes while the other slides the large tube open and shut. Through the interaction of the two chambers, a wide range of wonderful vocal sounds and animal and bird cries can be produced. In play, it feels as if two different types of flutes are joined, with the sound coming out one side or the other. See my article, *op. cit.* pg. 11 for a discussion of the pitch jump whistles. A sample of whiffle music is also included on the EMI cassette (Figure 28.6).

Figure 28.4 Flute Pile. 1994 6" x 6" x 6".

Space Flutes

The sounding tubes vary in length from about 4" to about 6". Decorated with various colors of engobes and glazes. The melody of these flutes is created by the production of strong combination tones. Combination tones are created when the sounds of the two pipes interact in our ears. I currently tune them to play perfect consonant intervals as well as thoroughly out ones. One pipe in each flute has three larger fingerholes; the other four very small holes. These sounds are difficult to record and play back with any fidelity (Figure 28.7).

Polyglobular Trumpet

Dark brown with white terra sigillatas, glaze, and gold trim. The design is inspired by drawings as found in Izikowitz, *Musical Instruments of the South American Indians* (originally published 1934; republished in 1970 by S.R. Publishers; drawings on pages 222 and 240). A powerful, breathy, altogether wonderful sound.

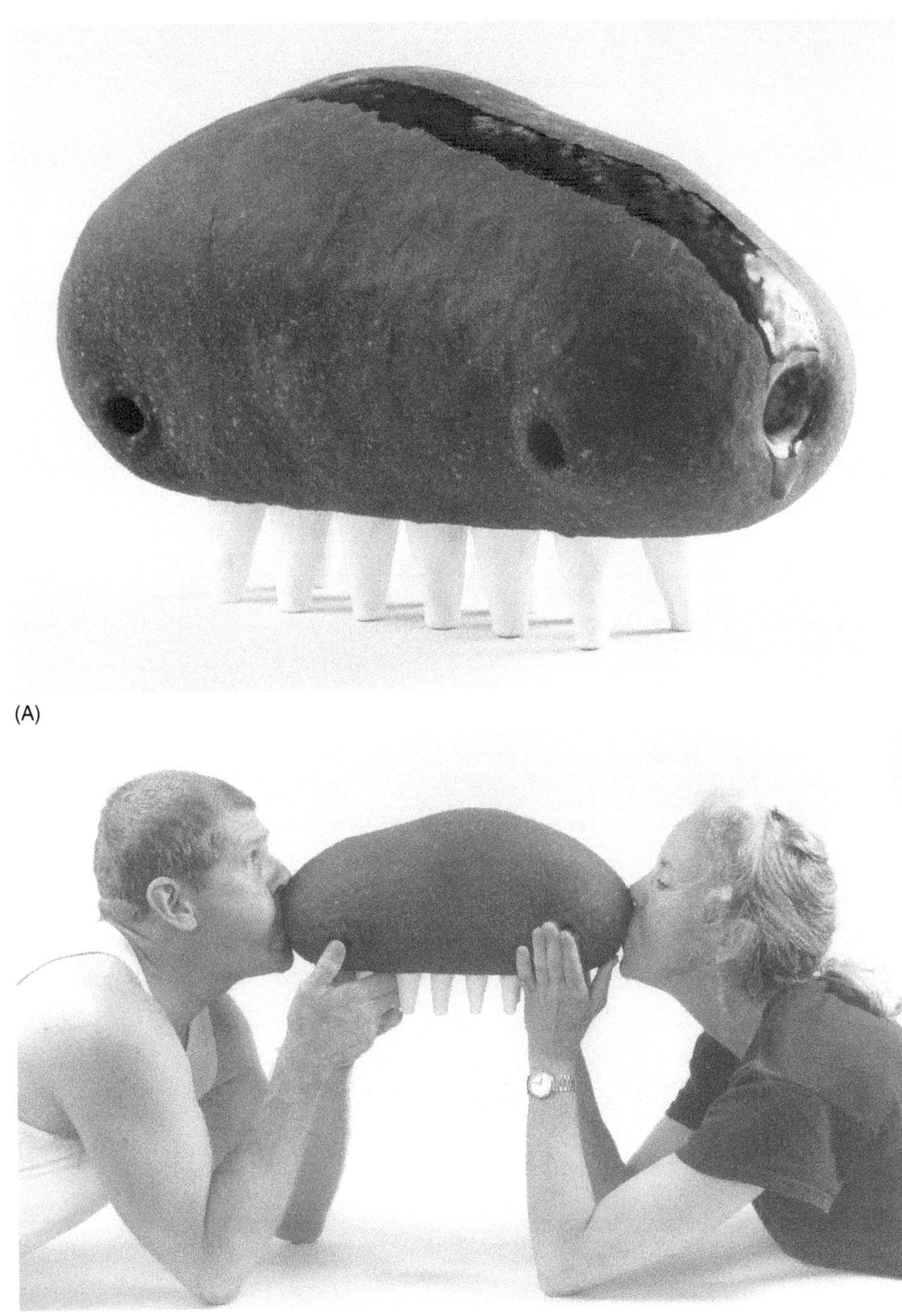

(A)

(B)

Figure 28.5 A and B. See Beastie. 1994 Ceramic 9" x 16½" x 7".

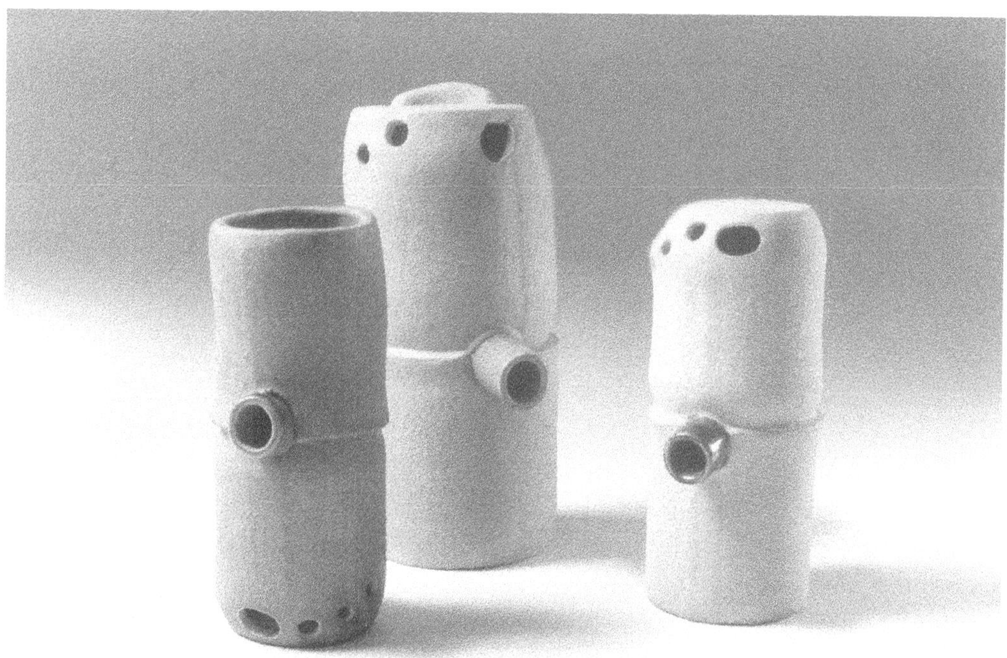

Figure 28.6 Whiffles. 1992 Ceramic 6½" x 2½" dia.; 8½" x 3½" dia.

Figure 28.7 Space Flutes. 1990-1995 Ceramic.

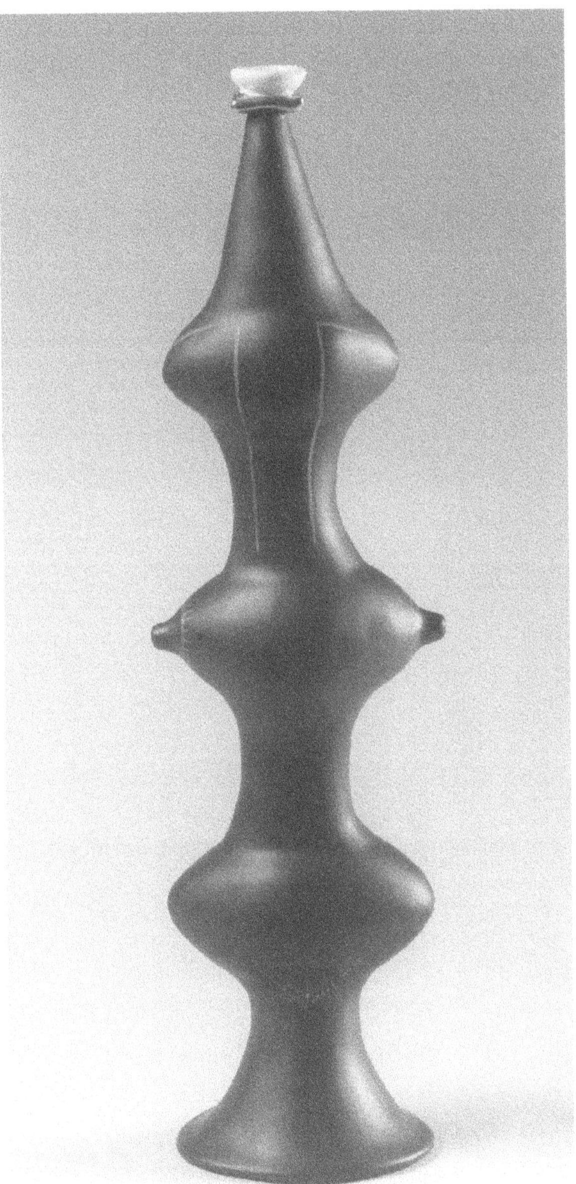

Figure 28.8 Polyglobular Trumpet. 1994 Ceramic 22" x 6" dia.

A listener called the instrument a "female didjeridu," which delighted me. Trumpet-type instruments are a new departure for me, an outgrowth of my years of work with the didjeridu (Figure 28.8).

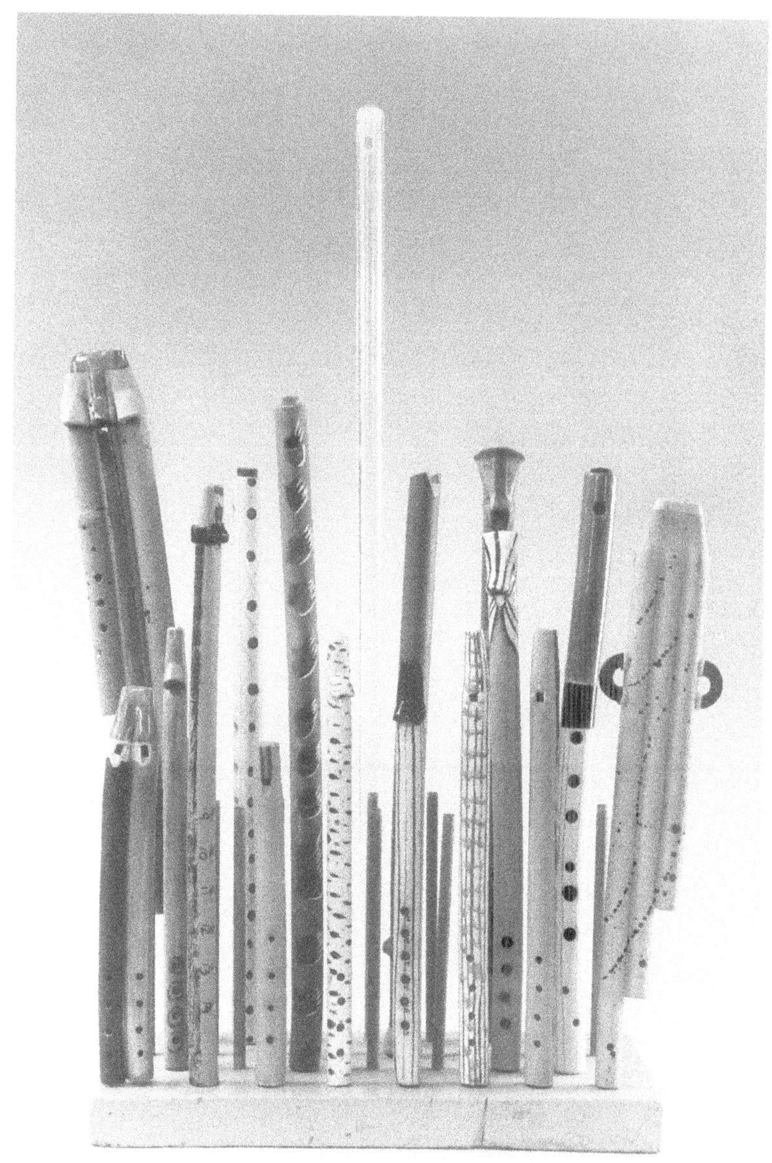

Figure 28.9 Rack of Flutes. Varied dates and sizes; ceramic, wood, nylon.

Rack of Flutes

Decorated with engobes, underglazes, glazes, and ceramic pencils. A small selection of my flutes from over the years, including double & triple pipes, hooded pipes, end-blown and transverse flutes, one harmonic flute, and flutes with pre-Columbian, sub-harmonic and diatonic scales. See my article, *op.cit.* pg. 9-10 for a discussion of tubular flutes. The EMI cassette includes the music of a hooded pipe and a triple pipe (Figure 28.9).

Related articles

Other articles by Susan Rawcliffe which appeared in *Experimental Musical Instruments* are listed below, followed by articles pertaining to ceramic wind instruments. These articles can be accessed through the Internet Archive at https://archive.org/details/emi_archive/.

"The Ceramic Whistles, Flutes, Ocarinas and Mirlitons of Susan Rawcliffe" by Bart Hopkin in EMI Vol. 1 #6

"Complex Acoustics in Pre-Columbian Flute Systems" by Susan Rawcliffe in EMI Vol. 8 #2

"The Flutes and Sound Sculptures of Susan Rawcliffe" by Susan Rawcliffe in EMI Vol. 11 #2

"Sharon Rowell's Clay Ocarinas" by Sharon Rowell in EMI Vol. 1 #2

"Sounds In Clay" by Ward Hartenstein in EMI Vol. 4 #6

"Udu Drum: Voice of the Ancestors": Frank Giorgini in EMI Vol. 5 #5

"Earthsounds" by Ragnar Naess in EMI Vol. 7 #2

"Globular Horns" by Barry Hall in EMI Vol. 14 #4

"Journey Through Sand and Flame; A Ceramic Musical Instrument Maker" by Brian Ransom is included in the current collection. It originally appeared in EMI Vol. 14 #4

29

Some Basics on Shell Trumpets and Some Very Basics on How to Make Them

Mitchell Clark

This article originally appeared in *Experimental Musical Instruments* Volume 12 #1, September 1996.

At the request of the editor of *Experimental Musical Instruments*, to whom I once casually mentioned that I had made a few shell trumpets, I will write something about the process of making such an instrument. But, to the possible disappointment of the editor, there's not an awful lot for me to say about their construction, as the simple forms of shell trumpets are quite easy to make. So, in the style of an entry in a cookbook where the author gives lots of history, lore, and anecdotes, and then finally gets down to the recipe, somewhere in what follows are some basic instructions for making shell trumpets. Endnotes – often referring to illustrations which may be consulted in other sources – are included and contribute additional texture.

I'll start by saying that when I was young, I knew about shell trumpets but obviously did not quite understand the principle of how they worked. I thought that no alteration was made to a conch's shell, which I thought was very beautiful and that it would be a shame to deface it. Rather, it seemed that getting the shell to sound was a matter simply of blowing *very, very, very* hard. Fortunately, I did not rupture any blood vessels trying out this theory.[1]

But the shell trumpet (an instrument in the domain of study of the *organologist*) has indeed been altered from the animal's natural shell (a natural object in the domain of study of the *conchologist*) in such a way that would make life uncomfortable for the actual mollusk itself (an animal in the domain of study of the *malacologist*) – that is, a hole's been poked in the shell. A shell trumpet will obviously have to made after the mollusk has (willingly or unwillingly) vacated.

There are two basic places this hole may be placed, and so there are two basic approaches that can be taken for making a conch shell into a shell trumpet. A hole is made either at the apex (the tip of the spire) of the shell, or, alternatively, in one of the whorls to the side of the spire. The mouth hole may be at the apex if the spire is shallow, as on a *Strombus gigas* ("queen conch" or "pink conch," common in the Caribbean),[2] *Cassis cornuta* ("horned helmet," found in the Indo-Pacific region), or

Cassis tuberosa ("king helmet," found in the Caribbean). The mouth hole may be on the side of the spire if the spire is more steep, as on a *Charonia tritonis* ("Triton's trumpet," distributed throughout most of the tropical Pacific and Indian Oceans). In some cases, the hole itself forms the mouth hole; in others, a mouthpiece is added. Mouthpieces seem to be a matter of what tradition has evolved, as sometimes the same species of shell may be found with or without a mouthpiece. For instance, a variety of approaches will be found with *Charonia tritonis*. In Polynesia, a mouth hole cut into the side of the spire is the norm.[3] Occasionally a side-blown *tritonis* will have a mouthpiece added, as found in the Marquesas Islands;[4] this appears to be a rare arrangement. Concerning end-blown *tritonis*, on the Hawaiian *pu*[5] and on the Korean *na*,[6] a mouth hole is cut into the apex. On the Japanese *hora*, the *tritonis* (called *horagai*) is given a mouthpiece, placed at the apex.[7] Other shells used for trumpets usually have the hole in the apex, with a mouthpiece or (perhaps more commonly) without.

The qualities of sounds which shell trumpets can produce are varied and also layered in the meanings and responses such sounds evoke. As children we learn of one of the poetic associations of shells – that if you hold a conch shell to your ear, you will hear (however far away from the coastline you may be) the sound of the sea.[8] Yes, perhaps it is indeed the air column enclosed by the shell filtering the ambient level of noise to create a faint roaring sound. But the association of shells with water, and the sea especially, is also at the basis of the many of the ceremonial uses of shell trumpets around the world. Shell trumpets have often been used at great distances from the sea, and this has contributed to the sacredness of their sounds. Thus, the hearing the of sea in a shell may be a vestige of these older, profound associations. Shell trumpets produce a profound sound in every sense of the word – there is a sense of the sound coming from the deep past. This is both true as regards the actual antiquity of the use of shell trumpets, which dates to the Neolithic era,[9] and in the very shell itself. The apex of a univalve gastropod such as a conch or a snail is the oldest part of the shell (the place where the young animal started growing): in blowing a shell trumpet the sound is passing from the oldest place to the youngest – from the past toward the present.

Concerning this antiquity of the use of shell trumpets, the etymologist Eric Partridge puts forth the idea that the word "conch" may be of echoic – that is, onomatopoeic – origin.[10] Echoic, I suppose, of the sound of the blast of a shell trumpet, and thus – given the early Greek roots of the word "conch" – indicating the great antiquity of their use. A common term applied in a number of parts of Polynesia to the shell trumpet – *pu* – would certainly also seem, in its own way, to be echoic.

The most common use of shell trumpets in many parts of the world – and they have a remarkably wide distribution – is as a signaling device. A shell trumpet may announce curfew in Samoa, or announce that fresh fish is for sale in Fiji, or may serve as a foghorn on the Mediterranean. The shell trumpet often has a magical role in relation to weather. It may on the one hand be used to calm rough seas, or on the other to summon wind when seas are becalmed.[11] Shell trumpets are also used in musical contexts, most often in conjunction with ritual. The Indian *shanka* has held a place in the Hindu religion for millennia. There it may be used as a ritual vessel as well as a trumpet.[12] The shanka is also of significance in Buddhism, where, besides its musical uses, it figures importantly into Buddhist iconography. Befitting their role in Tibetan ritual music, where they are called *dung-dkar*, shell trumpets made from *shanka* receive detailed decoration, with carving on the surface of the shell itself and with added ornamentation in metal and semi-precious stone.[13] Shell trumpets were also important ritual instruments in Pre-Columbian South and Central America and in Minoan Crete. In these latter areas, skeuomorphic reproductions ("the substitution of products of craftsmanship for components or objects of natural origin") of shell trumpets, in ceramic and stone, are found archaeologically. The details of their exact purposes remain a mystery.[14]

Generally, a shell trumpet is used to produce one note; harmonics are possible but seldom utilized. One exception is the Japanese *hora*, where three, sometimes even four, pitches of the harmonic series may be employed.[15] On the end-blown Fijian shell trumpet made from the Bursa *bubo* ("giant frog shell"), there is a fingerhole which will allow for a whole-tone change in pitch.[16] Shell trumpets with several fingerholes have also been explored.[17] Occasionally pitch is modified by the player inserting his or her hand into the aperture. Although shell trumpets would seem to lend themselves to being played in a musical context in homogenous ensembles, along the lines of ensembles of panpipes and stamping tubes in Oceania (particularly Melanesia), such an approach is actually very rare. Tonga (in Polynesia) is the only place where conch ensembles have been found, and then only in the more remote areas (some of the northern islands) and only in a few musical contexts (for recreation and for cricket matches).[18] In contemporary music and jazz, however, ensembles of shell trumpets have been used by trombonists Stuart Dempster and Steve Turre.[19]

Now, to get to work. I've made a few shell trumpets with the mouth-hole at the apex. A simple basic recipe is:

Ingredients:
The shell of a large univalve gastropod
A file
Jeweler's files for finishing work (optional)

Procedure:
File off the tip of the spire.
Smooth out the perimeter of the hole (optional).

That's it. But to be more specific: from my experience, for making a shell trumpet it seems that a conch of some size – something like seven inches or greater in length – is needed. My attempt at making an instrument with the shell of a young *Strombus gigas* (perhaps 5-6 inches long) did not work out: I just couldn't get a sound out of the thing. Perhaps a smaller shell such as that might work with a mouthpiece. I've made end-blown trumpets from *Cassis cornuta* (my shell of choice; see Figure 29.1), *Cassis tuberosa*, and adult *Strombus gigas*. My construction approach with the *Cassis* has been to file off the tip with an 8" mill bastard file and a lot of elbow grease, getting it to the point where the opening is about 5/8" in diameter. With the jeweler's files, I'll smooth down the insides of the opening. For a *Strombus gigas*, which has a steeper spire, I first cut off an inch or so of the tip with a saw, and then proceeded as with the *Cassis*.

It is certainly possible to get the job done more quickly. A friend once made a trumpet from a *Strombus gigas* by forcibly breaking off the tip – he's a percussionist – with little or no filing. In this case, it appears that the irregularities of the edges of the mouth-hole allowed for a more pronounced array of upper partials to the shell trumpet's tone. To remove the tip of a *Strombus gigas*, D.Z. Crookes (describing the process in his "How to make a shelly hautbois") supported the shell's tip "on an anvil, and nipped it off with a cold chisel," later carving a "half-civilized" mouthpiece.[20] I suppose one could also use a power grinder or sander to quickly get through the early stage on a *Cassis*, for instance, but I think a couple of hours or so of manual filing is not too big a price to pay. Of course, being physically involved with the stages of the manufacture of a shell trumpet, as with any musical instrument, increases one's connection with the instrument and its sounds.

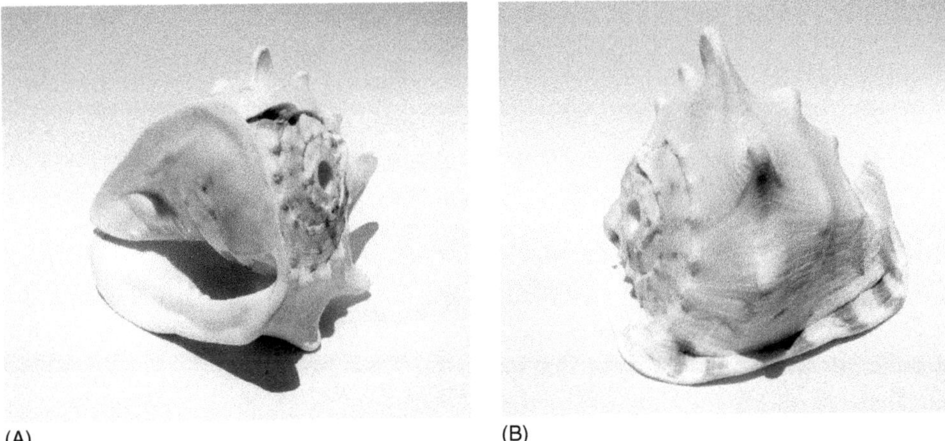

(A) (B)

Figure 29.1 A and B. Two views of an end-blown shell trumpet made by the author from a *Cassis cornuta* ("horned helmet"); length 8¼"; pitch B_3 (open) or A_3 (hand-stopped).

As regards side-blown shell trumpets, I've made one, from a *Charonia tritonis* (see photo below). For such a shell, a basic recipe could be:

Ingredients:
The shell of a large conch with a steep spire, especially a *Charonia tritonis*
A drill
Jeweler's files for expanding the hole and for finishing work

Procedure:
Drill a small hole into the side of the spire.
Expand the size of the hole and smooth out the edges.

Again, a little more detail. I placed the hole in the second whorl out from, and on the same side of the spire as, the aperture. With this arrangement, the aperture faces backward from the player when the trumpet is played. I used photographs of side-blown *Charonia tritonis* as my guide.[21] I used a drill bit of about 1/8" diameter to get the hole started and then followed with a 1/4" bit. I expanded the hole to about 5/8" with a half-round jeweler's file. A larger rat-tail file would also be possible (although one needs to be careful of a bulkier tool damaging the interior of the shell), before following up with the jeweler's file (Figure 29.2).

Although I've made a few shell trumpets, I have not yet made musical use of them in any *concerted* way. I do have a piece – forthcoming in my series of *Anthems* for ensembles of "peacefully co-existing" sustained sounds – for a plurality of shell

Figure 29.2 A side-blown shell trumpet made by the author from a *Charonia tritonis* ("Triton's trumpet"), length 11", pitch E_4 (open) or as low as C_4 (hand-stopped). (Photograph taken during the process of construction, i.e., the mouth hole is somewhat smaller than in the final instrument.).

trumpets and pre-recorded tape. Also, when you've got a shell trumpet around, blowing it every once in a while does impress neighbors and passers-by alike.

Again, these are the most basic of recipes. I look forward to other writers who have more background in the individual traditions of these instruments, and who are more acquainted with the acoustics and detailed construction,[22] to contribute further on the subject of these fascinating instruments.

Notes

1 Despite the fact that a large conch does need to be modified to make a trumpet, a small snail shell can be used, unmodified, as a whistle. An intact snail shell is essentially a stopped pipe, and if the aperture is of an appropriate size – so the player is able to create an embouchure – the shell can be an effective whistle. Unaltered large conch shells filled with water are used for their gurgling sounds by John Cage in his pieces *Inlets* (1977, which also makes use of a shell trumpet) and *Two*[3] (1991, which also includes a Japanese *shô* reed organ). A single such large water-filled conch was used by the present author in his *"concerning an aspect..."* (1988).

2 In general usage, the word "conch" is used to describe large spiral univalve gastropods even when it is not referring to what is, strictly speaking, a conch (the "true conchs" are members of the genus *Strombus*). This seems to be especially true in relation to shell trumpets, where the term "conch trumpet" is used quite freely.

3 Richard M. Moyle, *Polynesian Sound-producing Instruments* (Princes Risborough, England: Shire Publications, 1990), 39 and figure 25, which shows several side-blown *tritonis* being played in Tonga.

4 Richard M. Moyle, *Polynesian Sound-producing Instruments*, 39 and lower portion of figure 23.

5 Te Rangi Hiroa (Peter H. Buck), *Arts and Crafts of Hawaii, IX: Musical Instruments* (Honolulu: Bishop Museum Press, 1957, reprinted 1964), figure 256a.

6 Chang Sa-hun, *Uri yet Akki* ("Our Traditional Musical Instruments"; Seoul: Daewonsa, 1990), 31.

7 Hajime Fukui, "The *Hora* (Conch Trumpet) of Japan" in *Galpin Society Journal* 47 (1994): 47-62, where several photographs and a diagram of the mouthpiece are shown. One of the less-documented uses of the *hora* is in the Buddhist *Shunie* rite associated with the Tôdai-ji Temple in Nara, Japan (page 52). *O-Mizu-Tori* – a portion of this rite that includes the playing of shell trumpets – can be heard on the CD album *Harmony of Japanese Music* (King Records [Japan] KICH 2021). For a full-size color photograph of a *hora*, see Jane Fearer Safer and Frances McLaughlin Gill, *Spirals from the Sea: An Anthropological look at Shells* (New York: Clarkson N. Potter, Inc., 1982), 174–5.

8 Note that terminology relating to the human ear is rich in shell imagery. The *cochlea* (a Latin word derived from the Greek *kokhlos*, land snail) is the spiral, shell-shaped portion of the inner ear which transmits the signals to the brain which are interpreted as sound. As a word referring to a shell-like structure, *concha* (from the Greek *konkhe* – a shell-bearing mollusk in general – which, via Latin, is the ancestral form of "conch") is a term used to describe the human external ear, also known as *pinna*. And *pinna*, from the Latin word for "wing" or "feather," is also the name for a genus of large – and wing- or feather-shaped – bivalve mollusks (family Pinnidae).

9 John M. Schechter and Mervyn McLean, "Conch-shell trumpet" in Stanley Sadie, ed., *The New Grove Dictionary of Musical Instruments* (London: Macmillan and New York: Grove's Dictionaries of Music, Inc., 1984), I:461. Note that it is conjectured that the earliest use of the instrument was as a voice modifier – a megaphone of sorts.

10 Eric Partridge, *Origins: A Short Etymological Dictionary of Modern English* (2nd edition, New York: MacMillan, 1959), 114. Note especially one Middle English spelling, *conk*.

11 A recorded example of the former use, from Chuuk, in Micronesia, is included on the album *Spirit of Micronesia* (Saydisk CD-SDL 414), which includes a *conche* (note this alternative spelling) used for warding off storm clouds (track 22). Though brief, this recording beautifully captures, against a backdrop of storm waves, the shell trumpet's evocative qualities. (A photo on page 20 of the booklet shows a player of a trumpet made from *a Cassis* species.) The latter use – the summoning of

winds – is mentioned in the entry for the shell trumpet *ntuantuangi*, of the Poso Toradja of Celebes, in Sibyl Marcuse, *Musical Instruments: A Comprehensive Dictionary* (2nd edition, New York & London: W.W. Norton & Co., 1975), 368.

12 Note that the Sanskrit word *shanka* (which may be romanized in various ways, with or without diacritics; the English common name for the shell is "chank") does share the same Indo-European root as *konkhe*, and ultimately, "conch." The Latin scientific name for the *shanka* is *Turbinella pyrum*.

13 See Safer and Gill, *Spirals from the Sea*, 176-7, for two views of a specimen dated 1400.

14 Jeremy Montagu, "The conch in prehistory: pottery, stone and natural" in *World Archaeology* 12/3 (1981): 273–9, which focuses on these shell-trumpet skeuomorphs.

15 Hajime Fukui "The *Hora* (Conch Trumpet) of Japan," 51–2.

16 Moyle, *Polynesian Sound-producing Instruments*, 39 and figure 24.

17 D.Z. Crookes, "How to make a shelly hautbois" in *FoMRHI Quarterly* 80 (July 1995): 43, where he experiments with up to seven (?) fingerholes on *Strombus gigas*.

18 Richard M. Moyle, "Conch Ensemble: Tonga's Unique Contribution to Polynesian Organology" in *Galpin Society Journal* 28 (1975): 98–106. Also, his *Polynesian Sound-producing Instruments*, 41-2 and figure 25. Ensembles of three to seven, or more, side-blown *Charonia tritonis* are used.

19 Stuart Dempster's *Undergound Overlays from the Cistern Chapel* (New Albion NA 076) and Steve Turre's *Sanctified Shells* (Antilles 314 514 186–2) include contemporary creative uses of shell trumpets in ensemble.

20 Crookes, "How to make a shelly hautbois," 43.

21 For instance, Eric Metzgar, *Arts of Micronesia* (Long Beach, Calif.: FHP Hippodrome Gallery, 1987 {exhibition catalogue}), figure G, and Safer and Gill, *Spirals from the Sea*, 168.

22 See Montagu, "The conch in prehistory: pottery, stone and natural," 274–5, for a brief discussion of shell-trumpet acoustics which outlines some of the basic issues. Concerning shell-trumpet construction, note that Hajime Fukui's "The *Hora* (Conch Trumpet) of Japan" goes into a great amount of detail concerning making this particular instrument.

Related articles

Other articles by Mitchell Clark which appeared in *Experimental Musical Instruments* are listed below. Numerous reviews of books, recordings, and events written by Mitchell Clark also appeared in the journal. These articles can be accessed through the Internet Archive at https://archive.org/details/emi_archive/.

"The Qing Lithophones of China" by Mitchell Clark in EMI Vol. 10 #1

"The Wind Enters The Strings: Poetry and Poetics of Aeolian Qin" by Mitchell Clark in EMI Vol. 10 #3

"Crow-Quill and 'Cat'-Gut: The Lautenwerk and Its Reconstruction" by Mitchell Clark in EMI Vol. 10 #4

"Music As Fragile As Its Material: The Classical Repertoire of the Glass Harmonica" by Mitchell Clark in EMI Vol. 11 #1

"The Essential Thing the Pipes Play: Piobaireachd and the Great Highland Bagpipes of Scotland" by Mitchell Clark in EMI Vol. 11 #2

"A Conversation with Rex Lawson, Pianolist Extraordinaire" by Mitchell Clark in EMI Vol. 11 #3

"The Helikon" by Mitchell Clark in EMI Vol. 12 #4

"Sounding Antiquity: Reconstructions of Ancient Greek Music" by Mitchell Clark in EMI Vol. 13 #2

"Aeolian-Bow Kites in China" by Mitchell Clark in EMI Vol. 14 #3

30

A Musical Instrument Workshop in Hanoi

Jason Gibbs

This article originally appeared in *Experimental Musical Instruments* Volume 12 #1, September 1996.

Tucked into a comer of the Hanoi Music Conservatory campus is a spare building that doubles as the workshop and living space for Tạ Thâm and his crew of musical instrument artisans. The front room is both a living room and an instrument showroom; the back room is where the work is done, evidenced by the wood trimmings, an assortment of bamboo tubes, and the various manual and power tools found throughout the room (see Figure 30.1). Upon entering the building, one cannot help but notice the variety of unusual instruments hanging from the walls. These are both traditional instruments of Vietnam and instruments of Tạ Thâm's invention.

Tạ Thâm was born in 1929. He came of age during the time Vietnam was fighting for its freedom from France and took part in the historic battle of Điện Biên Phủ in 1954. Soon after North Vietnam achieved its independence, he enrolled in the first class at the Hanoi Conservatory in 1956 where he studied both classical Western and traditional Vietnamese instruments. While studying the violin and the piano, he came to realize that these instruments were the result of continuous improvement and development overtime and asked himself why Vietnamese instruments should not also benefit from such improvement. Many traditional Vietnamese string instruments are soft in volume, better suited to playing in intimate chamber settings or personal contemplation than for the concert stage. He felt that instruments needed to be built with fuller resonance and louder volume. Another dream behind his work and research has been to create a Vietnamese orchestra made up of instruments with a degree of expression and precision and in a variety of tessitura like the Western orchestra.

In 1957 he set out to research Vietnamese traditional instruments in order to understand their history and construction with a view toward improving them. From that time he also began to design his own instruments. In a traditional society like that of Vietnam, going against the grain and venturing into new areas is often not encouraged and Tạ Thâm's visions were often neglected and at times even obstructed by the musical authorities. For this reason, he often had to work

Figure 30.1 Putting the finishing touches on a *đàn bầu* in Tạ Thâm's workshop.

independently. This independence, however, made it difficult to earn a living. Periods of poverty slowed the realization of his dreams. At times, he had to work in rice fields, as a laborer and even as a fisherman in order to support his research. Despite many years when he hid his work for fear of derision, he was rewarded in 1987 when he won several national prizes for his inventions. For the past several years, he has been building traditional instruments at the National Conservatory of Music. During their American tour in October 1995, I met several Vietnamese musicians touring with the Thăng Long Water Puppet group from Hanoi, some of whom used traditional instruments constructed in Tạ Thâm's workshop.

Tạ Thâm's invented instruments use the raw materials common to traditional instruments, such as hard and soft woods, various sizes of bamboo, gourds, etc... He bases his inventions not only on the instruments of the ethnic Vietnamese, but also upon those of Vietnam's many ethnic minorities. To further his research, he has traveled extensively in northern Vietnam's mountainous regions, searching out the instruments of the country's minority peoples. During my stay in Hanoi in1995, he took several trips to Hòa Bình, a city on the edge of the mountainous home of the Mường people.

Vietnam is best known for its string instruments, like the 16-string zither *đàn tranh* – a cousin of the Chinese *zheng*, the Japanese koto, and the Korean *kayagum*; the four-string lute *đàn tỳ bà* – a cousin of the Chinese *p'i p'a* and the Japanese *biwa*; the *đàn nguyệt*, the two-stringed moon-shaped lute; and the *đàn nhị*, the two-stringed bowed lute (similar to the *tro duong* from Cambodia). There are also a wide range of percussion instruments like drums (*trống*) , woodblocks (*mõ* and *phách*), bells (*chuông* and *quả nhạc*) and gongs (*cống* and *thanh la*), and castanets (*sinh*). Two string instruments unique to Vietnam are the *đàn bầu*, or monochord, and the *đàn đáy*, a three-string lute used primarily in *hát ả đào* music, a highly literary entertainment song form performed by professional singers. Vietnam's many minority peoples have also created a wide range of distinctive musical instruments.

Many of Tạ Thâm's instruments are hybrids of existing Vietnamese instruments. In Figure 30.2 on the left is a *đàn bầu* made of a mixture of wood and bamboo with two strings and necks allowing for two fundamentals and two harmonic series. On the right is his *nam tranh kép*, using design characteristics of two traditional instruments, the angular-shaped *đàn đáy* and the flower-shaped *đàn sến*. It has two necks, each with a string – one made of silk, one made of metal. The *trăng thau* (Figure 30.3) is based on the moon-shaped *đàn nguyệt*, that uses the heightened frets

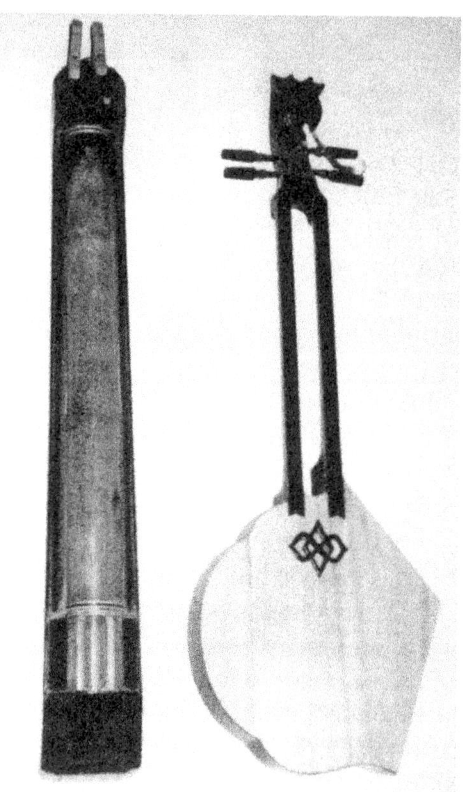

Figure 30.2 Left - Tạ Thâm's two-string variation on the *đàn bầu*. Right- *Nam tranh kép*.

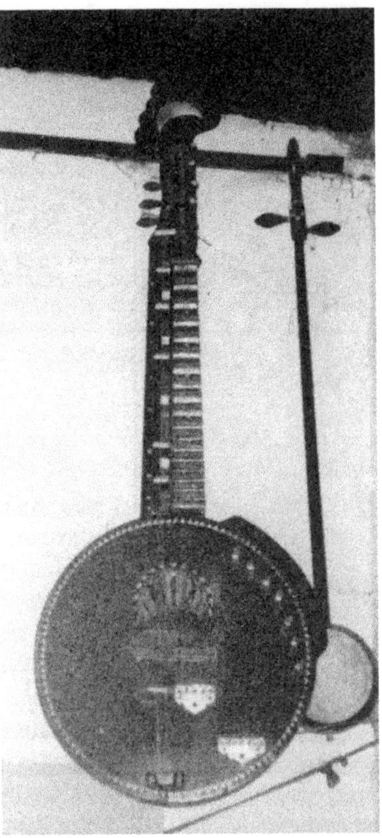

Figure 30.3 *Trăng thau*, a variation on the traditional *đàn nguyệt*.

common to Vietnamese lutes allowing for liberal pitch bending. An innovation of the trăng thau is the adoption of two different fret systems side by side. In Figure 30.4, Tạ Thâm is pictured with what he called, in French, a *contrabasse vietnamienne*, in fact modeled on the much smaller tính tẩu lute used by shamans of the minority Thái people of northwestern Vietnam. Many of Tạ Thâm's creations are percussion instruments. The *nhạc tiền* (Figure 30.5) is a variation on the traditional *song loan* and *sênh tiền*. It has a clapper connected to a woodblock like the former and a number of rattling coins, augmented by bells like the latter. The dàn mõ trâu (Figure 30.6) uses a series of tuned water buffalo-shaped cowbells that are rattled side to side.

In a traditional country like Vietnam, there is no concept of experimental music like we have in the West – there is no avant-garde to speak of. Vietnam's cultural policy encourages the unity of the nation's peoples and the development of culture

Figure 30.4 Tạ Thâm's contrabasse vietnamienne.

that serves them. An inventor of musical instruments like Tạ Thâm is less interested discovering new sounds or performance techniques than in working with the raw materials and musical system at hand. He is curious about activities outside Vietnam and is interested in meeting his colleagues in the world of musical instrument invention. You can probably find him in his workshop at the Hanoi Conservatory.

Đàn Bầu

If one sound had to be chosen to evoke Vietnam, for many it would be the sound of the *đàn bầu*, also known as the *đàn độc huyền* (single-string instrument). The word *bầu* means gourd and refers to the dried gourd fastened to the handle, surrounding the string at the point where it connects to the handle. In the past, this gourd may have served as a resonator, but today it survives as a decorative feature. Nowadays the *đàn cầu* is constructed using hardwood for a frame and softwood for the surface. The handle is made of flexible carved bamboo or water buffalo horn, and the string

Figure 30.5 *Nhạc tiền.*

Figure 30.6 *Dàn mõ trâu.*

Figure 30.7 *Đàn bầu.*

is made of metal. At the present time, it is almost always amplified (Figure 30.7, a *đàn bầu* side by side with a gourd). Historically the *đàn bầu* was played by blind street musicians or *xẩm*. The earlier *đàn bầu xẩm* (Figure 30.8) is constructed from a split bamboo tube. It used a silk string and occasionally substituted a half coconut shell for the dried gourd. In the days before amplification, a trunk could be placed under the instrument as a resonator.

Historical records trace the invention of the *đàn bầu* to 1770, but some scholars have claimed earlier origins and antecedents for the instrument. Some speculate that it originates from a string stretched from the teeth, others believe its antecedent is the *trống quân* – a "drum" consisting of a rope fastened to the ground at both ends stretched over pole that serves a bridge. This pole is positioned over a trunk or empty pit that serves as a resonator. Tạ Thâm believes it originates from the *tàn*

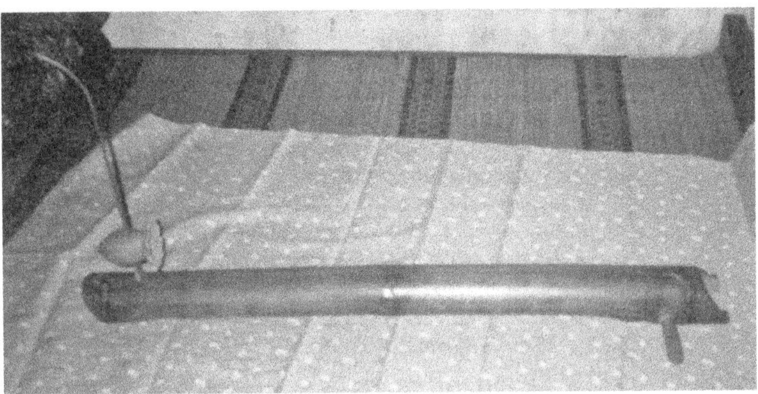

Figure 30.8 *Đàn bầu xẩm.*

máng, an instrument of the Mường minority constructed from a bamboo tube with a bamboo thread carved from out of it that is plucked like a string. None of these instruments, however, employ harmonics, the performance technique that makes the *đàn bầu* unique. It uses these harmonics exclusively, produced at nodes at 1/2, 1/3, 1/4, 1/5, and 1/6 the length of the string. A small bamboo plectrum held in the right hand plucks the string while the lower side of the hand stops the string at the appropriate node. The left hand moves the handle to bend the pitch downward by moving in the direction of the instrument, or upward by pushing the handle away from the instrument. The pitch can bend as much as a 4th or 5th in either direction. The left hand also produces a variety of vibratos, glissandos, and grace notes. The instrument's virtuosity and expressiveness are found in its left hand technique, which should have a subtlety that mimics the sound of the Vietnamese singing voice or declaimed poetry.

Traditionally the đàn bầu has been played by groups of blind musicians (*nhạc xẩm*), and in Vietnamese chamber music (*nhạc tài tử*). More recently it has also been used in the ensembles of *chèo* and *cải lương* theatrical music. In Vietnam today there is a growing virtuosic literature with solo works and concertos written for the đàn bầu.

Related articles

Further references to Tạ Thâm, and specifically to *đàn bu*, appeared in the Letters section of the following issues of *Experimental Musical Instruments*: Vol. 9 #4, Vol. 12 #4, Vol. 13 #2, and vol. 14 #1. These can be accessed through the Internet Archive at https://archive.org/details/emi_archive/.

31

Sirens

Bart Hopkin

This article originally appeared in two parts in *Experimental Musical Instruments* Volume 12 #4, June 1997, and Volume 13 #1, September 1997. Part One focuses on basics – what sirens are and how they work – as well as history. Part Two presents practical information on the making of sirens as musical instruments.

Part 1

Prior to the appearance of the jet engine, the man-made sound source with the greatest capability for sheer volume was the siren. Because of that characteristic, sirens have been used primarily as signaling and warning devices, such as air raid alarms and fire engine alerts. But sirens have a significant and colorful role in the history of acoustic theory. They also have unexploited potential for music-making, and they don't have to be piercingly loud.

As an acoustical device, a siren takes what would otherwise be a continuous flow of air and converts it into a series of pulses. These pulses, occurring at some frequency within the hearing range, propagate out into the surrounding atmosphere as sound. In its simplest form, the heart of the siren is a rotating disk. The disk is perforated around its perimeter with a ring of evenly spaced holes. A narrow air stream is directed against the disk through a tube in such a way that the holes pass as closely as possible in front of the tube's open end. In the moments when one of the holes passes in front of the tube, a puff of air passes through. Between the holes, the air flow is momentarily blocked by the surface of the disk. The puffing frequency corresponds to the rate at which holes pass in front of the air-supply tube. That frequency is the disk's number of rotations per second times the number of holes – for instance, a disk with a ring of 44 holes rotating at 10 revolutions per second will produce an audible tone at a frequency of 440 Hz, or A above middle C.

The siren can thus be seen as an air-gating device. In this respect, it is related to reed instruments. Consider the harmonica, which has metal reeds of the sort that are called free reeds. Within the harmonica, the reed at rest blocks a small air passageway. Under the pressure of the player's breath, the reed momentarily flexes out of the way to allow a puff of air through, then springs back to block the passage again, then flexes away again, and so forth, effectively converting the steady flow of the player's breath into a series of puffs at some audible frequency. Both the siren and the harmonica can be classified as "interruptive free aerophones"[1] – the "interruptive" referring to the air-gating approach, and the "free" part referring to the fact that they typically are open-air instruments, without flute-like resonant air tubes or chambers.

Models of the very simple siren just described (a rather quiet instrument, incidentally) were made and used by a number of 18th and 19th century researchers in acoustics. It is an engraving of this very instrument that holds the place of Figure #1 in the most important of 19th-century acoustics texts, Hermann von Helmholtz' famous *On the Sensations of Tone (Die Lehre von den Tonempfindungen,* 1877. See this article's Figure 31.1). But the basic idea – that of an air flow rendered as an audible series of pulses by means of a perforated rotating element – has over the years taken many forms. Let us step back now and review the history of sirens, insofar as it is known.

According to several secondary sources,[2] the invention of the siren mechanism can be credited to the 18th century Scottish natural philosopher John Robison (1739–1803). Helmholtz makes no mention of Robison in *Sensations of Tone* but describes the simple siren in his figure as being "after Seebeck," without further explanation. Presumably the reference is to the German physicist Thomas Johann Seebeck (1770–1831). In a slightly later acoustics text, J.A. Zahm also refers to a similar device as "Seebeck's Siren" (Rev. J.A. Zahm, C.S.C., *Sound and Music,* 1892 – a wonderful book).

The name *siren* (French *sirène*) seems to have been given by the French engineer Baron Charles Cagniard de la Tour (1777–1859). It was de la Tour who came up with the reconfiguration which converted the rather quiet instrument of Seebeck and Robison into a sound producer of great power. He did this by means of an enclosure and a double-disk system. We have fine illustrations of the idea from

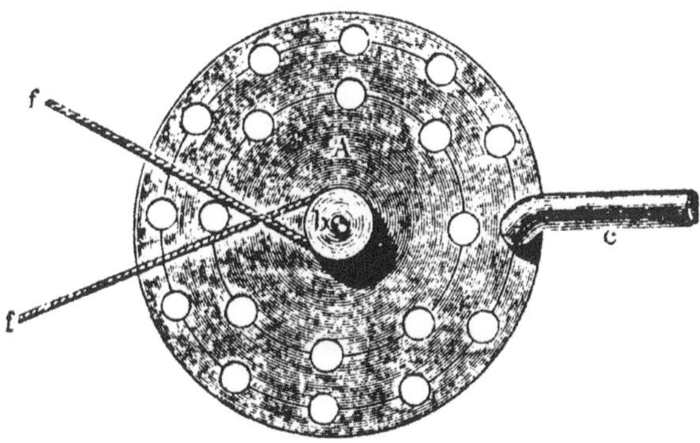

Figure 31.1 Simple siren "after Seebeck," from Hermann Helmholtz' *On the Sensations of Tone.* The part labelled *c* is an air tube.

both Helmholtz and Zahm (Figures 31.2 and 31.3). Here's how the de la Tour siren works: The air under pressure is directed into an enclosure. The front of this enclosure is something like a typical siren disk, with a ring of holes around the periphery, but this disk is fixed (not rotating). Directly in front of this is a second disk, with identical hole spacing. This second disk rotates. When in the course of rotation the holes in the two disks are aligned, then all holes simultaneously allow air to pass. When the holes move out of alignment, all the holes close and no air passes. To cause the outer disk to spin, its holes can be drilled obliquely, at something like a 45-degree angle. The air rushing out then drives the disk in a manner similar to a turbine or windmill. (The effect can be increased by drilling the holes on the stationary disk obliquely in the opposite direction.) With this arrangement, air pressure

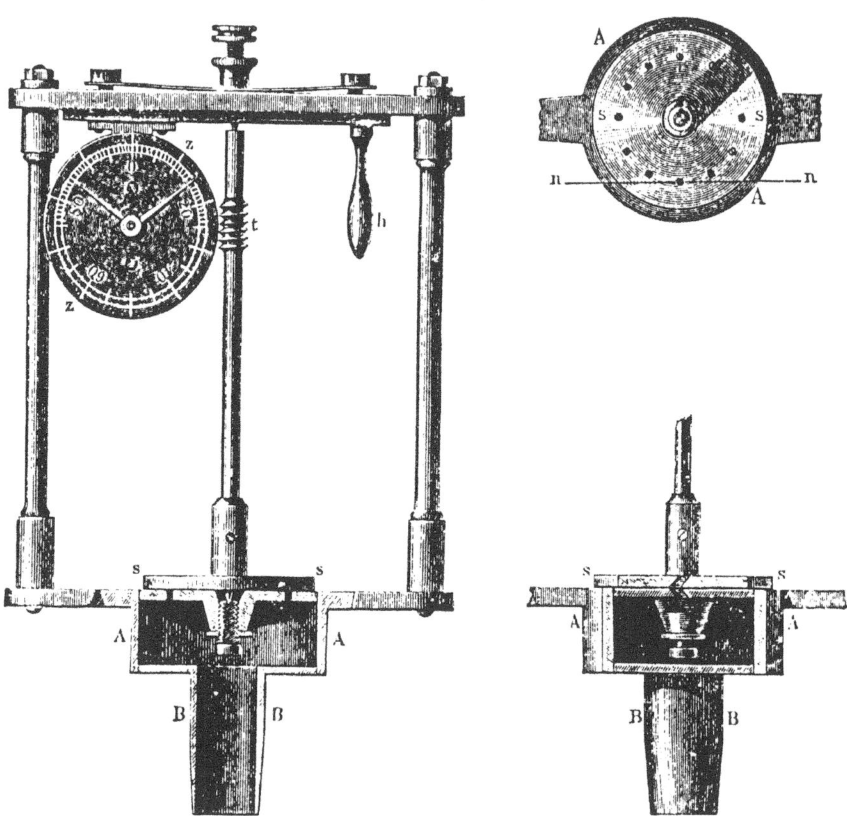

Figure 31.2 Three views of an enclosed siren with rotation-counter mechanism, from Helmholtz. Two of the views are cutaway views, showing the inner workings.

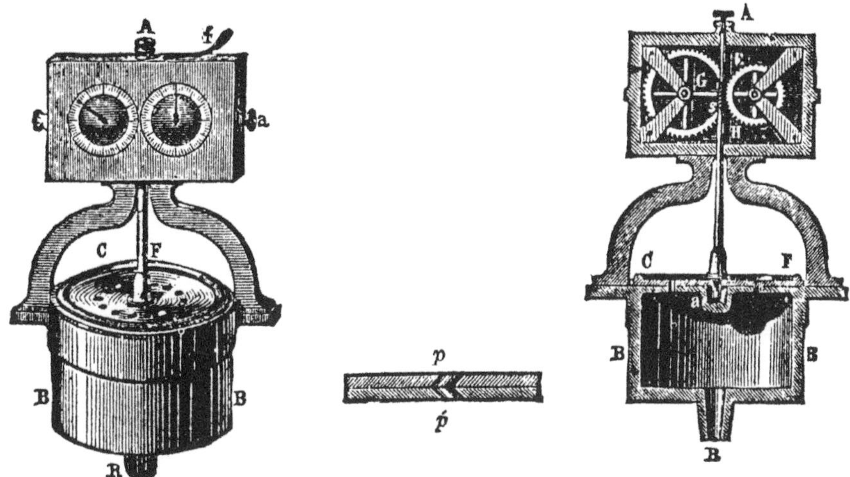

Figure 31.3 Another enclosed siren with rotation counter, depicted in Pietro Blaserna's The *Theory of Sound in its Relation* to *Music*.

alone is sufficient to operate the siren; no external driver for the disk is required. The mechanics of this air-drive system explain the rise and fall of the wailing sound we associate with sirens even today: as the air flow causes the disk gradually to accelerate, the audible frequency rises. This system, with its double disk and enclosure, is far louder than the simple siren for two reasons. First, it multiplies the effect by providing a large number of holes emitting pulses of air simultaneously instead of just one. Second, it eliminates cancellation problems that are inherent in the simpler design. This takes a bit more explaining, and I'll go into it more fully later on.

Another significant early innovation in siren design was the appearance of multiple rings of holes. Helmholtz credits this development to one Dove, apparently being the German meteorologist Heinrich Wilhelm Dove (1803–79). The idea is to have several concentric rings of holes, with different numbers of holes in each ring (Figures 31.4 and 31.5). Each ring will then provide a different pitch: while a ring of 44 holes in a disk rotating at 10 rps will produce a tone at 440 Hz, a smaller concentric ring of just 22 holes will yield a tone an octave lower at 220 Hz. In a de la Tour siren, all the rings typically sound simultaneously, producing a chord made up of intervals determined by the ratios of the numbers of holes in the rings. (Alternatively, each ring could be given the same number of holes, thus further reinforcing the single tone.) But in simple open siren with multiple rings of different numbers of holes, you can select which tone will sound by directing the air stream at different rings. In the early sirens, the number of rings, and thus the number of

Figure 31.4 A siren with many rings of holes, depicted in J.A. Zahm's *Sound and Music*.

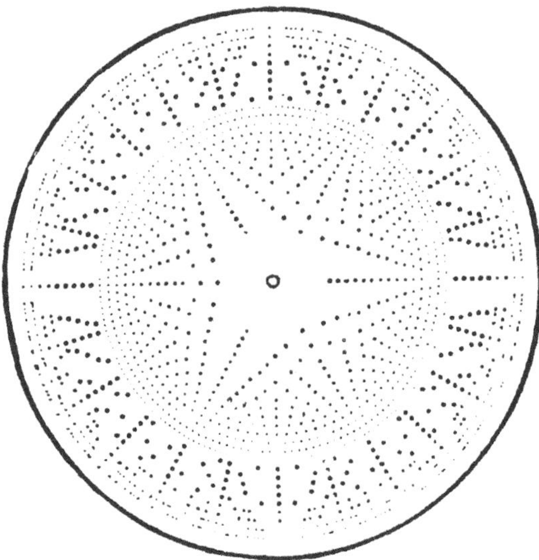

Figure 31.5 An unusually elaborate siren disk containing 24 rings of holes, depicted in Zahm and credited there to an acoustician named Oppelt.

pitches available, typically was two or three or four. But with enough rings of different numbers of holes, you could have an instrument capable of playing a more complete scale.

There was a reason that sirens played such a prominent role in Helmholtz' teaching, and in other experimental and pedagogical work that followed. The early sirens served physicists and professors as an effective tool for observing, demonstrating, conceptualizing, controlling, and measuring sound in the air. As a rudimentary example, what else could more convincingly illustrate the idea that sound is a matter of rapidly recurring pressure pulses in the atmosphere? Of particular importance was the siren's role as a frequency counter, providing researchers with a convenient method for counting vibrations per second, and clearly demonstrating the relationship between the physical phenomenon of frequency and the perceived phenomenon of pitch.[3] To work in this fashion, the researcher needed to know how many times the disk had rotated in a known time period. As shown in Figures 31.2 and 31.3, sirens depicted in early pedagogical books are often equipped with rotation counters operating by means of clockwork-like mechanisms – intriguing examples of 19th-century mechanical ingenuity.

Helmholtz took matters further when he had the double siren shown in Figure 31.6 built. The two sirens, which face one another, have identical disks and operate on a single drive shaft, so that they normally spin at the same rate and are capable of producing identical tones. The two siren casings and their disks can be aligned so that their rings of holes open and close at the same time, producing a strong sound. But the casing for the upper siren, including its fixed disk, can be slowly rotated in a controlled fashion by turning a crank. If the upper siren is rotated just the right amount and left in that position, it can be set so that, while operating at the same frequency, its holes close just as the lower siren's holes open, and vice versa. The two sirens are then out of phase; the instant of maximum pressure for each pulse from one coincides with the instant of minimum pressure from the other. They cancel, and the sound becomes weak. Thus, another essential concept in acoustics was made observable, demonstrable, controllable, and measurable – the concept of phase relationships. And more: by turning the crank continually so that the fixed disk on the upper siren rotated slowly even as its rotating disk spun rapidly below it, Helmholtz was able to slightly alter the frequency of the upper siren. This situation, with the upper siren sounding slightly detuned relative to the lower one, gave rise to the phenomenon known as beating – the wavering rise and fall in volume that comes about when two sources of near but not identical frequency sound together. The relationship between beating and phase relationships was thus convincingly demonstrated in a manner easily observed and conceptualized.

Using the double siren, Helmholtz also demonstrated underlying principles of just intonation. Central to the study of just intonation is the idea that for two or more tones sounding together, consonant harmonies arise when the frequencies of

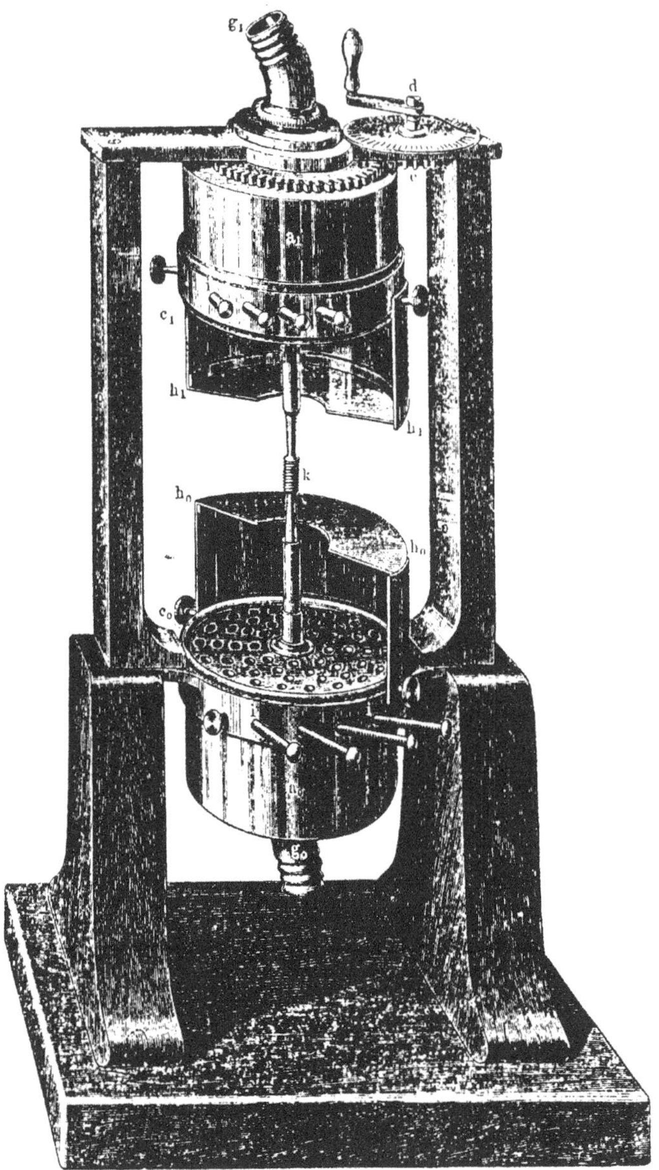

Figure 31.6 Helmholtz' double siren. Notice that there is an outer casing (cutaway in this drawing) over the two disks. This was to create an air resonance chamber to enhance the lower frequencies and make the upper partials less strident. Notice also the four bolt-like components protruding from the side of each siren. These were the manual controls for an elaborate internal mechanism that allowed the sounding of one ring of holes at a time by selectively blocking or un-blocking the air flow through each of the individual rings.

the tones form simple proportions – for example, frequency ratios such as 2:1, 3:2, or 5:4 tend to strike the ear as consonant, while more irregular ratios involving larger numbers seem more dissonant. In the siren disk with its several rings of different numbers of holes, the proportional relationships are laid out for all to see. For example, whatever the speed of rotation, an outer ring with 44 holes will always produce twice the frequency of an inner ring of 22 holes on the same disk, for a frequency ratio of 2:1, corresponding to the musical interval of the octave. And, of course, one can hear the actual effects when the siren sounds.

Another area for early exploration had to do with the sizes of the holes within each ring. Karl Rudolph Koenig (Prussian acoustician, 1832–1901), created a disk of seven rings in which the holes were not uniform in size within each ring, but appeared in regular patterns of larger and smaller holes (Figure 31.7). With such a ring, one hears not only the frequency of the total number of holes within the ring, but also lower frequencies associated with the periodic appearance of larger holes. The higher frequencies can then be heard as overtones of a lower-frequency fundamental. It would be impractical, but taking the idea further, it's not hard to imagine extraordinarily elaborate disks using gradations in hole size to create something analogous to what today might be called digital wave forms.

Following this line of thought, Koenig went on to develop another variation, the wave siren. Several of these appear in Figure 31.8. The wave siren design is not well suited to de la Tour's enclosed, double-disk approach; it would typically take an open form like the simple disk sirens and would probably, like them, produce a weak tone. In the wave siren, there is no ring of holes; rather, the outer edge of the disk has a wavy shape. Air flow from a tube is directed against this edge. As the waves rise and fall in front of the tube, they alternately block and unblock the flow through the tube opening much as the holes in other siren disks do. An advantage of the wave

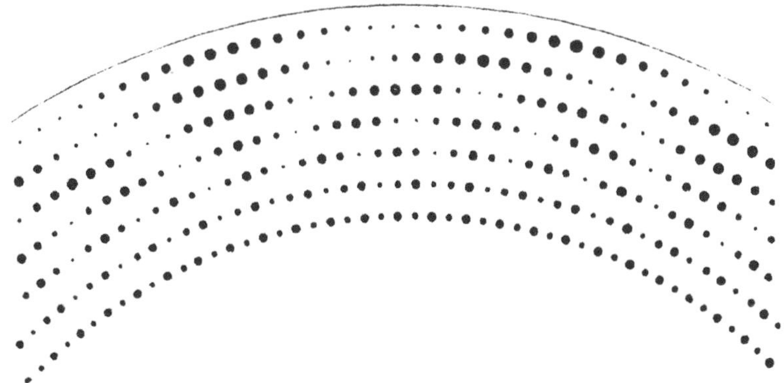

Figure 31.7 A portion of a siren disk created by Koenig with varying hole sizes within each ring, creating the effect of waveforms programmed into the rings.

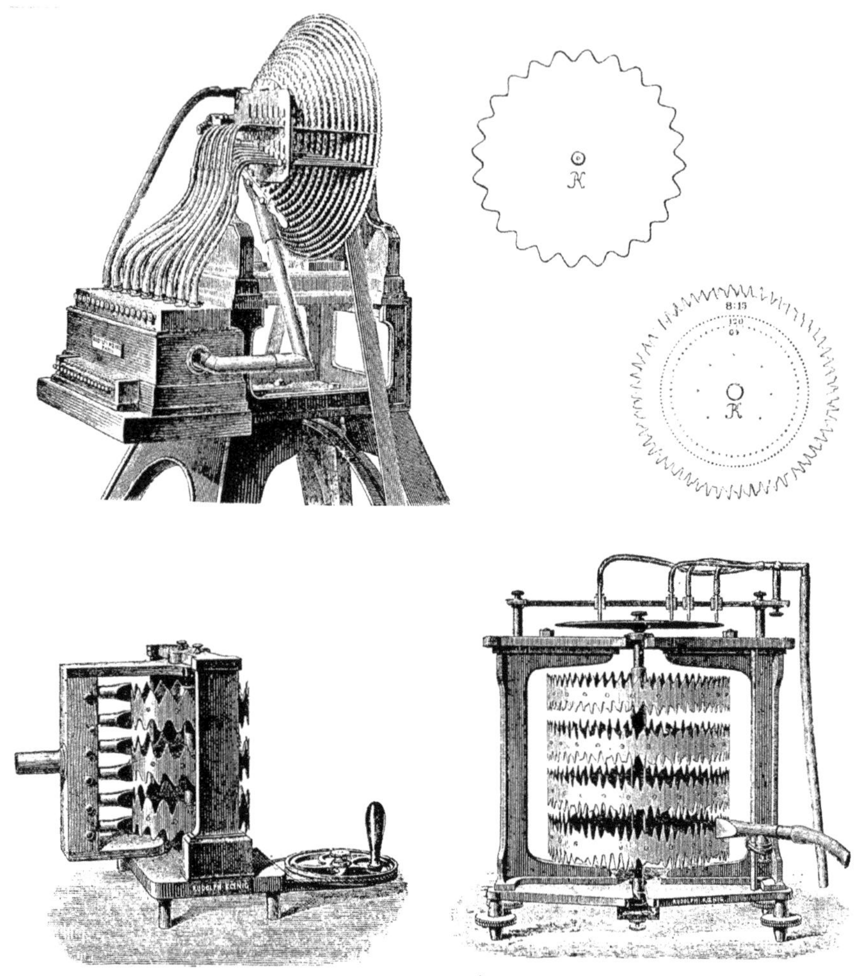

Figure 31.8 Several variations on wave sirens and siren disks by Koenig, all depicted in Zahm's *Sound and Music.* In a wave siren it's not possible to have multiple tones in a single disk. At the upper left you can see a solution to this limitation: an elaborate wave siren with a total of 15 disks, providing a scale of some sort. It's provided with stationary tubes directed at the edges of the disks and a keyboard-like control console. The disk edges appear to be shaped in something close to a sine wave form. The two drawings in the upper right show individual wave siren disks, one with a sine-wave shaped edge and the other with a more complex waveform. The two drawings below show wave sirens in a new configuration: they have wavy-edged cylinders in place of disks. (We will see more of the cylinders idea as this article progresses.) The instrument on the left has six possible tones – but notice that the cylinder edges all seem to have the same number of equally spaced waves, meaning that, spinning at the same rate, all will produce the same fundamental tone. Yet they have different wave forms – some pointed and some rounded. The instrument appears to have been designed to compare the tone qualities resulting from different wave forms at the same fundamental pitch. The drawing on the lower right shows a similar instrument with more complex waveforms. In addition to the cylinders, this instrument also has a regular siren disk mounted on top.

siren is that it is relatively easy to deliberately create different sorts of wave forms in the edge of the disk. The shape of the disk's wavy edge, in theory, corresponds to the wave form of the resulting acoustic signal. By cutting disks to different edge shapes it should be possible to create a siren sound with, for instance, what acoustics people call a sine wave, or a square wave, or a sawtooth wave. For a variety of reasons, I suspect that the correspondence between the disk shape and the actual resulting acoustic wave forms would not be close as one might wish. But it should at least be true that by experimenting with different disk edge shapes, you could come up with different tone qualities from the siren.

I mentioned earlier the turbine-like system by which the force of the air alone can cause the siren disk to tum, eliminating the need for a separate rotational driver. It turns out that this situation can be reversed as well. If a siren has a motor to drive the disk, then the rotation of the disk can be used to propel the air, eliminating the need for a separate air source. One way to do this is to add tilted blades or baffles to the surface of the rotating disk, extending over the holes, to propel the air like the blades of a fan. Helmholtz made brief reference to this trick in the first appendix to *Sensations of Tone.* (In the same appendix he also describes a primitive electric motor with speed control for driving the disk.)

In this century, many warning sirens have been made with rotating cylinders rather than rotating disks. In this configuration, a perforated cylinder rotates within a stationary housing provided with a matching set of perforations. Their operation is much like disk sirens: they can be driven by compressed air, or they can employ motor-driven rotational motion to force the air. Typically, however, the openings are not small holes, but windows made rather larger, since the larger opening area contributes to a louder sound.

I managed to get a look at one of these cylindrical warning sirens at a local fire station. It was mounted behind the bumper of one of the big trucks, and I had to lie on my back under the truck to view it, but I was able to see it well enough to confirm that it did conform to the principles of siren design I had been reading about. It consisted of an electric motor in a housing about 8" long and 4" in diameter, driving a rotating element of about 8" diameter and 4" long, set within a cylindrical outer housing with the required window-shaped openings. The rotating element was designed (so I surmised) to drive the air not by means of baffles, but by centrifugal force, in effect whirling air out through the windows in the outer casing.

Traditional acoustic warning sirens such as this, referred to as "growlers" one fireman told me, are considered old fashioned by people in fire-prevention and law enforcement. In recent years, most of them have been replaced by purely electronic sound makers. This was confirmed for me when I got hold of a sales catalog entitled *Audio Alerts and Alarms* from a company called Projects Unlimited. It contains fifty pages of audio warning devices for industry and law enforcement, and every one of them is entirely electronic in operation. Most of them employ piezo-electric transducers as the noise-producing element. Some are programmed to produce an electronic version of the familiar siren's wail.

True acoustic sirens are still made today, but on a more humble scale. Small sirens appear occasionally in children's toys, for instance. My son at one point had a small toy airplane with a tiny mouth-blown siren housed in it. It was simply and roughly made, but it worked, with the familiar rising wail. A couple of people have spoken to me of a type of siren that was made, years ago, to go on a child's bicycle. It had a drive mechanism that pressed against the bicycle's rear wheel as it turned, thus activating the disk and stimulating the air flow. As an attention-getter, I'm told, it was quite effective.

In vaudeville and silent film days, a mouth-blown siren was one of the essential items in the sound effects person's trick bag. The small, high-quality mouth-blown sirens are still made by the Acme company in England, which produces a variety of other fun sound effects as well. The Acme Siren takes the form of a cylindrical stainless steel enclosure, two inches long and a little under an inch in diameter. Inside is a stationary disk with a small, freely rotating disk directly in front of it, each containing six holes. The holes in the rotating disk (which is about ⅛" thick) are drilled obliquely so that the air flow activates the spinning. It works great – the sound is unmistakably siren-like, and respectably loud for something so small, though, of course, nowhere near as loud as a full-sized siren (Figure 31.9).

Another variant of the siren that has arisen in the 20th century is the light siren. Light sirens are not wind instruments and perhaps should not be considered true sirens, but the conceptual parallels are strong. A light siren is a sound instrument having a rotating disk set between a light source and a photoelectric cell. The disk may have rings of holes or slits to let light through, or it may made of transparent material printed with concentric patterns in opaque pigment. As the disk rotates, the holes or transparent portions let light through, while the opaque portions block the beam. The on-off-on-off of the periodically interrupted beam on the photoelectric cell creates an alternating current which can be sent to an amplifier and speakers. Hugh Davies,

Figure 31.9 The Acme Siren, as depicted on the box it comes in.

in his article under the heading "Electronic Instruments" in the *New Grove Dictionary of Musical Instruments*, lists quite a number of such instruments that have been built since the early part of the century, some of them quite sophisticated. The most exciting work today comes from the French maker Jacques Dudon, whose Photosonic Disk is a marvel of engineering and art: complex, mathematically derived patterns of opaque ink computer-printed on transparent disks give rise to music of extraordinary richness and subtlety. The player controls the sound output simply by moving a hand-held light so as to bring different parts of the disk into play. [See the article by Jacques Dudon in this collection for the full description].

If we can step back again to the 19th century, I'll describe one more peculiar siren variation, the heat siren. This is another of the exotic scientific devices described by J.A. Zahm in his 1892 treatise, *Sound and Music.* Zahm gives it the name *radiophon* and credits it to M. Mercadier. The device, shown in Figure 31.10, employs a gas jet or electric lamp serving as a radiant heat source. With the aid of a concave reflector, the heat is directed at a typical rotating siren disk with concentric rings of holes. On the other side of the disk is a small brass tube covered with lamp black. Zahm explains: "Through the perforated disk on the rotator, intermittent flashes of heat... are allowed to impinge on the soot-covered brass tube. This, by producing rapid changes in temperature, causes corresponding expansions and contractions in the metal tube, and a continuous sound follows in consequence." In the version shown in the drawing, a horn, seemingly coming off of the top of the brass tube, reinforces and directs the sound, yielding "notes that can be perceived at a considerable distance from the instrument."

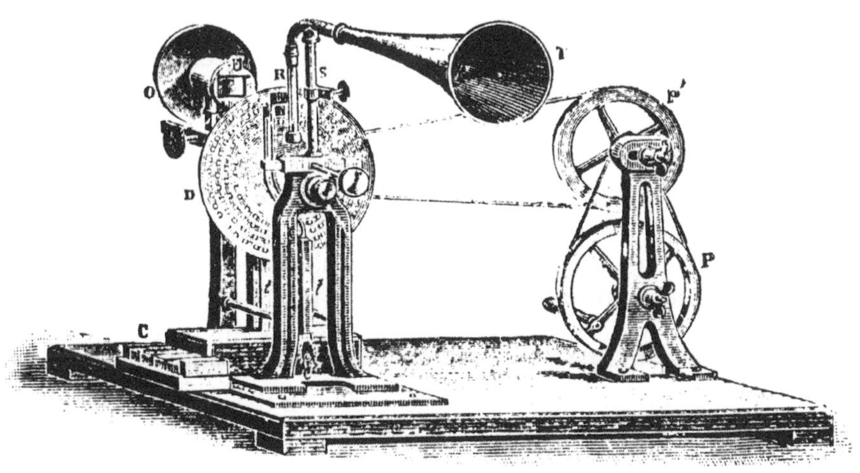

Figure 31.10 The radiophon, a heat siren credited to M. Mercadier in Zahm's *Sound and Music.*

In this discussion thus far, I've said relatively little about sirens as musical instruments. That's because sirens have only rarely been used for musical functions. In most cases when they have been used in music, the instrument called for has been a signaling siren been brought in for special effect. The idea that sirens could be designed specifically as musical instruments, with a range of pitches to be controlled by a player, has been explored only minimally.

So I took it upon myself, a few years ago, to make a siren expressly as a musical instrument. My design was based upon the simple sirens shown at the start of this article, with the single exposed rotating disk and a blow tube, with many rings of holes to provide a range of pitches. More recently I built another musical siren with a slightly more sophisticated design. I learned a lot in making the two of them, and in Part 2 of this article I'll try to pass some of that information on.

Part 2

At the close of the first part of this article, I commented that sirens have only rarely been designed and used as musical instruments. This is despite the fact that many of the intriguing historical sirens that we saw clearly suggested musical possibilities, with the capability to produce musical scales in a controlled and melodic fashion. With this in mind, a few years ago I decided to make a simple siren designed expressly as a musical instrument. Later I built another with a slightly more sophisticated design. In this part of the article, I'll describe those instruments and review some of the information gleaned in the process of making them. The instruments were not fancy at all, and they were flawed in many ways, so I'm writing now not to show off my work, but just to pass on a bit of what I learned – humbly hoping to contribute, in this way. to a brighter future for musical sirendom.

The first siren I made is shown in Figure 31.1. It follows closely the model of the simple single-disk sirens described in Part 1. To describe its basic elements: The heart of the instrument is a single disk with nine concentric rings of holes. The disk turns on a bearing mounted on a wooden base that can sit on a tabletop. There's a small electric motor mounted off to one side, with a belt drive to turn the disk. The player sounds the instrument by blowing through a flexible plastic tube of about two feet long. One end of the tube has a nozzle fashioned out of wood which concentrates the air stream. Mounted to the side on the base and extending out over the disk are two wooden arms that I call "fences," by analogy to the fence on a woodworker's lathe. To hold the nozzle steady and as close as possible to the surface of the spinning disk, the player can rest the nozzle on the edge of one of the fences (the exterior of the nozzle has a kind of shelf that makes this easy to do). You can slide the nozzle along the fence to position it over different rings of holes to produce different pitches (Figure 31.11).

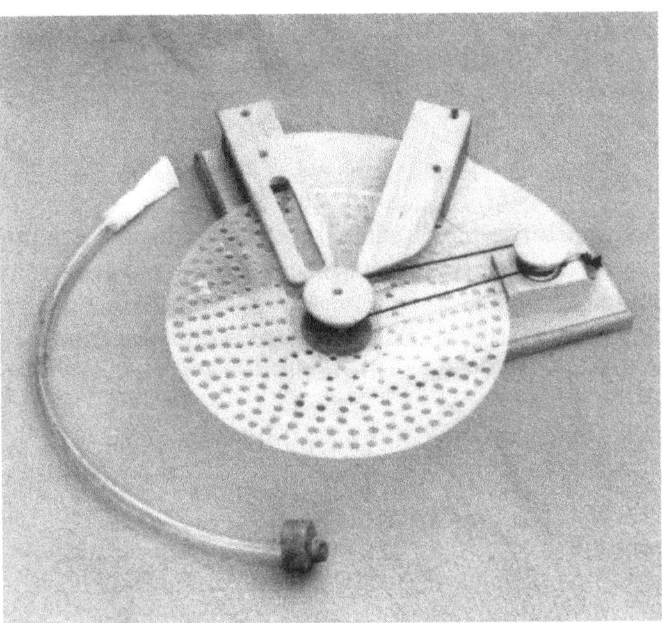

Figure 31.11 Simple open siren made by the author. The separate hand-held blow tube is on the left.

This siren's sound is very quiet. I do like the tone quality, though. It is a little nasal, with an added element of white noise resulting from air turbulence at the nozzle and around the holes. A slight warpage in the disk gives the tone a bit of a waver as the surface of the disk alternately approaches the nozzle more closely and then moves back as it rotates. The playing technique – moving the nozzle from place to place as you blow through the tube – is easy to master. The effect of sliding the nozzle along the fence to produce a glissando is very pretty. In part because of the scale of pitches that I happened to make available on the disk, the instrument seems to like to play a kind of mournful cowboy music.

It turns out that the number-one consideration for this sort of simple siren is: you must get the nozzle of the blow tube as close as possible to the surface of the spinning disk. And no matter how close you get it, it's still not close enough. If there's any gap between the mouth of the tube and the surface of the disk, then the air flow isn't blocked in the between-holes intervals as well as it should be. Instead, it keeps flowing, simply spreading out to the sides through the gap between tube and disk. The overall picture, then, is that when a hole passes, air rushes through and creates a pressure increase and resulting wave from on the far side of the disk. In the between-hole interval, air escapes through the gap between the tube end and the disk surface, and creates a similar pressure increase and resulting wave front on the near side of the disk. Thus, oscillating pressure variations are being set up on opposite sides of the disk, precisely out of phase with one another. Inevitably, as the waves spread, they cancel. That's one of the reasons why this sort of siren is so quiet. The de la Tour siren described in Part 1, with its air chamber enclosure, is louder in part because this sort of cancellation is minimized.

If you try to bring the nozzle too close to the disk surface and the nozzle accidentally touches the disk, the result is a disastrous clatter. Touching also tends to reduce the speed of disk rotation, which makes the pitch go flat, especially if the motor is weak. In short, you can never get the nozzle close enough, yet you can't afford to get too close. In making my simple siren, my response this dilemma was to decide to be happy with a very quiet instrument (and also to try my hand later at building something more like a de la Tour siren, which I'll describe in a moment). I mentioned earlier that my simple siren produces a lot of un-pitched white noise. This comes from air turbulence at the nozzle and around the hole edges. In addition, I suspect that a part of the pitched sound may come not from pulses of air through the holes, but from periodic changes in turbulence patterns, or from simple reflection, associated with air hitting the hole edges or the flat sections in between as they whiz by.

For such a quiet instrument, it's good to have a nice, quiet drive mechanism. On my simple siren I used an old reel-to-reel tape recorder motor to drive the disk. It's admirably quiet, but it has turned out to be too weak to do its job well. Someday I'll get around to replacing it with a stronger motor. The weakness is apparent in the fact the disk comes up to speed slowly and even then is prone to speed variation, resulting in unwanted shifts in pitch. A great boon for a siren like this would be a motor-speed control, which would allow tuning adjustments. Someday maybe I'll get around to that, too. For the disk's bearing, incidentally, I used one of the assemblies on which the tape reels turned from the same old tape deck – also nice and quiet.

For the disk itself I used a disk of clear plexiglass a foot in diameter which came pre-cut from a plastics outlet. It was light and convenient, and it drilled well for easy hole-making without splintering, but as I mentioned earlier, it proved to be warped. And what about tuning? How did I determine how many holes to put in each row to get a particular scale? The explanation gets a little involved, so rather than crowding it into the main text of the article here, I've placed it in the appendix at the end of this article headed "Scale-Making on the Siren Disk."

I built my second siren with the idea of creating a louder instrument (Figure 31.12). I wanted to test the hypothesis that an enclosure in the style of de la Tour would help increase volume by reducing the turbulence and cancellation effects described earlier. But I didn't want to make a double-disk siren in which all the rings of holes open and closed simultaneously, causing all the notes to speak at once. I wanted something on which I could play melodies, one note at a time. So I designed an instrument with a single siren disk spinning in an air-tight enclosure. Directly over the spinning disk, positioned as close as possible, is the flat lid of the

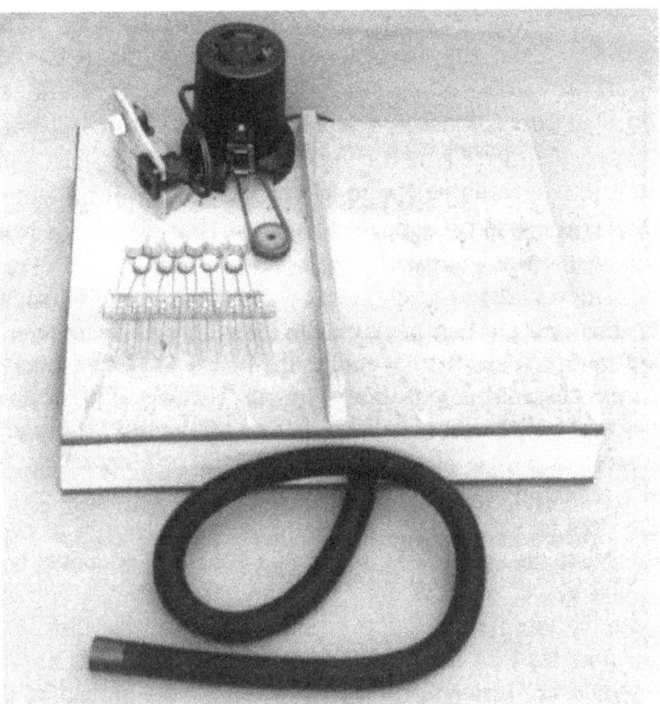

Figure 31.12 Enclosed siren made by the author. The blow tube is in the lower part of the photo. The power cord (not shown) includes a foot-pedal speed-control from a sewing machine.

enclosure. In this lid is a set of single holes – eleven holes altogether, each positioned over one of the eleven rings of holes in the spinning disk below. Mounted on the top are simple lever keys with key-heads and pads that come down over the eleven holes, so that the holes are normally closed. Pressing one of the keys opens one of the holes, allowing the corresponding ring in the siren disk to sound.

The instrument is built around an old record player. The original aluminum turntable, now full of holes, serves as the siren disk. The record player is good for the purpose, because the disk is perfectly flat, the bearing extremely stable, and the operation quiet (all standard requirements for a phonograph). I built the air-tight enclosure out of wood, right around the original record player casing. I put a thick, soft gasket of sponge rubber weather strip along the tops of the sides of the enclosure, and the lid screws down over this. This arrangement makes an air-tight seal. Further, it allows me to adjust the position of the top over the turntable by screwing in and squashing down the weather strip to variable degrees. By fine adjustment I can bring the lid as close to the disk as possible without touching. I drilled a hole in the side of the housing and inserted a flexible blow tube. The original phonograph motor is too weak and slow for the purpose, being designed for a maximum speed of 45 rpm. That's less than one revolution per second, while the siren calls for a minimum of several revolutions per second. So I mounted an old sewing machine motor on the lid. The idea of using a sewing machine motor seemed like an ideal solution: it has a speed control (the foot pedal), it's stronger than a tape recorder or turntable motor, and it's available cheap ($5 or $10 for an old sewing machine from a thrift store). But it too has turned out to be less than ideal: it's noisy and doesn't have the required steadiness of speed.

As I had hoped and expected, the enclosed siren is definitely louder than my earlier simple siren, though still not nearly as loud as a true de la Tour siren in which many holes open and close simultaneously. It takes a lot of wind to operate it, but the instrument sounds equally well on the intake as on the out-blow. (As far as the atmospheric vibration is concerned, a negative puff of air is as good as a positive one). As a result, you can play strongly and more or less continuously, if you don't mind hyperventilating. The valving system works OK, allowing you to play simple keyboard-like melodies with one hand.

However, after all that work, I find that I prefer the playing-feel and playing-motion of my earlier simple siren. I also prefer the greater expressiveness and variety of tone color that its non-keyboard technique affords. If I were to make yet another musical siren – something I would like to do someday – I'd think in terms of an improved version of the simple siren. I'd want a bigger disk for more pitches and greater range, and a good, strong, quiet speed-controlled motor with dependably steady rotational speed. To improve the volume, I'd want to have larger holes, and I might use an outside air source capable of moving more air than my lungs can manage. As for the white noise of air turbulence in the open-air siren, I wouldn't try to eliminate it. I've come to like it; it's part of the rough-edged personality of the instrument.

Appendix: Scale-Making on the Siren Disk

The musical intervals available from a siren disk will be determined by the relationships between the numbers of holes in each ring. But the numbers of holes will not determine the actual pitches, because pitch depends on the disk's rotational speed. After the disk is made and the siren is in operation, you can tune all the notes together up or down by adjusting disk speed, but you can't change the pitch relationships between them.

This appendix describes how to go about laying out the holes on a musical siren disk to produce particular a scale. I'll start the explanation with some very minimal background for those who may not be familiar with the necessary underlying musical scale theory. The human ear seems naturally to hear musical intervals in accordance with the ratios of the frequencies of the tones involved. For example, for any two tones for which the frequency of one is twice that of the other – that is, any two tones having a frequency ratio of 2:1 – the ear will hear the interval between them as an octave. It doesn't matter what the actual frequencies are; if the frequency ratio between them is 2:1, they'll be heard as an octave apart. The same goes for other intervals: for a perfect fifth, the frequency ratio is 3:2, for a major third it's 5:4, and so forth.

By their very nature, sirens are conceptually friendly for those who want to think about musical intervals in terms of ratios in this way. On a siren disk with multiple concentric rings of holes, the ratios of the pitches available will correspond to the ratios of the numbers of holes in each ring. For instance, if one ring has twice as many holes as another, then for each rotation of the disk, it will give twice as many puffs of air. The frequency ratio between the two rings will be two to one, and their tones will sound an octave apart. When the disk spins relatively rapidly, both frequencies will be relatively high; if it spins slowly, they will be lower; but for any given rotation speed, their tones will still always be an octave apart. The trick, then, is to make a disk with many concentric rings of holes such that the ratios of the numbers of holes in each ring correspond to the ratios for the intervals of the scale you want to hear. Here's a simple example: Suppose you want a six-note siren producing a basic major pentatonic scale. The intervals of this scale, identified both in musical terms and in terms of ratios, are:

Musical Interval	*Root*	*Major 2nd*	*Major 3rd*	*Perfect 5th*	*Major 6th*	*Octave*
Frequency Ratio Relative to Root	1:1	9:8	5:4	3:2	5:3	2:1

To achieve this set of ratios, you can make the six rings with the following numbers of evenly spaced holes:

Number of Holes	24	27	30	36	40	48

To produce a second octave of the same set of intervals, make an additional set of rings with the hole numbers doubled. See? Simple!

Except it's not always so simple. Suppose you want to create the same scale, but you don't want to drill so many holes. You decide to start with just 12 holes in the first ring. Then the second ring, producing the major second, should have 12 × 9/8 = ... um ... thirteen and a half holes! The half hole is a problem. How can you have thirteen and a half equal spaces between? It's a conceptual anomaly. To avoid it, the numbers of holes in the rings needs to be whole numbers.[4]

The job of scale making on a siren disk, then, involves the following steps:

1) Decide what sort of scale you want and define it in ratios.[5]
2) Find a set of whole numbers having those ratios, but preferably not so large as to require an impractically large disk or an excessive amount of drilling when it comes to actually making the holes. Those who remember the arithmetic of their school days will recognize that this calls for finding the least common denominator for the set of ratios involved.
3) Lay out the hole positions on the disk. To do this: Establish the center of the disk and, using a compass, draw the appropriate number of equally spaced concentric rings. Then mark the holes positions on each of the rings, starting with the innermost ring/smallest-number-of-holes/lowest-intended-pitch. To figure the equal hole spacing, divide the intended number of holes into the ring into the 360 degrees of the circle, and use a protractor to space the holes the resulting number of degrees apart. (Simple example: for eight holes, 360°/8 = 45°; thus, the eight holes should be evenly spaced 45 degrees apart around the circle.)
4) Drill the holes. If you have access to a drill press this job will be much quicker and easier than if you drill by hand. You can set up a jig such that the disk rotates on a spindle under the drill. Position the jig for each ring of holes and then drill the holes in succession, rotating the disk the suitable amount for each new hole.

Regarding the size of the holes, there are a couple of considerations. On one hand, for maximum volume and minimum turbulence, the holes should be as large as space allows. On the other hand, for a smooth wave form and, again, good volume, the hole diameter would ideally be about half of the distance between holes within the ring. (This assumes a round blow tube of a diameter similar to the hole diameter.) This is something to aim for in a general way, but not something to try to achieve at all costs, because the geometry and arithmetic of the situation make it unrealistic to try to achieve it consistently in a set of many rings.

Notice that while the major pentatonic scale given above is a rather simple scale, it still demanded altogether a large number of holes. More complicated scales, with more intervals and ratios involving larger numbers, tend to require still greater numbers of holes – often *far* greater. Lots of drilling! Now you can see why the early siren makers rarely included more than four rings of holes in their disks.

In theory you can produce any just intonation scale on a siren, although the number of holes required might become absurdly large. (A scale in just intonation would be one in which the intervals are defined as frequency ratios, as we've been discussing.) Tempered scales, and in particular, the 12-tone equal temperament which is standard in most western music, are another matter. In 12-equal, the frequency relationships are derived by a different sort of mathematical logic, and the resulting frequency ratios turn out to be irrational numbers. They can't be expressed as ratios, and so you can't realize such scales as varying numbers of holes in the concentric rings of a siren disk. But the situation is not as hopeless as it sounds. Remember that 12-equal was originally designed as a convenient approximation to just scales which were themselves actually preferred, so a return to the ideal of just intonation might not be seen as a loss. Furthermore, the ear is more forgiving than our friend the compulsively precise scale theorist might have you believe. Rounding off the ideal number of holes in a ring to the nearest whole number will cause some detuning relative to the ideal, but it might not be so bad as to be unlivable. A key consideration here is the number of holes in the ring. Where the number is small, one hole more or less makes a big change in pitch. Where the number is quite large, one hole more or less makes a smaller difference. So it all comes down to this: if you're willing to make a large disk with large numbers of holes in each ring and a huge number of holes altogether, you have a pretty good chance of being able to achieve least a fair approximation to any tuning, just or not, that you seek. If you choose to keep the overall number of holes smaller, then your tuning options will be quite a bit more limited. The fewer-holes option, to my mind, isn't necessarily bad: there's a kind of organic quality to music and musical systems which arise from the natural form and requirements of the instrument rather than being conceived and imposed from the outside. My quiet little nine-hole simple siren, with its scale of just four notes per octave, is an example of this.

By the way, the scale on that siren disk is:

Ring #	*1*	*2*	*3*	*4*	*5*	*6*	*7*	*8*	*9*
No. of Holes	15	18	20	24	30	36	40	48	60
Ratios	3:4	9:10	1:1	6:5	3:2	9:5	2:1	12:5	3:1
Intervals	P5	m7	Root	m3	P5	M7	Root	m3	P5
Sample Scale	E	G	A	c	e	g	a	c'	e'

(Notice that I'm calling the 3rd ring the root tone. One could as well call the 4th or some other ring the root tone. The letter pitches are identified as a "sample scale" because, while the intervals of the scale are fixed, the actual pitches vary according to disk speed.)

And the scale on my 11-ring enclosed siren is:

Ring	*1*	*2*	*3*	*4*	*5*	*6*	*7*	*8*	*9*	*10*	*11*
No. of Holes	8	9	10	12	14	16	18	20	24	28	32
Ratios	1:1	9:8	5:4	3:2	7:4	2:1	9:4	5:2	3:1	7:2	4:1
Intervals	Root	M2	M3	P5	Sept7	8ve	M2	M3	P5	Sept7	2-8ves
Sample Scale	C	D	E	G	↓B♭	c	d	e	g	↓b♭	c'

That's two octaves of a pentatonic scale including the septimal seventh (7:4), which is a little flatter than the minor seventh in the familiar 12-tone equal temperament.

Notes

1 This terminology comes from the widely used Sachs-Hornbostel system for the classification of musical instruments.

2 Sources for historical information on sirens are few. Most of the information in these paragraphs is gleaned directly from Helmholtz and Zahm, augmented by Hugh Davies' excellent article on sirens in *The New* Grove *Dictionary of Musical Instruments* (London: MacMillan Press Ltd. and New York: Grove's Dictionaries of Music, 1984) and, to a lesser extent, some other general-knowledge sources.

3 In this connection, Helmholtz notes that, prior to the use of sirens, "The exact determination of the pitch number [frequency] for such elastic bodies as produce audible tones, presents considerable difficulty, and physicists had to contrive many comparatively complicated processes in order to solve this problem for each particular case."· (From Alexander Ellis' 1885 translation of *On the Sensations of Tone,* Dover Publications, Inc., New York, 1954.)

4 Actually, it is possible to have something like a fractional number of holes in a ring, but it involves some compromise. If you space the holes as you would in order to have 13½ holes (or some other non-whole number), after completing the ring of evenly spaced holes, you end up with a ½-sized space (or, optionally, 1½-sized space) between the last hole and the first. As the disk spins and the siren sounds, the puffs of air will pass through the holes at the desired frequency, except at the half-space point there will be an instantaneous irregularity, amounting to a sudden phase shift. The ear will hear the intended frequency, but the peculiar phase change will be heard as a kind of glitch occurring once per rotation. I've never heard this effect on a true siren, but I did have the opportunity to hear it, many years ago, in one of Jacques Dudon's optical sirens (described in Part 1 of this article).

5 Unless you really like working by trial and error, this requires some familiarity with the desired intervals and their corresponding ratios, probably beyond what I've been able to offer in this article.

Bibliography for the primary sources

Pietro Blaserna, *The Theory of Sound in its Relation to Music* (London: Kegan Paul, Trench & Co., 1883)

Hermann L. F. Helmholtz; translated by Alexander J. Ellis, *On the Sensations of Tone as a Physiological Basis for the Theory of Music* (New York, Dover Publications Inc., 1954; originally published by Longmans & Co., 1885)

J. A. Zahm, (The Rev) *Sound and Music* (Chicago: A.C. McClurg & Co., 1892)

Special thanks to Reed Ghazala for bringing the Zahm book to my attention and making it available to me, and likewise to Craig Tucker for the Blaserna book.

Related articles

Other articles pertaining to sirens which appeared in *Experimental Musical Instruments* are listed below. These articles can be accessed through the Internet Archive at https://archive.org/details/emi_archive/.

In this collection, Jacques Dudon's article "The Photosonic Disk" discusses light sirens, which are instruments operating on a principal closely analogous to the sirens discussed in these articles, but based on periodic interruption of a beam of light falling on a photocell. Dudon's work, including light sirens and other instrument types, is also covered in "Jacque Dudon's Music of Water and Light" by Tom Nunn in *Experimental Musical Instruments* Volume 3 #5

"Deeper into Fleshtone; Sound Energy within the Human Body" by Monte Thrasher, in *Experimental Musical Instruments* Volume 14 #3, describes a fantastic imagined siren-base robotic sound sculpture

32

Slate

Will Menter

This article originally appeared in *Experimental Musical Instruments* Volume 13 #1, September 1997.

For many years I lived on a farm called Watercatch, near Bristol in the west of England. In the field behind it was an enormous underground tank, built early this century as a water supply. The sound was great after rain – drips from the roof and a long, long reverb. I made some beautiful recordings down there. This farm brought together lots of my musical interests. I had done several performances with dripping water before I moved there, and developed it further while I was there. Next to the farm were woods with lots of dead sycamore in, which kept us warm in winter and provided the material for making rope-ladder-style log xylophones in summer.

The field in front of the farm was called "Slate." I don't know why, but it pointed to another direction, because in 1986, I spent three months working 200 miles away in the slate quarrying area of North Wales around Mount Snowdon. A beautiful area – high mountains and steep valleys, remote sheep farms, and in the distance, the Irish Sea, the sandy beaches of Ynys Mon (the island of Anglesey) and on the Llyn Peninsula, the Whistling Sands of Oer. The sands whistle, says the tourist brochure. I was convinced it must be something to do with the wind or the sea and kept listening harder and harder (if that's possible) but could hear nothing apart from waves, wind, and a distant diesel generator. Maybe the weather wasn't right for sand whistling. Then a group of children ran past, and the sand was squeaking under their feet! So I ran too and the sand squeaked for me. Or creaked, I'm not sure which, but that must have been it, the whistling. After all, I reasoned, the "Squeaking Sands of Oer" wouldn't do much mileage as a tourist spot.

But the sands were a diversion which I'd explored on one of my days off. The main thing was that here I was at the very heart of the slate industry, surrounded by unlimited supplies of this wonderful material. During the 18th and 19th centuries slate from Llanberis, Bethesda and Blaenau Ffestiniog was shipped all over the world for use as a roofing material – as far as Valparaiso in Chile for example. The quarries employed thousands of craftsmen and laborers and were responsible for roofing the factories of the early industrial revolution in Manchester and the north of England. This was part of the complex jigsaw of social and geographical conditions that led England to industrialize its production earlier than the rest of the world.

The stone is gray. But each quarry has its own distinctive shade – green, blue, purple, and even red. The red is unusual but has such a lovely warmth to it. It's a sedimentary rock, formed from layers of mud at the bottom of the ocean, then metamorphosed through heat and pressure. And it has such a strong horizontal grain structure that it can be easily split with a chisel into sheets 5 mm or less thick which are still strong enough to withstand the battering of wind, rain, and hailstones on roofs for a hundred years or more, and the battering of wood and rubber percussion beaters for – well, for how long? I don't yet know, but the first instruments I made with slate are now over ten years old and are going strong.

What instruments? Slate marimbas, llechiphones. Although I don't speak Welsh it seemed right to give them a name based on the Welsh word for slate (*llechen*, singular; *llechi*, plural). The form is similar to a wooden marimba. Slate bars about 5 mm thick are supported on rubber tubing at their nodal points. Underneath each bar is a plastic resonating pipe. Beaters are rubber balls on wood or leather-bound wood. Tuning is by grinding underneath with a file or (quicker but noisier) an angle grinder.

Figure 32.1 Dinorwic Quarry.

What do they sound like? Probably much mellower than you'd expect. More like an African marimba than an orchestral one, something like a metallophone but not such a long ring. A unique sound, and I'm surprised it hasn't already been used much more. Some people have compared it to a harp too, but while I can hear a similarity I think it is also to do with cultural associations since the harp is an important instrument in Welsh folk music.

Figure 32.2 Slate Beach, Porth Penrhyn, near Bangor.

Figure 32.3 Archive photo of a bugler signaling blasting.

What tunings and ranges? I've made them from C_1 (two lines below the bass clef) up to G_5 (an octave above the Treble clef), but the tone is strongest, and most distinctively slate, between C_2 and G_4. The tunings have been strongly influenced by the first one I made. Seven slates split from one rock, all the same length but different thicknesses, turned out by chance to be a five-tone scale: D_2, E_2, G_2, $B\flat_2$, C_3, D_3, E_3. I love this scale. It's very encouraging for workshops because it is so harmonious, the notes forming a ninth chord if played together, and at the same time, because of the tritone between E and B*b*, it's not as bland as the more common pentatonic C D E G A. One of the reasons I responded positively to this initial accident of tuning was that I'd heard this scale used in the beautiful music of Hukwe Zawose from Tanzania. So I've made a lot of llechiphones on this scale. I've also made many fully chromatic models, and many less usual tunings. My particular favorite at the moment is an equipentatonic tuning, where the interval

Figure 32.4 Slate fences above Penrhyn Quarry.

between each key is 2.4 semitones. For a while I made a standard range of llechiphones which I made available for sale, but now I just make to commission, and try to respond to individual needs.

In the area around the quarries, slate is used not just for roofing, but all sorts of other things – floors, walls, work surfaces, doorsteps. Driving down the narrow mountain road to Penrhyn Quarry in Bethesda, a twenty-foot-high slate wall suddenly borders the road – rich dark gray – and at Porth Penrhyn near Bangor the whole beach is made up of thin shale-like slates stacked on their sides making a wonderful accidental sculpture. Back on the mountainside, many of the fences between the narrow fields are made with huge slabs of slate sunk into the ground and tied together with wire, born of practicality and necessity, but giving the area a unique sculptural landscape. Even more dramatic is the town of Blaenau Ffestiniog. Ten miles inland and south of Mount Snowdon, it seems that all you can see is slate. Here the slate is mined underground rather than quarried, but huge man-made mountains of waste slate surround the town and merge with the distant green hills. The scrap makes a great sound too. You sometimes hear wild goats running over it. Walking over it yourself is a musical experience either listening to your footsteps or selecting pieces that you like the sound of and pocketing them.

And yet... this is a romantic view. I struck up a conversation with a young woman in a pub in Blaenau Ffestiniog. "I don't know what you like it for," she said. "There's nothing here except slate and sheep and rain." Slate has become a relatively expensive roofing material so most of the quarries are now closed, and it's an area of high unemployment with few opportunities. Traditionally quarry work was hard and dangerous, and many local people regard the tips as a monument to the grueling labors of their fathers and the lives lost through accidents and industrial disease. It's only a minority that would prefer the tips to be grassed over and forgotten about. While this young woman was desperate for some excitement in her life which she knew she was never going to get from slate, she'd never managed to travel further than Colwyn Bay, a coastal resort about thirty miles away.

Slate also symbolizes the exploitation of Wales by England – or rather the English bourgeoisie. Many of the quarry owners and bosses were English, while the laborers and craftsmen were Welsh. The craft of working slate can't even be properly discussed in English, because there are many technical terms in Welsh that have no English equivalent.

I started my experiments with slate as a musical material in 1986 during a residency at the slate museum in Llanberis. I was ignorant of the dangers. Tuning with an angle grinder in a public corridor, slate dust filled the air but I was quickly

Figure 32.5 Five llechiphones.

Figure 32.6 Detail of chromatic llechiphone.

Figure 32.7 Will Menter with llechiphones.

informed that once the dust gets in your lungs it stays there, so a face mask and good ventilation were essential for working with slate. During my residency I soon came to recognize the feeble walk and posture of some of the older men in the area whose lungs had been ruined by long exposure to the dust. In the 19th century, some quarry owners had even claimed that slate dust was good for you. Of course, the quarry workers all knew it wasn't, but it took until the 1970s for official recognition and compensation to be given for industrial disease caused by the dust. All the stewards working in the slate museum were former quarry workers and I learnt much from them and struck up friendships with a couple of them.

I came to love the area and after the residency was finished, started visiting regularly. Like a lot of instrument makers, ultimately I am more interested in music than instruments – or rather I see instrument making as being part of the creative process of music more than an end in itself. Some of the other parts of this process are well known – composing, performing – but sometimes overlooked are listening and, perhaps most important of all, creating a physical and social context for all these activities to happen. My approach to music making is to be involved with as much of this process as possible.

I gradually developed the idea that a performance that was about slate, that used slate to make the sounds and that was performed in the heart of the slate quarrying area would be particularly resonant (not just acoustically) and I set my heart on achieving this. The musical idea was expanded to include other art forms. I commissioned six poems from Gwyn Thomas and set them to music to be sung by Sianed Jones and five other musicians. The music was built up around the llechiphones, but also incorporated traditional instruments – sax, violin, trombone, bass viol. Sculptors Andy Hazell and Lucy Casson made a set out of scrap and industrial materials that echoed the land- scape of the quarries. Textile artists Annie Menter and Barbara Disney made batik backdrops that drew attention to the colors and shapes in slate, and also drew on the local tradition of engraved slate fireplaces. Andy also took many slides of the slate area, edited some archive film we found of work in the quarries and manipulated a complicated three-screen projection system while the musicians played. All this eventually came together in the form of a cross-art-form performance called *Cân Y Graig – Slate Voices.*

We performed *Cân Y Graig – Slate Voices* fifteen times in six different locations in 1990. Two of them had this special resonance that I was looking for. The first was the slate museum itself in Llanberis. The space we used had been the foundry for the second-biggest quarry of all and still had the original furnace and industrial saws and planes powered by an enormous water-wheel. Here we had the most complete exposition of our material. But more exciting for the audience and acoustically more resonant were our underground performances at the Llechwedd slate mine in Blaenau Ffestiniog. Four hundred feet underground, the audience had to wind their way down low-roofed tunnels carved out of the rock to reach the enormous cavern which had once been a solid, silent mass of slate but was now an equally silent void beside an underground lake. Each person present at the performances could only guess where the particular slate whose space their own body now occupied had ended up – as far away as Rio de Janeiro, or as near as the waste tip by the mine entrance. Our performance lasted for about an hour and at the end we moved beside the lake to read out a litany of the names of quarry workers who had died in accidents. For evoking the life and work of slate workers over the years, this context couldn't be surpassed and many of the audience said afterward they were deeply moved by the experience.

Cân Y Graig – Slate Voices was definitely the high point of my work with slate so far but I have gone on making slate instruments and using them in lots of different music. For example, I combined them with Zimbabwean mbiras in a more recent project called *Strong Winds and Soft Earth Landings*. The tuned slate marimbas I suppose are my main slate instruments, but just as interesting are simple hanging chimes made from random pieces of slate ranging from a few centimeters across up to almost a meter. As with most percussion instruments of this type, the bigger ones are dominated by overtones rather than the fundamental pitch and give very rich individual tones. I've set them up as wind-blown instruments and also in arrangements as mobiles where they collide with each other producing more random rhythms and series of sounds.

Figure 32.8 Performance of *Cân Y Graig – Slate Voices* in Welsh Slate Museum.

Figure 32.9 Performance of *Cân Y Graig – Slate Voices* in Llechwedd Slate Mine.

At the moment I'm working on a project to bring together my interests in water and slate. I took five of my llechiphones outside and placed them underneath a large oak tree, so I could stand on one of the branches and take a photograph looking down. When it started to rain a few minutes later, my first thought was to rush the instruments back inside, but before I had a chance to do this, I noticed what a great sound the big drops of water falling off the tree were making as they hit the slate. So instead of taking them in, I brought my tape recorder out and recorded it. Later I used this tape in a piece with two singers, but I also had a prototype design for a new instrument – the dripping llechiphone – which should be finished this summer. My plan is to use this in an installation with a development of another water instrument of mine, the gurgler. This one consists of a set of plastic drainpipes standing in a bowl of water with air bubbling up inside them through a hose sprinkler. The sprinkler turns slowly and delivers bubbles under each tube in turn, making a deep pitched gurgling sound. The idea was inspired by listening to bath water gurgling as it emptied. The sound I was going for was more of a glugging than a gurgling but the instrument has turned out to sound nothing like this! All this should come together in some performances and installations in mid-1998.

Related articles

Other articles pertaining to lithophones which appeared in *Experimental Musical Instruments* are listed below. These articles can be accessed through the Internet Archive at https://archive.org/details/emi_archive/.

"The Till Family Rock Band" by D. A M Till in EMI Vol. 7 #5

"Musical Pillars Commentary" by Matthieu Croset in EMI Vol. 8 #1

"Elemental Mallet Instruments" by Jim Doble in EMI Vol. 9 #3

"The Qing Lithophones of China" by Mitchell Clark in EMI Vol. 10 #1

"More on The Deagan Chimes and My Father's Stone Chimes, Too" by Ellen Schultze in EMI Vol. 10 #1

"The Icelandic Lithophone": Elias Davidsson in EMI Vol. 14 #1

33

Mechanical Speech Synthesis

Martin Riches

The article presented here is made up of two articles on speech synthesis by Martin Riches which originally appeared in *Experimental Musical Instruments* Volume 13 #1 and Volume 13 #2, September and December 1997. In Part 1 the author provides a basic grounding in the mechanics of speech production, and then describes four important historical synthetic speech devices. Part 2 is devoted to the mechanical speech synthesizer that he himself created, called The Talking Machine.

Part 1: Speech Production and Four Historical Speech Synthesis Projects

What is speech synthesis? Well, a very short definition might be: Speech synthesis is the art of producing what sounds like the human voice but without using any form of recording.

Intelligible electronic speech synthesis first became possible in the late 1930s but the art of producing speech artificially is much older than that. Long before the first experiments in electronic speech synthesis, there were several successful attempts to reproduce the sound of the human voice using the techniques of the organ builder and instrument maker.

I will describe four historical projects which I read about while preparing to make a talking machine of my own; four projects where I found informative and credible drawings and texts. But first, since talking machines are analogs of the human speech apparatus, I shall first give some brief workman-like notes on acoustic theory and on the human production of English speech sounds.

Acoustics

The human vocal tract is a tube about 17 cm long, open at the lips and effectively closed at the other end at the vocal cords. If you take a piece of cylindrical tubing of this length, fit it with a reed in place of the vocal cords and blow it, you will get a strong resonance around 500 cycles per second with further peaks of strong resonance odd-number multiples of 500 – at 1500, 2500, 3500, 5500 cycles per second and so on. In fact, the insides of the mouth and the throat do not form a perfect cylinder and, what is more, the hollow shape is constantly changing as we talk. As a result, these resonance peaks are shifted around away from that original regular pattern. These strongly resonant frequencies give the various speech sounds their individual character. Just three, or even two, of the lower peaks are all we need to identify a speech sound. These peaks of strongly resonant frequencies are referred to in the speech synthesis field as "formants." The formants are also influenced by the continuously changing timbre of the vocal cords. If a frequency is absent from the original sound it will be absent from the final speech sound. However this effect is of secondary importance as far as speech production and recognition is concerned.

How to Speak English[1]

The human vocal apparatus consists of:

an air supply, the lungs;
an adjustable resonator, the mouth, containing a lump of flexible muscle of constant volume but variable shape, the tongue. The volume of the mouth cavity can be increased by lowering the jaw. This adjustable resonator is about 17 cm long and the cross-section can be varied from zero to about 20 sq.cm.
a non-adjustable resonator: the nose, joined to the mouth by a flap valve. It has a capacity of about 60 cc.
four noise-making devices: the vocal cords, tongue, teeth and lips.[2]

The name "vocal cords" is misleading; it seems to suggest that the voice is a string instrument. In fact it is more like a trumpet. The vocal cords are two strips of soft flesh – two internal lips – which form a constriction in the pipe leading up from the lungs. This constriction can be opened and closed at will. When it is open we breath freely. When the vocal cords are partly closed, the air passes through the constriction causing the vocal cords to vibrate making a farting noise - like the sound you can make by blowing between your closed fingers or the sound that trumpeters make with their lips. Like a trumpet player's lips, the tension of the vocal cords can be increased to raise the pitch.[3]

The vocal cords, slapping together about 50 times a second, produce a sound which is rich in overtones. To demonstrate how this sound is adapted by the mouth resonator to produce speech sounds, try the following experiment. Simulate the mouth cavity by forming an enclosed space between your two hands and then blow into it with a farting sound. Continue blowing while changing the shape of the cavity and open and close some fingers to simulate the lips. You should be able to make some convincing OO-AH vowel sounds and some semi-vowel sounds like WAH WAH.[4]

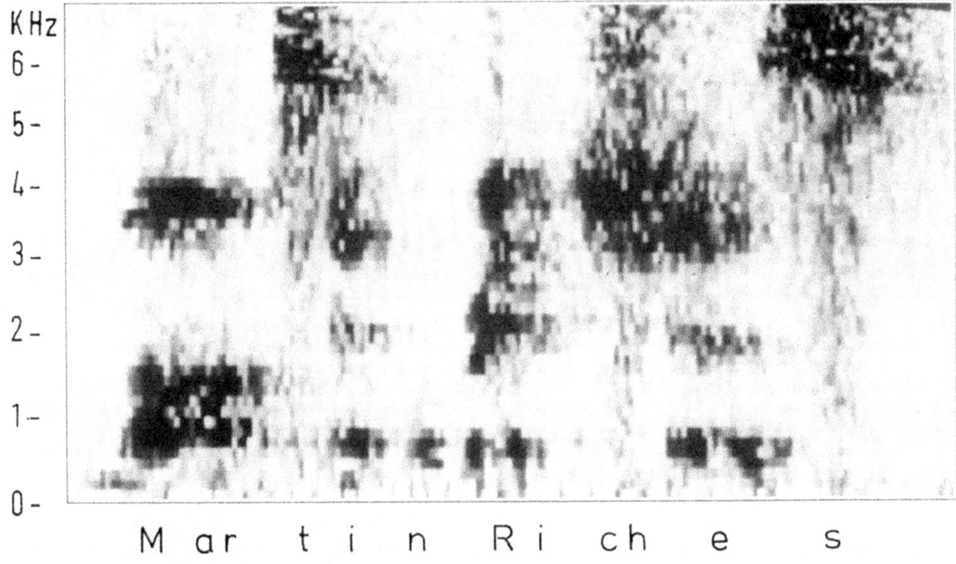

Figure 33.1 Spectrogram of the author's voice speaking his name. Note the three bursts of high frequency sound for the *T*, *CH*, and *S* sounds. The short horizontal bands for the AR, I and E sounds represent the formants of these sounds. The spectrogram was produced by Dipl. Ing. Klaus Hobohm of the Technical University, Berlin.

As you can hear, there is no essential difference between the production of vowels and the production of voiced consonants like W. The voiced consonants are just performed with the tongue moving faster. *L* and *R* sounds are produced by flapping movements of the tongue originating from two different points on the roof of the mouth.[5] *K* and *G* sounds are made by the tongue touching the roof of the mouth further back from the *L* and *R* positions and momentarily blocking and then releasing the air supply. The *K* sound is produced without the vocal cords vibrating.

If the mouth is closed or the mouth cavity is blocked off by the tongue and, at the same time, the flap valve leading to the nose is opened, humming "nasal" sounds are produced. The three English nasal sounds, *M*, *N* and *NG*, are distinguished according to the position of the tongue. An *M* is a humming sound with all the sound coming out of the nose. Try it. The tongue is low. Now move your tongue up, blocking off the front of the mouth behind your front teeth, and it changes to an *N* sound. Since you have blocked off the front of your mouth it makes no difference to your *N* sound whether you now have your lips open or closed. Now move your

Figure 33.2 Sections through the vocal tract, pronouncing various vowels. (Redrawn from: Gunnar Fant, *Acoustic Theory of speech production*. 's-Gravenhage, Mouton & Co, 1960).

tongue further back and you get an *NG* sound. Hold your nose and the sound stops. If you say "I siNG a soNG to the MaN iN the MooN" while holding your nose, you will hear that it is only the nasal sounds which get distorted.

As we have seen with *K*, some sounds are produced without using the vocal cords. If you speak while touching your Adam's apple (the outer casing of the vocal cords), you will feel that most of the time they are vibrating. But they can open and allow the air to pass freely to the front of the mouth to produce the "unvoiced" hissing sounds: *S*, *SH*, *TH*, *F* and *T*, the fricatives. These sounds are produced by the teeth working in combination with the tip of the tongue and lips to form whistles with different resonators formed by the lips. *T* is just a short explosive *S*. Sounds such *V* and *J* (as in JuDGe) are produced like *F* and *SH* but with the vocal cords "switched on". The *H* sound is produced by producing a hiss in the half-closed throat.

The lips have two functions. They modify the vowel sounds – compare the position of the lips when you say AH, EE and OO. The lips can also be closed, retaining a supply of compressed air which is released to produce the explosive *P* and *B* sounds. *B* is pronounced with the vocal cords vibrating; *P* is just the sound of the released air. (The explosive sounds *P-B*, *K-G*, *T-D* are known as "plosives" or "stops").

None of these speech sounds are absolute. Speech is a continuous process: sliding from one speech sound to the next one. Since the tongue and other speech organs have an appreciable mass, it takes a little time to shift them into the correct positions, so that each sound is influenced by the sound that came before and anticipates the next sound to come.

Furthermore, everyone speaks differently: even the short telephone message "Hi! It's me!" is sufficient to identify someone whom you have heard a few times before. Because there is such a wide variety of speech, we are highly skilled at recognizing the underlying patterns – but we also rely very heavily on context. Provided that the listener has some idea of what to expect, it is possible that even a comparatively simple device which can reproduce human speech sounds at the correct tempo, will be understood.

Four Historical Speech Synthesis Projects

Kratzenstein

In 1779, the Imperial Academy of St. Petersburg, for its annual scientific competition, offered a prize for an answer to the following two questions:

1. What is the nature and character of the sounds of the vowels a, e, i, o, u that make them so different from each other?
2. Can an instrument be constructed, like the vox humane pipes of an organ, which shall accurately express the sounds of the vowels?

The prize was won, two years later, in 1781, by Professor Christian Gottlieb Kratzenstein, a German living in Copenhagen, who constructed a set of acoustic resonators to speak the vowel sounds. To produce the sound of the vocal cords Kratzenstein used a "free reed", the type of organ reed which was later used in the mouth-organ, the accordion and similar instruments. Apparently he invented it for this purpose (Figure 33.3).

On top of the reed Kratzenstein placed a resonator – a wooden box – an analog of the inside of a human mouth. This box adapted the sound of the reed just as a real mouth adapts the sound of the vocal cords. The shape of the resonator was different for each vowel. Looking at the sections of Kratzenstein's pipes it seems

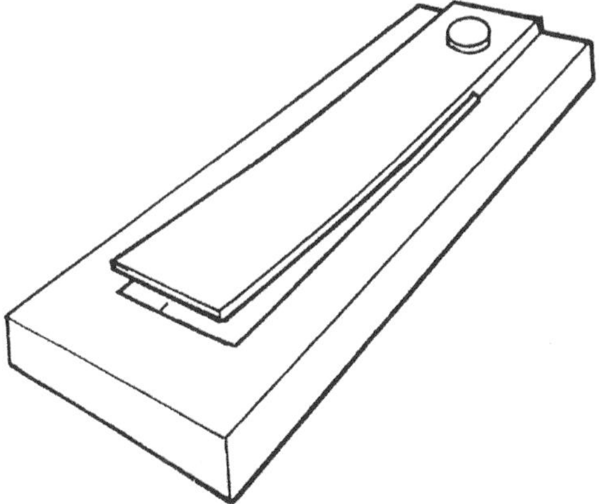

Figure 33.3 A free reed. The reed is just small enough to swing through the opening in the plate, rather than beating against it, thus producing a less harsh sound.

Figure 33.4 Kratzenstein's vowel resonators. (Redrawn from: James L. Flanagan, "The Synthesis of Speech" in *Scientific American*, Vol. 226, No.2, pp 48-58, February, 1972).

probable that they were designed on a trial-and-error basis without trying to precisely imitate the mouth positions. Evidently a wide variety of shapes will produce recognizable vowel sounds (Figure 33.4).[6]

Von Kempelen

It was a contemporary of Kratzenstein who is generally acclaimed as the Father of Speech Synthesis. In 1791, after some 20 years of research, Wolfgang von Kempelen published "The mechanism of human speech, together with a description of a Speaking Machine". Von Kempelen's Speaking Machine not only successfully imitated the vowels and many of the consonants but was also able to speak complete words and phrases. Air was supplied by pumping a bellows. The mouth was simulated by a flexible rubber resonator. Kempelen's left hand manipulated this mouth, while his right hand controlled the ivory "vocal cords," opened the "nose" tubes to provide *M* and *N* sounds and pressed keys to play the *T*, *S* and *SH* sounds.

At the time people were a little suspicious that there might be a trick involved – and not without reason. Kempelen had previously caused a sensation with his amazing chess-playing automaton. It turned out that the chess playing mechanism was operated by a concealed human chess player hidden inside. However his Speaking Machine was a completely legitimate device and quite well known in its time. Wolfgang von Goethe commented on it: "Kempelen's Speaking machine... is not very talkative but it pronounces various childish words and sounds quite nicely..."

One might suspect that Kempelen's Speaking Machine could say little more than "Mama" and "Papa" like a Victorian talking doll; indeed it did have a rather high-pitched and therefore childish voice. But according to von Kempelen's own account it could also say:

> Vous etes man ami – je vous aime de tout man coeur Leopoldus Secundus – Romanorum Imperator – Semper Augustus

It was not so good at Kempelen's native German, perhaps because it was difficult to reproduce the staccato quality of the language, the pure vowel sounds and the many stops, with sufficient accuracy. Kempelen sums up his operating instructions as follows:

1. The right hand should rest on the wind chest so that the index finger and the middle fingers just cover the *M* and *N* nose holes. The thumb comes over the lever or key *SCH* and the little finger over the *S*. The left hand covers the opening of the mouth.
2. The right elbow always rests on the bellows. To speak a letter the bellows must be pressed down, sometimes hard, sometimes softly. This pressure has to continue until the word which is being spoken is finished otherwise the letters and syllables will not join together with each other. If one lifts the elbow, the voice is silent.
3. The nose must remain closed for all letters, except *M* and *N*.
4. For all unvoiced consonants and wind consonants the mouth must remain shut.
5. For all wind- and voice-consonants, the mouth should not be completely closed but kept open enough to allow the voice to sound as well. (Trans.: M.R.)

He then goes through the alphabet describing how to speak each letter.

Although his description of speech as "voice passing through restrictions" may be seen as a mechanistic interpretation of the filtering effects of resonance, he did not understand the fundamental acoustic processes taking place. Nevertheless his practical approach was sufficient to produce a machine which really did work. Perhaps his greatest achievement was his book "The Mechanism of Human Speech (…)" which is extremely readable, well illustrated, and has the reputation of being one of the finest physiological books of its time (Figure 33.5).

Faber's Euphonia

Professor Joseph Faber produced Euphonia in 1830 (or 1835) and it took 25 years to build. The "facade" was the head of a young woman mounted on a machine draped with women's clothes. A second version, Euphonis, took the form of a bearded Turk. The machine was played by sixteen keys connected by wires to the rubber speech components, similar to the organs of a human being. Air came from a double bellows. It could recite the alphabet, and could say "How do you do, Ladies and Gentlemen?" It asked and answered questions, whispered, laughed and sang. Since it was built by a German speaker it spoke English with a German accent. There follows a rather patronizing description by a professor of the faculty of medicine, Paris.

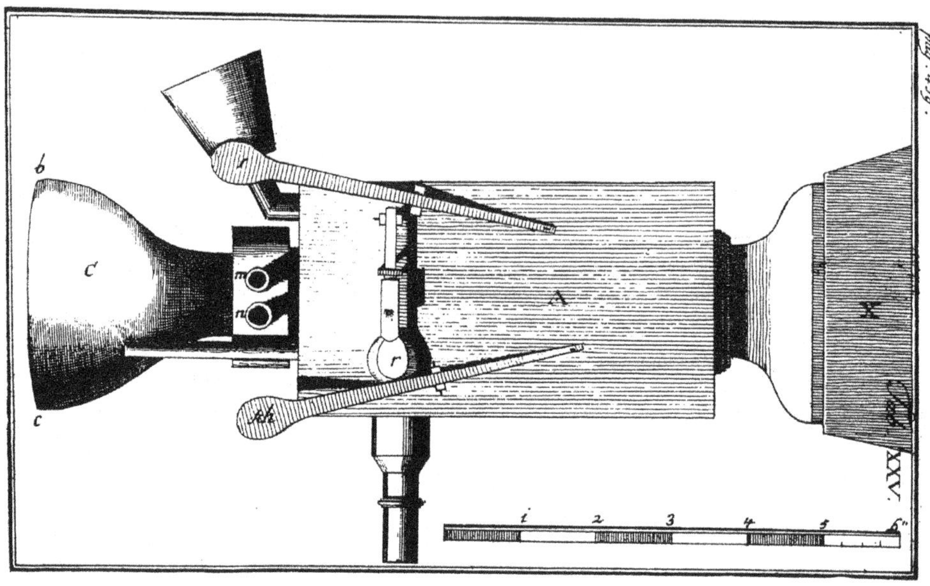

Figure 33.5 Von Kempelen's Sprechende Maschine. (From: Wolfgang von Kempelen, *The mechanism of human speech together with a description of a speaking machine.* Wren: Degen 1791.) A. Box containing an ivory reed. Manipulated by left hand: b-c – mouth; C – rubber resonator; Underneath C - a small auxiliary bellows, not visible. Operated by right elbow; X – Part of the bellows. about 16" long. Operated by fingers of right hand: S – S key. Under it, the S pipe; Sch – Sh key. Next to it the Sh pipe m, n - Nostrils. normally closed by fingers; r – Reed key. When depressed it stops the reed vibrating. The scale is in inches.

> Machine Parlante de M.Faber par M.Gariel
>
> Monsieur Faber has set out to construct a machine which really talks, that is to say produces the sounds and articulations and to arrive at this result he has imitated, at least in a general way, the organs of phonation.

The machine is composed essentially of three parts:

1. the bellows
2. the sound-producing device
3. the articulating device

1. We do not have anything particular to say about the bellows which is intended to send a current of air to the larynx.
2. The device which produces the sound, the larynx, is a reed made of ivory, the length of which can be varied, within certain limits, so as to change the pitch of the sound it produces. It is regrettable that Monsieur Faber did not attempt to use a system of membranous reeds which would have had the advantage of bringing the machine closer to reality.[7]

3. The articulatory device consists of a part which produces the vowels and a part for the consonants. The vowels are produced by the passage of air through openings of various shapes cut in partitions which are successively placed in the path of the current of air by the operation of levers driven by the keys; furthermore there is a special cavity which can be connected to the previous one which is intended to produce the nasal sounds, the connection being made at will by means of a special lever. The consonants are produced by parts which function in a very similar way to the lips, the teeth and the tongue. A special rotor produces the roll of the R. All these parts and organs are set in motion by fourteen keys which are most ingeniously laid out so they can drive the organs with suitable intensity and in the correct order to produce a syllable. Fourteen keys suffice because, with the assistance of auxiliary keys the character of a consonant can be changed from unvoiced to voiced, etc.

   The voice of the machine is, of necessity, monotonous and, it must be added, it is not perfect. Some sounds produce a better effect than others. However, in general one can understand the words and phrases pronounced although one would certainly not think of comparing the sounds it produces with the varied intonations of the human voice. Despite the improvements which need to be made to it, this machine is nevertheless interesting because it clearly shows the mechanism of phonation, which can be reproduced artificially in this way, and which in consequence can be shown to exactly obey the laws of acoustics.

– *Journal de Physique, théoretique et appliquée*, Paris, 1879, pp 274, 275.
(Trans.: M.R.)

My information about Faber's Euphonia is confused because of conflicting dates which I have not been able to resolve. Professor Joseph Faber was born circa 1800 in Freiburg im Breisgau and it was here that he built Euphonia. He later moved to Vienna. Euphonia was first exhibited in 1840. He traveled with it, showing it at the Egyptian Hall in Piccadilly in 1848. Contemporary reports say that Euphonia was far superior to von Kempelen's machine and anything that came before. However, unlike the self-confident von Kempelen, Faber was a fanatical loner who built his machine under the greatest privations. He never became rich nor was he satisfied with his creation and finally he killed himself. He is said to have destroyed Euphonia in desperation (Figures 33.6 A,B).

Surprisingly we find that shortly before this, in 1853, Phineas Taylor Barnum rented Euphonia for $20,000 for a half year ("an exorbitant price," as he says in his memoirs). He showed it for a time in his American Museum. The picture, circa 1873, shows Euphonia as it was exhibited by Barnum.

(A)

(B)

Figure 33.6 a. Faber's Euphonia, from a contemporary poster. (Reproduced in: Alfred Chapuis & Edmond Droz, *Les Automates*, Neuchatel, Éditions du Griffon 1949.) This tantalizing drawing does little to explain how Euphonia actually worked. Part of the mechanism seems to be dedicated to animating the mask. If any reader has more information about Euphonia, perhaps they could share it with us. b. Euphonius, Professor Faber at the controls. (Reproduced in: Jasia Reichardt, *Robots: Fact, Fiction & Prediction*. Thames & Hudson, 1978.)

Chapuis (see bibliography) mentions that in another article (substantially the same as above, in *Magasin pilloresque*) Gariel refers to Faber as an engineer and suggests that perhaps Gariel is referring to Joseph Faber's son and to an improved machine. This would resolve the conflicting dates. If any reader has information about Euphonia's career in America, perhaps he or she can share it with us.

The Voder

In the 1920s, experiments began with electronic speech synthesis. Just as in the 18th century, they started off by trying to synthesize vowel sounds and then progressed to complete speech. In the mid-1930s Bell Telephone Laboratories started a major project called the Voder (Voice Operation DEmonstratoR), intended as an exhibit for the World's Fair which was to be held in New York in 1939 (Figure 33.7 A–C).

The Voder had ten main keys. Each key, connected to a potentiometer, operated a band-pass filter controlling one section of the sound spectrum up to 7500 cycles per second. A wrist switch changed over from the voiced to the unvoiced sound source. Three additional keys triggered the *T-D*, *P-B* and *K-G* plosives. There was a pedal to regulate the pitch so that the Voder could speak with a natural intonation.

Like von Kempelen's Speaking Machine and Faber's Euphonia, the Voder required considerable skill and manual dexterity to operate it. Bell Telephone adopted a methodical but extravagant approach to the problem of operating the machine which is in marked contrast to the lonely efforts of the pioneers.

Bell Telephone first selected 320 New York Telephone Company and AT&T switchboard operators and put them through an aptitude test: basically, sitting them in front of the keyboard and seeing how they got on making a simple sentence. Thirty of the candidates were selected for the instruction course. After a year of individual tuition on ten identical machines they had narrowed the field to a pool of twenty-four girls who could play the Voder keyboard and make it talk almost without thinking about it.[8] They worked in shifts operating the machine at the New York World's Fair and later in the Golden Gate Exposition in San Francisco.

The intelligibility of the Bell Telephone Voder was apparently quite good but to make quite sure everyone could understand it they had a narrator with a microphone who would talk with the machine to provide "cues" and so prepare the listeners for what they were about to hear. The performance, constantly repeated in front of a standing audience, lasted five or six minutes.

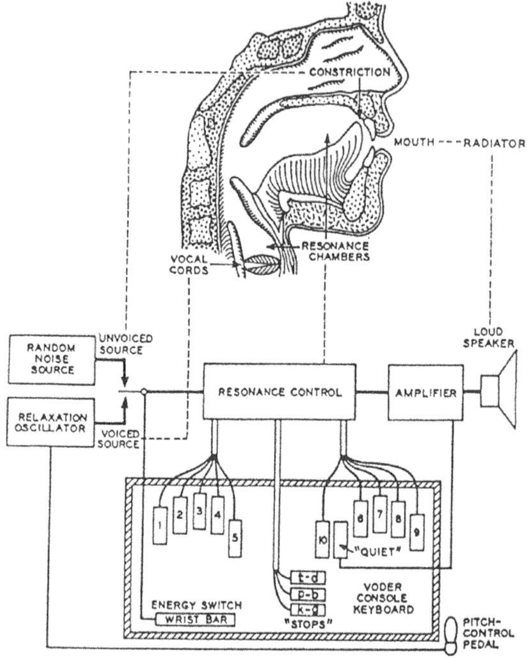

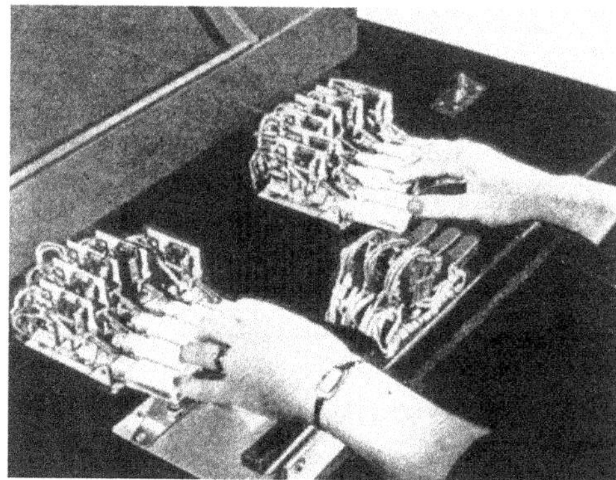

Figure 33.7 a. The essential parts of the Voder. (This and b and c are all from: H. Dudley, R.R. Reisz, S.A.A. Watkins, "A Synthetic Speaker", in *Journal of The Franklin Institute*, Vol. 227, June 1939, No. 6 pp. 739-764. b. The Voder and operator. c. The Voder keyboard.

Part 2: The Talking Machine

While I was voicing some pipes I had made for a computer-controlled organ, I noticed that when they were badly adjusted – playing with too much attack or "chiff" – they sometimes made sounds quite similar to human speech. I wondered what would happen if I replaced all these organ pipes with pipes that were especially designed to make speech sounds and whether this new organ could be taught to speak (Figure 33.8).

With this at the back of my mind, in 1990 I applied for a grant from a local arts authority to make a talking machine. It was something of a shock when I learnt

Figure 33.8 Martin Riches with resonator pipes for the Talking Machine.

that I had secured the grant; I was now under considerable pressure to produce a working mechanical speech synthesizer.

At first I had hoped to make a single mouth with a mobile tongue and lips which would speak all the vowels, semi-vowels, nasals and glides. Those consonants which need air-tight closures, the plosives, *P-B*, *T-D*, and *K-G*, could, if necessary, be spoken by additional instruments. I made a rough, hand-operated model which spoke 12 English vowels. But, being under pressure to succeed, this single mouth approach seemed too risky for a first attempt.[9]

Since I was starting the project in a state of utter ignorance, I followed the strategy of making a separate instrument for every single speech sound. This seemed to be a cautious step-by-step method which might be expected to produce at least partial success. I already had a suitable air supply and control system available – my computer-driven pipe organ – so all that needed to be done was to make the instruments themselves, fit them into the organ in place of the organ pipes and rewrite the software to deal with words instead of music. As it turned out, this approach worked quite well.

This machine has a larger repertoire of vowel sounds than Professor Kratzenstein's five resonators (described in the preceding article on historical talking machines), but although I used a different kind of reed, the vowel sounds are produced in much the same way. In fact, apart from the electrics and electronics, the machine could have been built in the 18th century. They would probably have programmed it using a pinned cylinder like that in a mechanical organ.

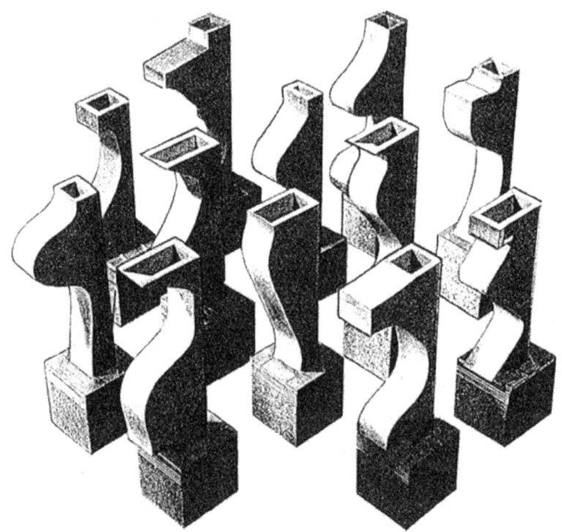

Figure 33.9 Talking machine resonators.

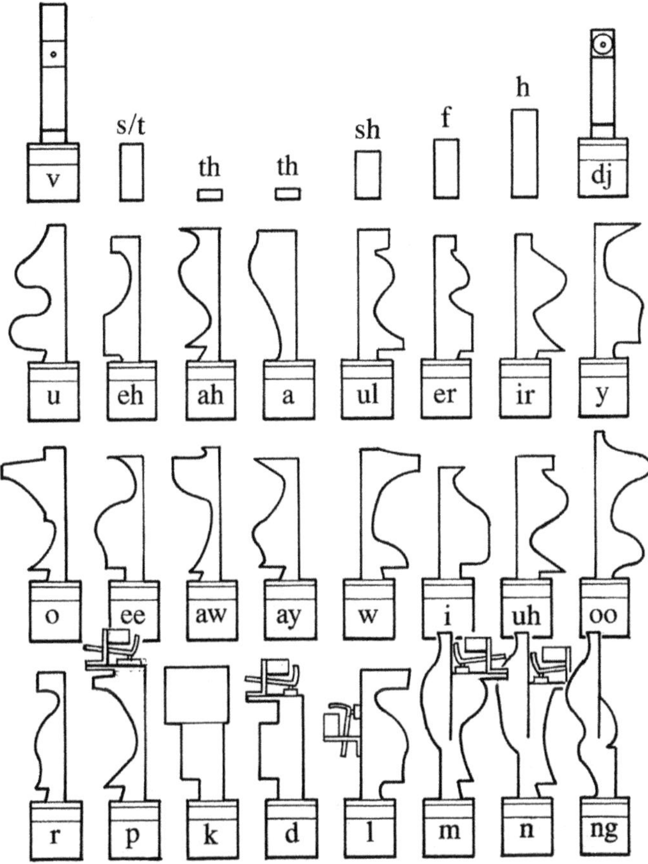

Figure 33.10 The complete set of resonators.

A vowel pipe consists of a simple organ reed corresponding to our vocal cords, and above it a wooden resonator that reproduces the spaces we make inside our mouth when we speak. The resonators are based on measurements of X-ray photographs taken of a person speaking (Figures 33.9, 33.10, and 33.11).

The machine's intelligibility is based on the ear's ability to accept jumps from one sound to the next in much the same way as the eye merges the jumps from one picture to the next when watching a film. I did not follow my "one-sound-per-instrument" principle consistently: a few instruments have an additional valve which performs the function of the lips and changes the sound, as in "m-er". Also, some of the fricatives *F*, *S*, *SR*, *CR*, *T* etc. are produced by combinations of instruments playing together (Figure 33.12).

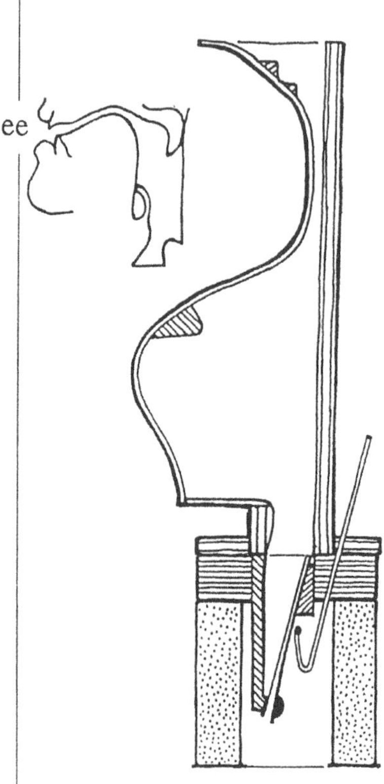

Figure 33.11 The EE pipe reproduces the narrow shape of the human mouth saying EE. On the left beside the pipe diagram: cross section of the vocal tract producing the same vowel EE.

Each instrument consists of two components: a noise maker (special whistles in the case of the fricatives and metal reeds for the rest) and a wooden resonator which filters this noise into the required sound. The resonators resemble the shapes made by the mouth and throat when speaking. Making the resonators required the accumulation and processing of a quantity of anatomical, acoustical and phonetic information. The following methods were used:

1. Reading up the subject and consulting experts. (See bibliography and acknowledgments).
2. Observations using a dentist's mirror and a pair of calipers.
3. X-ray data.
4. Calculation using acoustic formulae and tables.

Figure 33.12 *S/T* pipe. If this pipe is played continuously it will produce an S sound. Played short it will speak a light T. The small piece of metal fixed over one side of the opening changes the sound of the whistle to a hiss. The other fricative pipes are similar, except the *H* which is just an empty resonator.

5. Experiments with an adjustable resonator and a model mouth.
6. Interpolation between existing resonators.
7. Luck; a few resonators which would not make the hoped-for sound turned out to be just right for others.

My basic technique was to take a diagram of a cross-section of the human mouth based on X-ray data, photocopy it up to full size, tack it to two layers of 3-ply, cut them to shape on a bandsaw and use them as two sides of a resonator. The third side was a rectangular piece of plywood while the curvilinear fourth side was made of flexible 1mm. plywood. A few of these resonators worked perfectly first time. Most required a good deal of tweaking and tuning and rebuilding; I accumulated many rejects.

My main problem was that the work was tiring for the ears. It was difficult to remain objective; after a few hearings, it was all too easy to convince one's self that the sounds which emerged were correct. I had to keep the listening sessions very short and consciously try to maintain a positive and patient attitude. It was perhaps particularly frustrating because the resonators were undoubtedly producing human speech sounds – of a sort – and it was easy to imagine that one was confronted by a most intractable human personality. Fortunately, I was able to intersperse the tuning sessions with work on the software.

The resonators had to work together as a set – like the pipes of an organ – so finally I had to tune each reed pipe to the same pitch and balance all them all in loudness and timbre. Building the resonators took several months.

During the early stages of tuning the resonators I had a fascinating session with an elocutionist, Honor Kovacs. She treated the machine exactly like one of her human pupils. For example: she would ask the machine to say AW and I would then play that pipe. "Well now," she would say, "if you could just try moving the tip of your tongue up a little bit and say it again?" I would then quickly change the shape of the pipe using plasticene and she would say "Yes, not bad" or perhaps "No, too much." This session gave me many insights and brought the dry acoustic theory to life for me.

Fortunately. as we saw with Kratzenstein's pipes, it is not necessary to reproduce the complex shapes made by the mouth exactly. The proportionally correct cross-sectional area at each point along the vocal tract is decisive – but the shape of that cross-section is relatively unimportant.[10] The actual shape of the cross-section was largely decided by fabrication techniques and, for me – working in plywood with a circular saw and a band-saw – rectangular sectioned pipes were easiest to build. I had initially experimented with Plasticine (not permanent enough), high density styrofoam (good but equally impermanent), and polyester resin (difficult to change and I find it unpleasant to work with). Some resonators are fitted with internal details to reproduce the velum, teeth and epiglottis. Some are partially lined with felt to increase damping.

The reeds are made of hammered brass sheet, held in position over the spoon-like kelch with a wedge. This is the method used in reed organ pipes. However the rim of the kelch is lined with thin leather making the sound less harsh. Because of my lack of experience in reed-making, not all my reeds have exactly the same timbre. I exploited this deficiency by taking care to match reeds and resonators. For example, the EE resonator would be matched to an EE-like sounding reed. This EE-reed would be unsuitable for the AW resonator because the characteristic AW frequencies would be missing or weak.

What about the consonants? The *S*, F, *SH* and similar sounds – the fricatives – are produced by special hissing whistles. These whistles are prevented from making a full musical whistling sound by small obstructions added to divert the air flow.[11]

Other pipes have "noses" added to them and this enables them to speak the nasals: the *M*, *N* and *NG* sounds. The *M* and *N* pipes have additional valves, the equivalent of opening lips so that they can say "m-er" and "n-er".

The most obstinate sound was *K*. Finally, I had to resort to a percussive click, wood on leathered wood inside a resonator. I later learnt to "suggest" the softer varieties of *K* by using a small pause followed by a release of air. The *P* sound is produced by a soft rubber valve cover which slaps a hole in a resonator making a faint popping noise – the same effect as slapping the rounded lips with the fingers.

The problem with the plosives, *K* and *G*, *P* and *B*, was that in natural speech they require a higher pressure and this is not normally available from the organ blower. Indeed, the blower and its magazine bellows are designed to keep the pressure exactly constant. It would however be possible to instantly raise the pressure by adding a mechanism to close the regulating valve and press heavily down on the bellows.

Many of the other problems with consonants could be solved by the programming of the controlling computer.

Programming

The whistles which produce the unvoiced sounds, *F, T, S, SH, CH* etc., speak fast; the reed pipes for the voiced sounds are slower. Some of the reeds have to be blown for an appreciable amount of time before they start to speak. This applies especially to narrow vowels such as EE and OO which offer more resistance to the flow of air. The machine therefore has to start playing the more reluctant instruments well in advance so that they will start speaking on cue. If it plays too late, there is a gap. If it plays too soon there is an overlap with the previous pipe, producing a curious choric effect. Each pipe has its own program variable to tell the computer how much lead it requires before it speaks.

To help me program a word I ask the computer to produce a graphic display which, simplified, looks like Figure 33.13.

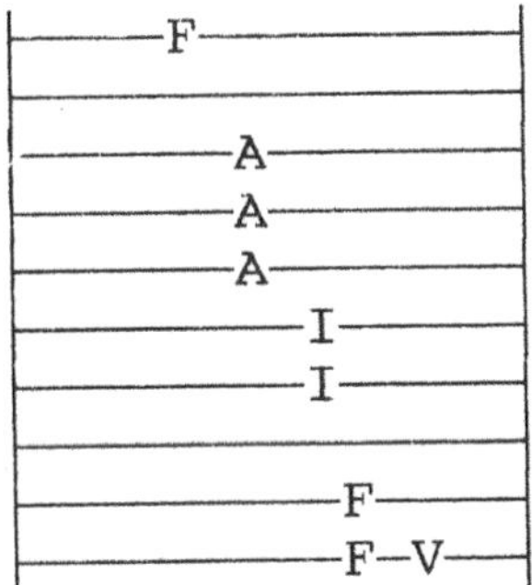

Figure 33.13 Graphic display for the word "five," showing the relative length of the sounds.

This is what "five" looks like. This graphic display helps me visualize the relative length of the sounds. Note the *AI* diphthong, the intentional gaps between the fricatives and the vowels and the simultaneous *F* and *V* sounds at the end. This diagram does not show how much the pipes have to be blown in advance. The computer takes care of that for me automatically.

This diagram is a little reminiscent of a Pianola music roll. Indeed the machine could also be driven mechanically (by a barrel organ or player-piano mechanism); the reason for using a computer is that it is far easier to make changes in the timing if the machine mispronounces a word and, in any case, a computer is far more flexible and reliable than a music-roll reader. To operate the machine the sentence to be spoken is typed into the computer. It then consults its pronunciation dictionary and if it finds the words there it will speak them. The actual pronunciation instructions for the word "five" as shown above would be: F1 1 A3 I2 1 F1 FV 1. Letters indicate pipes, numbers are durations or pauses.

I think that one of the reasons that the great mechanical talking machines of the past took such a very long time to complete – von Kempelen (1791) and Prof. Faber (1830) both worked for over 20 years on their machines – is that they did not have such a readily adaptable method of programming a machine, so they had to learn to play their machines by hand. A further problem: although no tactile sensation is more familiar to us than the feeling of the insides of our mouths, before the discovery of X-rays it must have been extraordinarily difficult to assess the volumes produced inside the mouth while speaking.

My machine always speaks at the same pitch: 220 Hz. – a tenor. I have not yet attempted to provide stress or intonation so the machine would not distinguish between the two sentences "You understand" and "You understand?" But it could be done by modulating the air pressure to change the pitch (and volume). It is also easy to change the emphasis by altering the tempo at which the words are spoken: "Y o u understand". The machine leaves a 1/5 of a second pause between all words and does not speak with mostofthewords joinedtogether aswedoinEnglish. Separating all the words certainly improves intelligibility but it also sounds rather deliberate.

Intelligibility is also dependent on context. For example, it is relatively easy to understand the machine when it is telling the time. We know that we can expect a vocabulary of 30 or so words arranged in a well-known format and rhythm: "It is Six... forty-five... and ten seconds." But this familiarity causes a problem when trying to improve the machine's pronunciation: I know what it is going to say next anyway. If it says "1, 2, 3, 4, 5..." I am going to hear the next word as "six" – even if it says "tigz" (which is, in fact, what it did say for a time until this was pointed out to me). During an elocution session I therefore get the machine to speak its repertoire of words at random. This helps, but does not overcome the problem that I have become thoroughly accustomed to all its mispronunciations and almost invariably recognize all words no matter how poorly they are spoken.[12]

Because of this problem of familiarity I feel that I have now gone as far as I can using my ears alone; criticisms by visitors with good analytical ears still bring improvements. I believe I could make further improvements by using a spectral analysis program to objectively compare the machine with my own voice.

The machine's mother tongue is British English. It can speak some 400 words, can recite the ABC, can count (forever, if necessary) and can speak a few words in other languages. In Japanese it can speak a few polite phrases and can count to 100. It can also count in German. I always maintain that it must be particularly adept at German because this language has more stops and purer vowel sounds than English and this corresponds much better to its style of speaking. My German friends disagree: the problem is that foreign languages are spoken with a British accent; the foreign sounds are missing. For example: it cannot not say "un bon vin" because it has no nasal vowels. Nor – sadly – can it speak Chinese where pitch is essential to meaning.

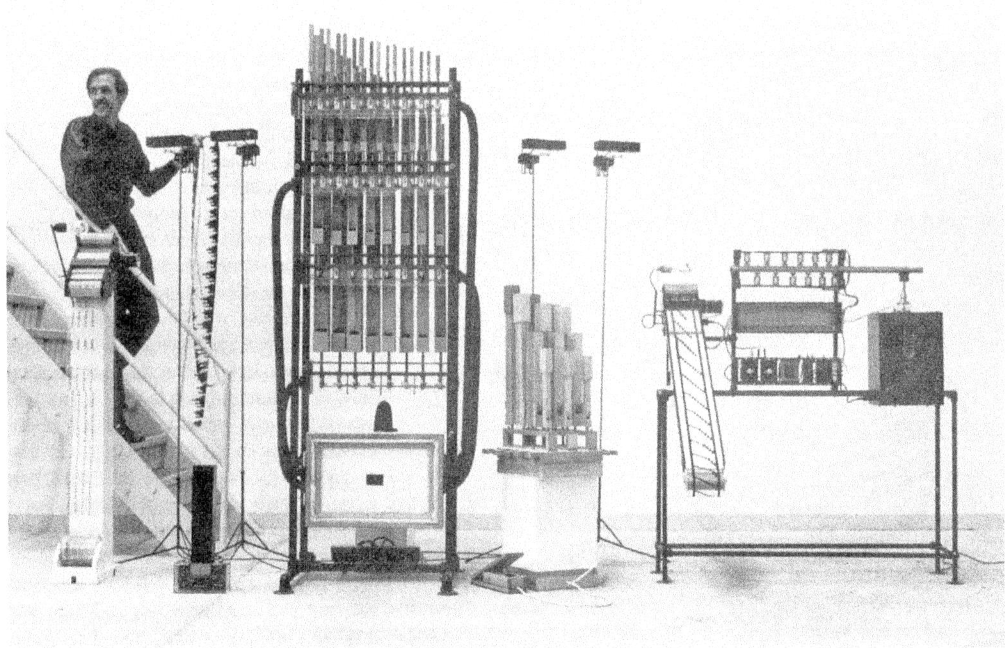

Figure 33.14 Martin Riches with some of his music machines. Left to right: A percussion machine, a mechanical organ, a small blower with 12 pipes, 4 stops and 3 pedals, and a flute-playing machine. The wind chests, blower and frame of the mechanical organ form the basis of the Talking Machine.

It has taken part in many concerts (reciting), and has been exhibited many times in Germany (where it lives), and also in Japan. It is quite entertaining to watch: all the instruments are visible, and the wind chests are transparent so that the valves which control the flow of air can be seen in action. The valves are fitted with LEDs so that even the very fast opening of the T valves can be recognized. It is particularly instructive to follow repetitive speech like counting.

The machine is sometimes exhibited in interactive mode. The public can use the keyboard to make it count and recite the alphabet and various poems and can also input sentences for it speak. The public seems particularly intrigued by its speaking "in stereo": each sound comes from a different place. This effect is of course particularly pronounced when one is standing very close to the machine.

The machine was featured in a radio program "A Talk about Talking" which I made with the American composer Tom Johnson. The talk includes a demonstration of the machine's capabilities and five duets for bassoon and Talking Machine. (The machine counts and recites the names of the notes). When the bassoonist arrived for the recording session I explained and demonstrated the machine for him at some length. When I had finished he nodded vigorously and said "Ja, Ja... So now show me: where is the loudspeaker?"

Acknowledgments

My warmest thanks go to the following for inspiration, information and helping in all sorts of ways:

> Prof. Manfred Krause, Dipl. lng. Folkmar Hein and Dipl. lng. Klaus Hobohm of the Technical University, Berlin; Jasia Reichardt, art organiser and writer, London; Peter Richards and the staff of the Exploratorium, San Francisco; the staff of the Instituut voor fonetische Wetenschap, Amsterdam; Renate Bonn, Berlin and Francisco Flores, Caracas; Prof. Remko Scha of the Institute for Computer linguistics, Amsterdam; Hanns Zischler, Berlin; George Sassoon, Loch Blue, Scotland.

Patient and critical listeners have included Yumiko Urae, Alex Veness, James Tenney and elocutionist Honor Kovacs.

The Talking Machine project was made possible by a grant from the Senator for Cultural Affairs, Berlin.

Notes

1 Languages vary; English has neither the click sounds of Xhosa and !Kung nor the nasal vowels of French.

2 But perfectly intelligible British English can be spoken while clenching a pipe between the teeth and not moving the jaws at all.

3 Men generally have lower pitched voices than women and children because their vocal cords are longer. The vocal cords have other functions. When you are about to lift a heavy weight you inflate your lungs and close the vocal cords to "hold your breath". This is braces the rib cage and helps transfer the load from the arms and shoulders to the spine and stomach muscles. The vocal cords also provide a last line of defense below the epiglottis to prevent food "going down the wrong way" into the lungs.

4 Sir Richard Paget used a similar technique, blowing a rubber reed, held between his thumbs, which enabled him to "talk with his hands". See bibliography.

5 Japanese and Chinese have a sound where the tongue starts from a point exactly between the English L and R positions. They often use this convenient central position for both Land R when speaking English. Thus, for English-speaking listeners, their Ls may sound like Rs and their Rs like Ls.

6 There is a superb modern version of Kratzenstein's vowel resonators at the Exploratorium in San Francisco called "Vocal Vowels". It was designed and built by Stephen Green using modern data derived from X-rays of the vocal tract.

7 Fr. *anches membraneuses*, presumably rubber reeds.

8 The training period was preceded by intensive study to optimize the controls, to develop a keyboard technique and to decide on the best methods of instruction. Each operator had six half-hour sessions a day, with lessons given every second day. This was found to be about as much as anyone could stand. Various refinements were added to the machines in the course of the training period as a result of experience gained.

9 I have in the meantime returned to the idea of a single fully motorized mouth. Its finest achievement so far is to speak, quite clearly, the vowels-only sentence "you weigh our air. We owe you our awe", (addressed to Torricelli, the inventor of the barometer). Examples of other long vowel-only sentences will be gratefully received. I am currently working on the consonants.

10 Having once built a square-sectioned flute I would venture to say that a flute is recognizable as a flute whether the tube is round or square.

11 The whistles are based on Venezuelan duck-calls. This was a fortunate misunderstanding: I had asked a friend in Caracas for a duck-call. I expected to be sent something with a reed but -to my surprise - received a paper bag full of little whistles labeled ~Pitos para patos~: whistles for ducks. These are quite different from the familiar mallard duck-calls-they are just pairs of dished aluminum washers glued together. Presumably they attract non-quacking varieties of duck such as pintail or widgeon; my local mallards ignore them. The dimensions are not critical; I was told that red-blooded Venezuelan males hand-craft their own whistles out of the metal ends of spent shotgun cartridges.

The whistles are based on Venezuelan duck-calls. This was a fortunate misunderstanding: I had asked a friend in Caracas for a duck-call. I expected to be sent something with a reed but – to my surprise – received a paper bag full of little whistles labeled "Pitos para patos": whistles for ducks. These are quite different from the familiar mallard duck-calls – they are just pairs of dished aluminum washers glued together. Presumably they attract non-quacking varieties of duck such as pintail or widgeon; my local mallards ignore them. The dimensions are not critical; I was told that red-blooded Venezuelan males hand-craft their own whistles out of the metal ends of spent shotgun cartridges.

12 This is the situation of the mother of a very young child – the only person who understands what her offspring is trying to say: "Gimimohnook." "Oh, you would like me to give you some more milk, would you?" "Yurg." Speech recognition is a far more complex skill than speech production; the listener does most of the work.

Bibliography

Richard Paget Bart: *Human Speech.* Routledge & Keagan Paul, London, 1930. (Reprinted 1963). A lively description of the mechanism of speech and resonance, and of his many experiments with resonators.

Alfred Chapuis & Edmond Droz: *Les Automates*, Éditions du Griffon, Neuchatel, 1949. Authoritative general history of automata, French. Some technical detail. Numerous illustrations, of mixed quality. The section "Androïdes Parlantes" pp. 329–334, mentions the works of Kratzenstein, Kempelen and Faber and a talking doll.

Peter B. Denes & Elliot N. Pinson: *The Speech Chain: The Physics and Biology of Spoken Language.* Bell Telephone Laboratories, Anchor Books. Anchor Press / Doubleday. Garden City, New York, 1973. Deals with acoustics, speaking, hearing and speech recognition. A good introduction.

H. Dudley, R. R. Reisz, & S. A. A. Watkins: "A Synthetic Speaker." In *Journal of the Franklin Institute*, Vol. 227, No. 6, pp. 739–764, June 1939. All about the Voder.

Gunnar Fant: *Acoustic Theory of speech production.* Mouton, Den Haag, 1960. X-ray data and acoustic theory. Provided most of the essential technical data for making resonators.

James l. Flanagan: *Speech Analysis, Synthesis & Perception.* Springer Verlag, Berlin, Heidelberg, NY, 1965. State of the art 1965. A very useful technical reference and good history.

James l. Flanagan: "The Synthesis of Speech." In *Scientific American*, Vol. 226, No. 2, pp. 48–58, February, 1972. State of the art 1972, for the general reader. Some history. Well illustrated.

C. Gimson: *An Introduction to the Pronunciation of English.* Edward Arnold, London, 1962, 3rd Ed., 1980. Detailed. Includes a history.

H. Kaplan: *Anatomy and Physiology of Speech.* McGraw-Hili, New York NY, 1960. Detailed anatomy of the vocal tract.

Wolfgang von Kempelen (1734–1804): *Ober den Mechanismus der menschlichen Sprache nebst Beschreibung einer sprechenden Maschine (The mechanism of human speech together with a description of a speaking machine).* Degen, Wien, 1791. Facsimile reprint with an introduction by H. E. Brekle and W. Wildgen: Stuttgart: Fromann-holzboog 1970. Worth looking at just for the fine technical drawings.

Kopp Potter & H.C. Green: *Visible Speech*, D. Van Nostrand Co., New York NY. 1947. How to read spectrograms. Good as background.

Jasia Reichardt: *Robots. Fact, Fiction & Prediction.* Thames & Hudson, London, 1978. Excellent history, illustrations and narrative. A continuing inspiration.

Radio Shack: "Technical Data Sheet for the SP0256-AL2 Speech Processor" (early 80s). An electronic equivalent of the Talking Machine. Provides a set of 64 speech sounds, including long and short versions of the same sound to be concatenated into words. Many useful practical hints.

Appendix

An appendix listing the speech sounds of standard British English appeared with the earlier edition of this article. This appendix can be accessed through the online archive posting of the complete article at https://archive.org/details/emi_archive/.

Related articles

"Incantors" by Qubais Reed Ghazala, in *Experimental Musical Instruments* Vol. 8 #4, also addressed the topic of speech synthesis. It can be accessed through the Internet Archive at https://archive.org/details/emi_archive/.

34

Beyond the Shaker

Experimental Instruments and the New Educational Initiatives

John Bertles

This article originally appeared in *Experimental Musical Instruments* Volume 14 #4, June 1999.

Simple musical instruments have long been a staple of the elementary school classroom. Many teachers have a class unit where the students create easy instruments, such as shakers or rubber-band boxes. But student-built musical instruments can go much further than those simple old warhorses. Furthermore, the evolving state of education now presents a more positive atmosphere for instrument builders to work in the classroom. This article will explore some of the new initiatives in education and the implications for teaching artists, as well as some of the theory and practice of building instruments in the classroom – admittedly from the viewpoint of an elementary school teaching artist in the New York City area.

Arts Education in the Late 90s

Arts education in the late 90s seems to be dominated by three major initiatives: The return of arts education in the schools; the new educational standards; and a cross-curricular approach to learning, with an emphasis on literacy.

The late 80s and early 90s were dark days for the arts in American education. Black Monday; the recession leading to a subsequent decimation of school budgets; and a relentless assault on the arts in America by some prominent politicians and other public figures combined to virtually eliminate the arts in many schools. Certainly the inner-city schools were the hardest hit, with many schools having no program whatsoever, but even in the most posh suburbs, classes in the arts were sustained only by parents and PTAs raising private funds to hire teaching artists and programs.

For those of us who were working to bring arts programs to schools at that time it seemed as though art was dead in America, that we were bringing up a generation of citizens who had had no experience with art in any form, or perhaps worse, an extremely superficial exposure. And yet, here in New York, the state mandates that

Figure 34.1 The author and the kids.

elementary students spend 15% of their time on the "arts." How that was interpreted was up to the city, district, and school. In one school I visited I saw an auditorium full of kids watching a Disney film. A disgusted assistant principal told me that that was part of their state-mandated 15%. The school had no funds to hire teachers or visiting artists, and the inner-city neighborhood couldn't raise funds to bring in outside arts programs.

It is truly tragic, but the kids who were in elementary schools during this period will really be the "generation without art," with all that that implies. A high-school music teacher recently complained to me that although he had enough instruments to create a wind band (thanks to new funding coming in), the students who came to him had not even the slightest knowledge of the rudiments of music, such as reading music or even knowing what the notes were.

However, there are indications that things are changing.

The major impetus for change has been the booming economy of the mid to late 90s. The Wall Street-fueled boom has finally started to trickle funds down into the coffers of education departments and schools. Board of Education arts coordinators, who for years had watched arts education, become the almost exclusive province of large arts institutions (who have their own agenda, after all), are suddenly finding that they have funding to hire new teachers, buy or rebuild arts supplies and materials, book programs, and otherwise revitalize their moribund curriculum.

Another reason for this change was a series of medical studies which showed a clear correlation between the arts and brain development. Most telling were the reports that showed that children who played instruments at an early age turned out to be better at math and science. The connections between math, music and science seem self-evident, but had not previously been documented in a scientific fashion. What was discovered was that the "pathways" in the brain between the areas which are the centers of math, music, and science were more developed – there were millions more neuronic connections in the brain of the child who had taken music. Studies like these began to creep through the sluggish bloodstream of the educational bureaucracies around the mid-90s.

Curricular Connections

A third force for change has been the cross-curricular initiatives (especially targeting literacy) which have finally begun to take a solid foothold in education, especially in the new Educational Standards (more about those later). When I was in school in the 60s and 70s, it seems to me that all of the subjects were taught as if they were exclusive from each other, with no connections between them, or very limited connections. For example, studies in history took no notice of climatology, and math and science seemed to be two separate realms, only distantly useful to each other.

Teaching across the curriculum and making connections to other subjects is nothing new in education. In various guises (the Whole Language concept of the late 80s is one), it has been in and out of favor for the last decade. Only in the last few years has it really begun to take hold.

For teaching artists, it is a two-edged sword. Part of the devil's bargain that the arts have had to make to get themselves back into the schools is an abandonment of "art-for-art's-sake" and an adoption of "teach-to-the-core-curricula-through-the-arts." For so many years as the arts were under siege, its detractors accused the arts of being elitist, irrelevant to the three Rs (writing, reading, 'rithmatic), not able to contribute to the national economy, not able to help young people get jobs, etc. One result of this assault was that to get funded, the arts were forced to show that they had relevance to the core curricula of reading and writing, math, science, and cultural studies.

I'm not going to get into a discussion here of the merits of "art-for-art's-sake" versus "teaching the core curricula through the arts", other than to say that there is room for both in an ideal world; but in the current funding atmosphere the latter will get funded and the former will not – almost exclusively. For the moment, anyway, it is best for those who wish to build instruments in the classrooms to make the connections (we'll take a look at some possibilities further in the article). Perhaps later, if the boom continues, we can return to teaching the arts for itself.

Educational Standards

The new Educational Standards are also taking a cross-curricular stand, giving a boost to those of us who have been using this method for years.

Standards have been part of the educational scene for as long as there have been bureaucracies, but the last few years have seen some major changes. Educational standards try to set goals for students and teachers to reach so that efforts to overhaul education can have a unified goal rather than a scattered one. To appeal to the wide variety of cultural, economic, and life-style diversity that makes up our US, these standards have previously been rather vague, not to say foggy. In the past they were also watered-down by politics and grandstanding to the point where they were unusable.

But starting about five years ago, there was a new drive to put in place a set of specific, understandable, and relevant standards. In my area, these standards have been put in place in a staggered fashion. New Jersey started their new standards several years ago. New York State has the new standards on art and literacy in place this year, with math, science, and cultural studies set in previous years. Even the federal government has gotten in the act, with a set of National Standards that is supposed to be coming out within the next few years (you can bet there's going to be a lot of anguished polemical hand-wringing in Congress over that one).

A lot of people – myself included – have a philosophical problem with standards in art. How can you judge the arts? It's easy enough to say that we all know bad art when we see it; but is that really a good way to judge? After all, we have our own biases, cultural and economical and political and religious, etc. . . . But the reality of the matter is that the people who have control of the money in education are not really interested in those matters. They want to know exactly what benefit arts education will have on children so that they can sell their programs by showing the grant-givers that on this pie chart it shows that the students have a 15.7 percent increase in attention-span and a 17.7 percent increase in attendance when arts programs are in their classroom blah blah blah. . . Arts educators today have to swallow hard, grit their teeth, and try to work within the system. One positive thing that has come out of the establishment of art standards is that there is now a solid basis for assessing arts programs – as long as the assessment is done with integrity and intelligence. One-size-fits-all assessments do not work with the arts.

One thing that the new standards seem to have is an emphasis on literacy – that is reading, writing, comprehension, and logic. Even math problems have been put into word form (which is perhaps better for real-life applications). Science and cultural studies are also expected to have bearing on literacy. Finally, the arts as well must have relevance to literacy. And here is where we get to the real meat of this article (after a rather long-winded introduction). Experimental musical instruments are ideal for making cross-curricular connections in all kinds of directions, including math and science, literacy, and cultural studies. In the next section, we will examine some of those connections and how they can be made.

Figure 34.2 Skip La Plante and Carina Piaggio join John Bertles on classroom-buildable wind instruments.

Making the Connection – Experimental Instruments Across the Curriculum

The study of mankind is a constant and ever-expanding web, nodes of knowledge connecting not only across the globe but across time. Students can 'surf' these connections by thinking on a deeper level, moving away from superficial knowledge and searching for catalytic factors that move in widening ripples away from the source. This sounds like gobbledygook, but actually it is just a statement of a philosophical goal for curricular connections. Using these connections, students can discover not just a concrete fact, like a historical migration of a tribe, but the range of affecting factors, such as climatological causes (perhaps a drought caused the migration); cultural repercussions (what happened when the tribe moved into the territory of another – war or integration? What happened to the arts? Were they mixed to form a new synthesis?). The students should then be able to draw conclusions (how does this compare to what has happened in history; what are the implications for us in our world?) and use those conclusions to further their understanding of the world. Let's use an example to illustrate the process.

The World According to Finger Pianos

Say a class is going to build finger pianos with me. I build my finger pianos from a small chunk of plywood with bobby pins broken in half stapled onto the wood. Each pin is stapled in a different place, thereby producing different pitches. Depending on their ages, much of the work will be done in the classroom by the students, except the cutting of the wood and the stapling.

Already we have a science unit right there. The reason the pins are producing sound is because of vibrations. Again, depending on the age of the class, we could go right off and explore the world of sound waves and vibrations. Another natural step would be an examination of energy transfer that causes the vibrations – take it right across the electromagnetic spectrum – sound, radiation, heat, the various colors of light. The next step would be to examine why the pins make different pitches – because the length of the vibrating pin determines the pitch.

As we build the finger pianos, we have to bend the pins up to let them vibrate freely. Another effect of bending the pins is to change the pitch as we change their rigidity. In general, the more bent the pin, the higher the pitch.

So we have already determined two basic ways to change pitch: length of the vibrating object, and rigidity of or tension on the vibrating object (of course it all comes down to the speed of the wave, but we are talking about observations that we can get from the finger pianos).

The next step is to make the finger pianos louder. Rather than use the traditional methods of building a sound box or resonator onto the bottom of the instruments I have the students explore the classroom as resonator. They go around the class finding out which surface makes their instrument louder. We make a list of what makes the sound a lot louder, a little louder, or not at all louder. After a while we take a look at the list to see what worked best. Of course, the things that have a thin rigid surface with air adjacent to it are going to make the best sound, like the glass on the door, or a hamster tank, or a corrugated cardboard box. The best of all is a big Styrofoam box.

In this experiment, the students have determined the ideal way to create a resonator. From here, it is possible to discover more interesting things about resonators, including why string instruments have boxes on them, or why it sounds so good to sing in the bathroom.

The next step is to look at the cultural implications of the finger piano. One way to do this is by resources. Our resources were wood and metal. People around the world build these instruments from those materials. But what if they didn't have metal? What could they use? Were these a people who had the resources around to discover how to make metal – the ore deposits, the proper fire, and forge-making materials like brick and hardwood and a bellows – or did they get their metal by trade – and from whom? What were the trade routes that were followed, and how did exposure other cultures influence them? (Think of how Hindu religion travelled to Indonesia, followed by Islam – through sea-borne trade routes that stretched from the Middle East down to India and from there down to the curve of the Indonesian archipelago).

Considering that the finger piano is extremely popular in Africa, we would next look at some of the names for the finger piano there, the kalimba, sensei, mbira. Why are there so many different names for the same instrument? This could lead to a discussion of languages in Africa. There are thousands of languages and dialects. What does that tell us about the social structure there, and the history of Africa?

We can look at where the finger piano has gone. Some of the Caribbean countries have taken the finger piano to an extreme, especially in the rumba box – basically a gargantuan finger piano on steroids. What does the presence of finger pianos in the Caribbean mean? How and why did the finger piano travel there, and who brought it?

Math connections could be made by figuring out a way to create a non-traditional notation. For example, if a finger piano has six notes:

1155665_
4433221_
5544332_
5544332_
1155665_
4433221_

The above is "Twinkle Twinkle Little Star". Of course, who knows what pitches will actually come out of those little pins, but the rhythm is there and so is the pitch contour. For really young kids, this is a great number recognition game. For older kids, it is an introduction to numerical matrices. Taking it further, using this simple notation (perhaps augmenting it with a simple rhythm system), students could create their own compositions and have them performed.

Literacy connections could also be engineered in. Each of the cultural concepts that we touched on could be the subject of a short research paper. One of the cornerstones of third and fourth grade education in New York City is the study of cultures, and this finger piano unit would be ideal to complement this portion of the curriculum.

Furthermore, the above study satisfies many of the Educational Standards. To be specific, it satisfies the following New York State Learning Standards for the Arts:

Standard I – Creating, Performing and Participating in the Arts (elementary level)
Music

1a. create short pieces consisting of sounds from a variety of traditional and nontraditional sound sources
2b. sing songs and play instruments, maintaining tone quality, pitch, rhythm, tempo, and dynamics; perform the music expressively; and sing or play simple repeated patterns with familiar songs, rounds, partner songs and harmonizing parts
1e. identify and use, in individual and group experiences, some of the roles, processes, and actions used in performing and composing music of their own and others.

So we made quite an excursion of exploration. And it all started with building the humble finger piano. But this is just one example of the possibilities of cross-curricular connections.

Making Connections – With Integrity

For the last eleven years I have been using experimental instruments (mostly made from recycled materials) in classrooms all up and down the East Coast. I have found that just about any kind of connection can be made, as long as it is done with integrity.

Following are some possibilities that experimental instruments offer:

Environmental Studies

When I started teaching in the late 80s, there was a big drive to rebuild the moribund recycling effort. Schools used these programs as a way to raise the environmental consciousness of the children, and hopefully through them, their parents. With recycling now firmly entrenched in the northeast, the demand for this kind of program has dropped, although around Earth Day it picks up again.

A typical program using an environmental emphasis might begin with an introductory assembly program to show students some sample instruments, get across concepts in recycling and reusing, as well as basic concepts of sound, acoustics, and vibration. This is followed by in-class workshops in which the students actually build simple instruments. The students are given a list of possible materials to collect, that might include (depending on the age) cardboard tubes, cans, plastic bottles, boxes, rubber bands, bobby pins. short metal pipes, and so on. Then in the workshop, the class builds their instruments in an orgy of experimentation. The advantages of this kind of workshop lie in the fact that the kids have collected the materials themselves, experimented with the concepts and realities of sound and vibration, and created their own orchestra of instruments. This kind of hands-on approach is infinitely better than just showing students instruments and explaining how they work.

Math

Music and math are so closely interrelated that you simply cannot have one without the other. Simple instruments can be useful in many kinds of math-related activities. For example, elementary school teachers often complain to me that their kids don't "get" fractions. My feeling about this is that teaching fractions as a system without relation to real-life applications is basically worthless. It is vital to show how fractions relate to the real world, and music is one example. Musical notes – whole notes, half notes, quarter notes, and so on – are of course fraction-based. A class, using their instruments that they have constructed can create and perform a piece based on the fraction values of musical notes, thus showing one real-life application of fractions.

Another math and music connection is in base numbers and matrices. Our standard base number is base 10 – because we have ten fingers. In other words, once we have reached 10, we start over, with 10+1 (11), 10+2 (12), and so on. But some of our systems are not base 10, for example, time is measured in base 60 (60 seconds, 60 minutes, etc.). Music is often in base 4 *(4/4* time) or other base numbers – base 6, base 5, and so on. Using the concept of base 8 (eight eighth notes), students can create rhythms by using a matrix:

Each box in the matrix represents an eighth note. Boxes with something in them represent a played note, empty boxes a rest. So, assuming we are using shakers and horns, a simple ostinato might be:

S	S	H		S	SS	H	

Two 8th-note shakes, one horn, one rest, one shake, two 16th-note shakes, one 8th horn, one rest.

Note the flexibility of this system – it is possible to subdivide the 8ths to 16th (as in the sixth box) or even to 16th note triplets.

One of the complaints about using simple instruments in the classroom is that elementary school students do not know musical notation. By using a simple system such as the rhythm matrix it is possible to create useful ostinatos. But what about pitch?

Another way to show pitch is by using bar graphs. Bar graphs are part of the elementary school curriculum starting usually around grade 3. Here is "Old McDonald" using a bar graph:

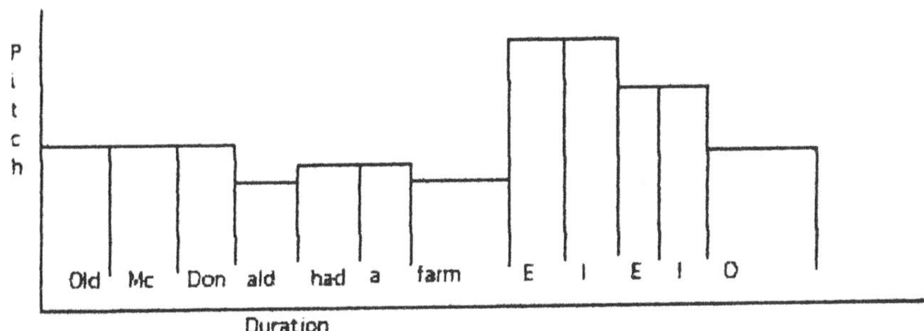

This simple form of notation can solve the problem of pitch. Note that the thickness of each bar represents the duration of the note.

Depending on the kind of simple instrument you build in the classroom, exact pitches may or may not exist, but at least you can create a "pitch contour" using this kind of notation. Using the simple notational ideas above, it is possible for students to create and compose musical pieces using their simple instruments, and in the process, make connections to math.

Science

Creating musical instruments has so much to do with science that it seems to be duh-simple. After all, energy transfer, lengths of vibrating objects, waves, and so forth are just part of the process of building instruments.

The science of acoustics has a major translation problem, however, and that is that acoustics basically deals with frequency of waves – a subject that is guaranteed to glaze over the eyes of most elementary school kids. However, the observable phenomena of acoustics are just perfect for students of that age. These phenomena can be boiled down to the following:

Pitch:
longer = lower and shorter = higher
looser = lower and tighter = higher
less dense = lower and denser = higher

Volume:
Strings need a resonator to sound louder
Winds (except those with toneholes) sound louder with a funnel

Of course, this is a vastly simplified set of observations and missing many nuances. But all these are easily shown with simple musical instruments:

Pitch:
Buzz your lips through a long tube and the pitch is lower; shorter is higher
Pluck a loose string and the pitch is lower; tighter is higher
Hit a piece of softwood (maple) and the pitch is lower; hardwood (cherry) is higher

Volume:
Play a rubber band stretched across the fingers and the sound is soft; wrap it around a box and the sound is louder
Buzz your lips through a cylindrical tube and the sound is softer; buzz through a conical bore (funnel shape) and the sound is louder

These are the kinds of things that will appeal to an elementary school student and are easily done using simple instruments. Furthermore, when that elementary student moves on and takes physics in high school or middle school, those observations will continue to help them understand the more difficult concepts introduced there.

Following is a concrete example of a simple instrument that also relates to science, math and music:

Diatonic Straw panpipes: Cut a piece of straw, block off the bottom with your finger and blow over the top. Check the resulting pitch on a piano and trim the straw to the desired pitch. Use a hot glue gun to put a dab of glue into the bottom of the straw to close the end. Cut other straws by using the whole step/half step ratio to get the correct pitches for a diatonic scale. Following are possible straw lengths to create a straw panpipe:

7 inches, 6 and 6/16ths, 5 and 12/16ths, 5 and 8/16ths, 4 and 14/16ths, 4 and 5/16ths, 3 and 12/16ths, 3 and 8/16ths

I started with 7 inches merely because typical straws do not go much longer than that. Place the cut and hot-glued straws on a piece of masking tape with the top ends even (for easier blowing) and you have your straw panpipe.

Another benefit of this instrument is that it gives students a way to clearly see the whole step/half step construction of a scale in a visual rather than aural way.

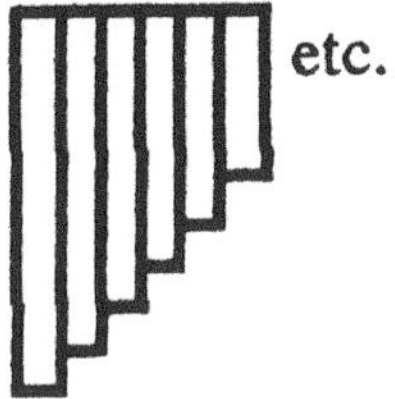

Cultural Studies

At several points in elementary school curricula, there are opportunities to study different cultures, usually first as communities, and then several grades later as distinct cultures. Simple classroom instruments can help give to the students a deeper understanding of how and why other cultures function.

A simple way of looking at the art of another culture would be to listen to music from that culture and build similar instruments. But to take it further would be to examine the resources that a particular culture has as reflected by what they build their instruments out of. For example, a culture that lives in a rain forest quite probably would build some instruments out of that all-purpose plant – the bamboo. Another culture living in a semi-arid place might be prone to building instruments from clay, or from animal remains (bones, horns, etc.).

Another way to look at cultures is by mere orchestration. For example, two of the major musical styles of the Caribbean are salsa and merengue. If you trace the roots of the musical instruments used in these styles, you find the three-fold roots of much of the Latin Caribbean cultures – European, African, and Native American (such as the Taino Indians of Puerto Rico). Keyboards, brass, and guitars come from Europe, drums, maracas, guiros, and so on from Africa and Native American cultures. Right there you have a lesson on how the Europeans came into the new world and introduced African slaves. The resulting mix is a large factor in the shaping of Latin American culture.

One thing that never fails to impress me is that through history, when cultures merge (through invasion, famine, disaster, etc.), the military and political organization of one tends to annihilate the other; the arts of each usually meld to form a new kind of art.

Literacy

When I first started teaching 11 years ago, the first cultural institution that I worked with was dedicated to helping students read better through the arts. This approach has over the years been embraced by many, many other groups. There is such a huge push on literacy these days that it seems to overshadow all the other core curricula.

There are many ways to further literacy using classroom instruments, but one of my favorite approaches is to tell stories by using the sounds of instruments.

In a typical such program, in the first session, the students are introduced to the concepts of building instruments from trash. The second session (and perhaps part of a third) are devoted to actually building the instruments. In the third session, we also make a catalog of sounds of the musical instruments: "thump", "clang", "boing", "shake", etc. Then we see what those sounds could represent. For example thumping could be sound of elephants stampeding, or simple footsteps, or thunder. Shaking sounds could be the sound of a rattlesnake, rain. And so forth.

At that point the students will write a story using the sounds of those musical instruments to help to tell the story. In the next session, some of the students would read their stories as the rest of the class performs the sounds with their instruments. All of this could eventually lead to a stage performance where the class could demonstrate their stories and instruments.

The real benefit of this is that many students get very excited when they find they can integrate music into their stories. I feel that so many kids are stimulated by television and movies where many arts are folded into the storylines – music, visual, acting, and so forth – that they find their own written stories to be somewhat dull.

I have had some fantastic results from these workshops. I have had teachers tell me that some students have jumped to write things (and volunteer to read them aloud!) who have never written enthusiastically before. I feel that this is something that the arts offer – a way to reach kids who just don't react to rote learning. Maybe they are not being stimulated in the way that they need.

Conclusions

The process of building musical instruments is one in which there is a confluence of ideas. To build the instrument, you must have materials. You must have knowledge of how sound works, and how instruments work. You have to have the forethought to design and plan your instrument. You have to have the tools to build your instrument. A certain amount of 'spirit' (for lack of a better word) goes into the building of the instrument. Finally, you must have the ambition and love and ability to play that instrument.

I think of education the same way. The same kinds of processes that go into building a musical instrument also go into building a student and citizen. We, as teachers, can help that building process, and hope that we can impart some of our fears and wishes. But in the final analysis it is only the student's ambition and love and ability that will truly shape the instrument that they will become.

Related articles and resources

Bash the Trash is the organization through which author John Bertles has continued to carry on the work described in this article. Website: https://www.bashthetrash.com/.

Other articles pertaining to children's instruments, instruments in the classroom, and pedagogy which appeared in *Experimental Musical Instruments* are listed below. These articles can be accessed through the Internet Archive at https://archive.org/details/emi_archive/.

"Teaching with Homemade Instruments: The Work of Robin Goodfellow" by Bart Hopkin, in EMI Vol. 2 #1

A series of seven articles by Robin Goodfellow, appearing in each issue of EMI from Vol. 13 #1 through Vol. 14 #3

"A Bibliography for Available-Material Instrument Making: With an Emphasis on Children's books and Teaching Materials" by Tony Pizzo, in EMI Vol. 2 #1

"Pedagogy, Santa Fe Research: Some of Their Work" by Marcia Mikulak, in EMI Vol. 3 #3

"A Children's Instruments Workshop" by Bob Philips, in EMI Vol. 4 #4

"John Maluda's Instruments for the Montessori Classroom" by Bart Hopkin & John Maluda, in EMI Vol. 5 #3

"Experimental Musicians: The Next Generation" by Joan Epstein, in EMI Vol. 5 #5

"Two Generations of *Experimental Musical Instruments*": Tilman Kuntzel and Margrit Kuntzel-Hansen, in EMI Vol. 12 #1

References to classroom instruments and children's instruments making also appear in the *Experimental Musical Instruments* Letters section in EMI Vol. 4 #5, Vol 12 #3, and Vol 14 #4

35

The Photosonic Disk

Jacques Dudon

This article originally appeared in *Experimental Musical Instruments* Volume 14 #4, June 1999.

A Brief History

In 1972, looking at the ceiling fan at my singing teacher's home in Benares,[1] I happened to have a strange idea: "what would happen if I fixed a ring on the blades, pierced with holes, in order to pulse the light sent to a photoelectric cell behind it, and plugged into the stereo?" Surely, I thought, if I could manage to space the holes at different regular intervals chosen to produce the right chords, that machine could be used as a tambura – that is, as an instrument for drone accompaniment. Then, I asked some friend to send me a photoelectric cell, which never came, and this is why this "optical tambura" never saw the light before I came back to France. It wasn't until 12 years later, on the 22nd of February1984, that I was able to experience the idea. What did I do during 12 years? Collect butterflies? No, I simply made other kinds of instruments, playing with water – but that's another story.[2]

So, on this particular day in February 1984, I was thinking about making an instrument that would be able to sing vowels – and then came this idea again – but this time I had a photocell. I rapidly cut some holes in a small cardboard disk attached to a tiny DC motor ... which instantly sang a few notes with a melodious, harmonica-like sound. Oh, the happiness that awakes with simple things!

Actually it didn't work at once in imitating the human voice, but on the other hand, after sketching a few disks, I was able to produce quite a good recreation of the sound of an Indian tambura. Anyway, from that day my interest went more towards creating a completely experimental instrument, and this is how it turned out to be. At first it took the shape of a platform, rapidly assembled with "Meccano" ("Erector Set") parts, able to receive interchangeable disks, up to four at a time, turning at various speeds. I called it the lumiphone (Figure 35.1, with the same machine seen in action with its cardboard disks in Figure 35.2).

After creating different versions of it and enjoying it more and more, I went to the patent office, proud of my discovery. I found there that electro-optical instruments had been already invented, probably since the use of optical sound tracks in

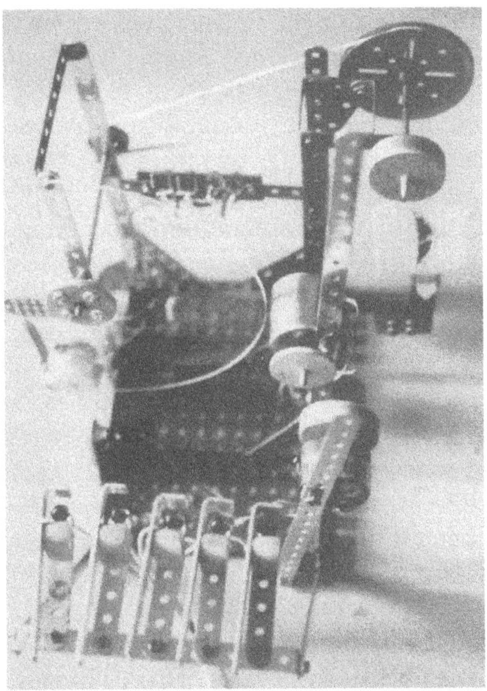

Figure 35.1 Lumiphone (1984).

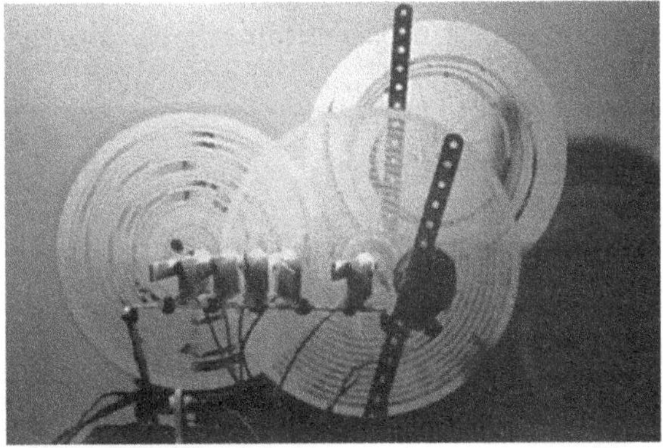

Figure 35.2 Lumiphone with cardboard disks.

the movie industry. But no conflicting claim was found by the patent research office with any of those previous instruments, I guess all of them being of the keyboard harmonium type, while my project was describing freely moving control elements (light sources, optic filters, sensors) with a vast array of interchangeable disks. As our organization applied for a grant to continue its researches, I was asked to register a patent, and this is what I did, not in the form of one precise instrument, but as a "new process for sound creation" – leaving it open to further evolution.

My intuition was right, because the instrument, following the progress I made in its gestural controls and in my compositional ideas, never stopped evolving. From a complex apparatus at the beginning, to rather discrete systems today, passing occasionally through various particular applications (playing with the sun, playing automatically, etc.), it became more and more centered around the disk itself. Thus, the instrument as a whole, now named the "photosonic disk"[3], gradually became identified with the central vibration-causing element.

After my successful experiment of February '84 with a disk of two inches in diameter, I wanted to explore all kinds of disk patterns, and what they sounded like. It was an absolute thrill, and still is today, to be able to hear sounds that you design visually. To understand better what was happening in the process, I needed to test as many disks as I wanted as fast as I could. My first ideas were cut with a knife in the most common A4 cardboard sheets available (21 cm wide). Up to now this is still my most standard diameter. Figure 35.3 shows myself manually cutting this type of cardboard disk. A hundred of those were made between '84 and '86 and served for many concerts in Europe.

In 1986, Daniel Arfib, computer music researcher in Marseille, and I began to design a software program for using a plotter to draw our disks in India ink on tracing paper. For every 40 disks plotted, I spent one full night developing negative films from them through photochemical processes. They were then left to dry, later to be cut in the shape of the final disks. With the patterns now appearing as opaque areas on a disk of otherwise transparent film, the cutting of holes was no longer

Figure 35.3 Cutting cardboard disks.

necessary – the light would be blocked in the opaque areas while simply shining through the transparent areas as the disk rotated. Between1986 and 1990, 400 different disks were designed this way – among them we find all the disks played in my CD *Lumieres audibles.* This ink technique allowed neither grayscales nor very fine lines, but still allowed many different waveforms in all types of scales. In 1990 however, after several repairs, our plotter became irremediably out of order, and that was the photosonic disks' momentaneous end. Recently, however, Patrick Sanchez, another computer researcher from Marseille, started to transpose Daniel Arfib's software into PostScript[4] language. We are still working on that tool today. It will allow in the coming years a real renaissance of the photosonic disks, with a quality level never attained so far. Among other improvements, besides a much finer precision and access to the best of electronic printing systems, this third generation includes grayscale possibilities and new graphic operations.

The Photosonic Process

How does a photosonic disk produce sounds? We know light is able to transmit signals of audible frequencies. This is how sound recorded on the tracks of the ancient "optical sound" movies is made audible (this is also how astronomers receive some of their information from the outer space, such as pulsar's revolutions, etc...). In the photosonic process, sound is generated in a similar way, by communicating audible frequency modulations to light beams, having them pass through a rotating disk which is alternately opaque or transparent. These pulsated light beams are received by a photoelectric sensor, which transforms their variations of luminosity into variations of electric current, directly amplified by an ordinary audio amplifier connected to a speaker.

The Figure 35.4 below retraces the path of the light, from the light source (L) to the photocell (C), passing by the disk (D) and an optic filter (F).

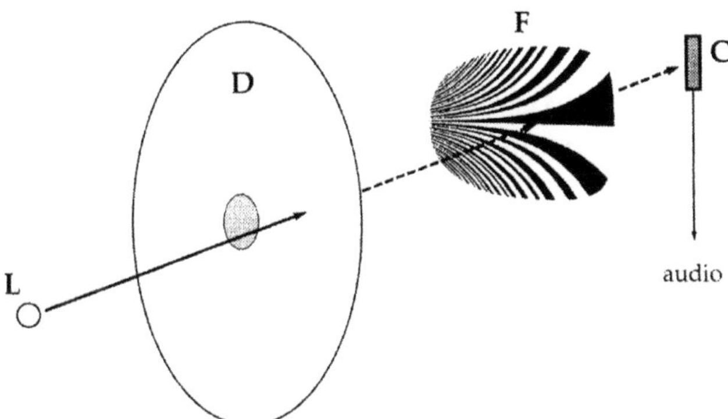

Figure 35.4 The path of the light beam.

The photosonic disk then works very much like a siren that uses light instead of air pressure, except that photosonic instruments are necessarily electric, needing amplification and speakers. On the other hand, while pressured air rapidly diffuses and loses its energy in open space, light does not, allowing all the elements used in the photosonic process (light sources, filters, sequencers or other additional disks, even sometimes the photocell itself) to be moved freely in the three dimensions, offering many sound controls with a maximum of precision.

Again, seeing that so many people have not believed it, it should be emphasized that these instruments are not connected to any kind of sampler or synthesizer that would make them sound. They proceed by a purely optical generation of sound and are simply amplified by electric means. But each photosonic disk can itself be seen as kind of a synthesizer program, where imaginary sounds are condensed in a graphic form.

Knowing this, you may ask: what are the sounds you can produce by this process?

Here we have to start from the vibration-inducing element, the disk in itself, and to consider three different aspects always present in any disk:

1) the waveforms used, that will determine some of the spectral qualities (the timbral dimension) of the sounds;
2) the number of repetitions applied to these waveforms, that will determine the musical scale of a disk;
3) the way different sounds are spread on the surface of a disk. As we shall see, these three aspects are often strongly related to each other.

Waveform Geometry

Types of waveforms I experienced in my disks are of a considerable number, and their description would exceed the length of this presentation. They range from "single hole" waves to fractal patterns that auto-develop themselves from geometric laws of progression, as well as waveforms designed to resynthesize external acoustical sources.

Only waveforms using two zones, like white and black (or in case of sirens, one hole per period), can be simply modeled. Those are determined by their cyclic ratio, that expresses the length of the opening (or the transparent section in disks printed on films) over the total wavelength. There is an acoustic law associated with those waves, which says that it shows a lack in its harmonics close to the inverse value of the cyclic ratio, or multiples by 2, 3, 4, etc. of this inverse value. For example, a "square wave" traced on a disk is supposed to cut harmonics 2, 4, 6, 8, 10, etc., in other words it will be richer in odd harmonics; a wave of an opening five times smaller than its length will have low overtones 5, 10, 15, etc. Waveforms using more than two zones (e.g., 2 holes per period, etc.), even for

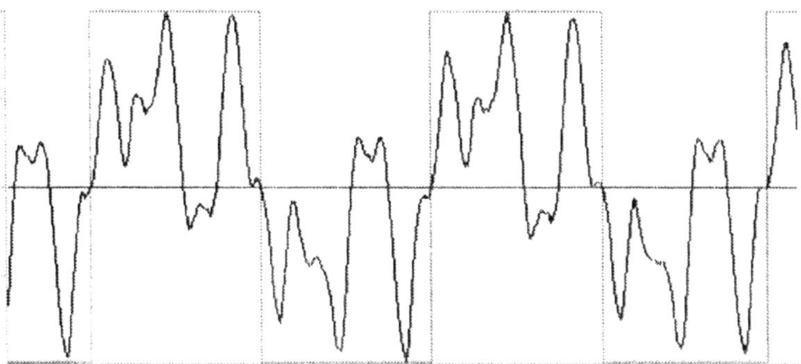

Figure 35.5 Electrical response to a simple waveform.

simple archetypes, enter at once into much more complex acoustical models, not to mention when those zones have more than two different gray scale values or worse, when they use gradients.

It is important to specify that what is drawn on a disk and what comes out of the photoelectric sensor are two very different things, making it quite impossible to predict what any given pattern is going to sound like. Figure 35.5 illustrates this with the simplest waveform one can imagine: a square signal that is alternating opacity and transparency in equal proportions. Superimposed over it in the figure is the waveform of the actual sound generated, which has very little in common, as you may notice, with the original pattern.

Furthermore, a disk never plays the same sounds either because its sounds totally depend on the way it is played. Speed, radius, light disk distances, and the presence or absence of optical filters will dramatically shape the output timbre.

Tunings

Portions of this section may serve as a complement to Bart Hopkin's appendix in his very well documented article on sirens.[5] In contrast to timbre, a disk's scale is totally predictable, the frequencies of the fundamental tones obtained being strictly proportional to the numbers of repetitions of the periodical waveforms printed on the disk. By simplification, I usually call these numbers of repetitions the "frequencies" of the disk (the frequencies heard are actually the product of these numbers and the rotation speed of the disk, in revolutions per second). For someone familiar with just intonation, intervals generated by a disk are easy to understand: their ratios are the same as the repetition numbers' ratios. Usually these are whole numbers, to avoid a noise created by an irregularity arising at every cycle of the disk, and this is why disks are perfectly suited for just intonation.

In pure theory, a disk can only produce overtones of its own speed rotation; but considering what we hear this is not so true.

Bart Hopkin mentions an early disk of mine bearing fractional repetition numbers. I made indeed such an experiment once, to hear how it would sound. I found three cases where this solution is of interest. The first case is when the disk turns so fast that the fundamental produced by the fractional irregularity passing once with each rotation becomes audible as a bass drone. The second is when you use this disk with an "optical sequencer" or in a photosonic Barbarie organ (described later): if the notes are short enough, you may not notice the noises. The third possibility is to use those disks as optical sequencers themselves, where they may serve specific rhythmic purposes.

About this question of fractional frequencies, later on I found different techniques, based on cyclic timbral modulations that make non-whole numbers acceptable. This can be useful because if it can be done in only one ring in a disk, it can help divide by two the total number of waveforms on the disk. In the case of a siren, this can save lots of drilling.

Figure 35.6's inner ring shows such a technique I call "grillon" (the cricket): between 1 and 2, the original line of the waveform splits in two, changing gradually into the octave waveform (in 2), which allows the integration of an additional line, increasing the fundamental frequency by 0.5 (from 7 to 7.5 in the example shown).

The outer ring of Figure 35.6 illustrates another technique I call "ours" (the bear), because of the gently groaning sound those waves usually generate: here is a wave composed of three dark lines of which the second and the third will slightly move forward on each period in such a way that, getting closer to the first line, after one full cycle they will simulate a dephased form of the original wave, being able to join with it at a different spot (line3 becoming line I here). As a result, the perceived repetition frequency is not 7 here, but 6.66.

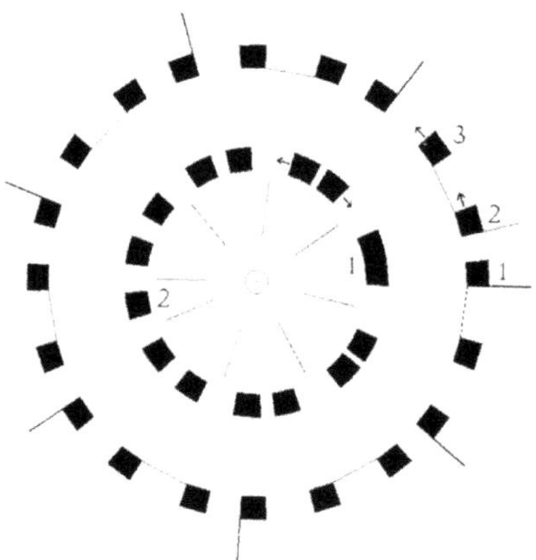

Figure 35.6 Fractional frequencies: *grillon* & *ours*.

That technique, and the *grillon*'s, could be used to create alternative tunings; actually, I have only used them so far to produce different timbres. Apart from these special cases, I only use whole numbers – which is enough to express practically any just intonation system.

In theory, audible frequencies drawn on a disk could range between 3 and 2000; in practice, the repetitions I use range approximately between 5 to 500, which covers almost seven octaves. Rather than being a limitation, although it forbids the use of tempered scales, I have found the rule of having tone frequencies proportional to whole numbers to be quite broad. By chance, it's also particularly instructive concerning the harmony of musical scales.

From Chords to Timbres

There are many ways to mix sounds on a disk, from pure additive synthesis to various graphic combinations that will result in the synthesis of totally new waveforms. Chords can also be created of course by playing with several lights. Different vibrato movements given to each light will then make the chord sound more like a polyphony, instead of a single composite sound.

The simplest way to produce chords on the same area of a single disk consists of printing small rings of different frequencies close to each other. The sensor that reads a certain portion of the surface of the disk then makes the sum of the frequencies of each ring. It is even possible in this manner to organize small sound spectrums.

Another way is what I call an "omission": by creating a periodical accident on a waveform; omitting one line out of every n lines, etc. If then f was the original tone frequency, we add to it a bass tone of frequency f/n (this can be seen in disks E & F on the following pages). n has to divide f a whole number of times, otherwise a noise of frequency 1 will be heard.

If not, the technique of omission can be applied to what I call "composed frequencies", of which Figure 35.7 second and third rings show two examples: just next to a frequency 48 (first ring from the center), the 2nd ring shows five groups of four lines each, while right next to it, the third ring selects five similar groups out of a frequency 53 (fourth ring).

Five is neither a divisor of 48 nor 53, but selection has been made in the best-balanced way to give the illusion of a regular frequency 5 in both cases. The result sounds nevertheless as a bass drone (frequency 5) with "inharmonics" 48 & 53, in a sound comparable to oboe or bassoon multiphonics. "Composed frequency" disks can also be used successfully as optical sequencers (discussed later with the instruments).

The outer ring of Figure 35.7 shows one last specimen of chords I call "intermodulations", generated by graphic superimposition of waveforms. We notice that on

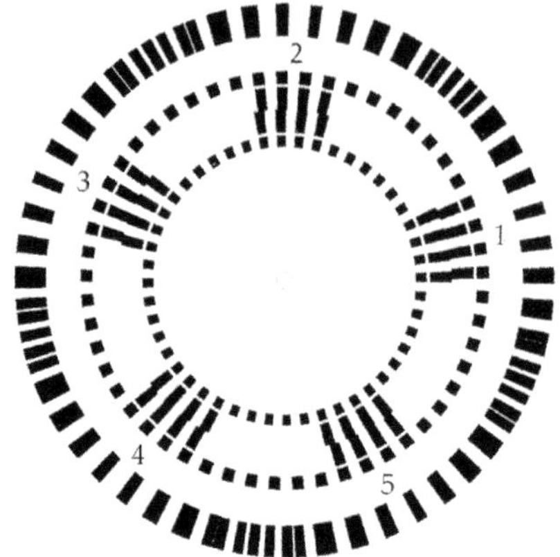

Figure 35.7 Composed frequencies & intermodulation.

five spots (1 2 3 4 5), the two frequencies (48 and 53 here again) are in phase, causing an increase of transparency, while in between, phase opposition darkens the resulting waveform. But the result, rather than being a simple sum of the two waveforms, is a complex signal which contains strong difference tones, such as a clearly audible frequency 5 here, the difference between 53 and 48.

Depending on the harmonies between the original frequencies, these intermodulations may sound from compact harmonic tones to dissonant or metallic "inharmonic" timbres. Here 53, 48, and 5 are a bit low, but with higher frequencies we are able to see the resultant difference tones through visible "moiré" effects (Figures 35.8 (h) and 35.19 on the following pages).

There are many ways I found to create intermodulations of various timbres, which is one of the interests of that type of chords, out of black & white waves, and even more with gray level waves, by integrating gray level values in the additive operations.

And, finally, having the same light beams passing through two or more successive disks will also create intermodulations, generating strong difference tones with powerful phasing effects.

Composing a Disk

When composing a disk, bass tones usually start from the center, and higher ones spread to the periphery, but everything is possible, such as special orders, arpeggios, glissandi, gradual blending of rings, etc.

A disk is basically the disposition, within the proper usable radius (between 16 and 100 mm in my usual format), of specific waveforms, repeated at specific frequencies. In complement, certain dephasing is often arranged by rotation in between rings in order to smooth transitions and to optimize chord timbres, if not for purely visual esthetic considerations.

So, what kind of music can be drawn inside a circular image of 20 cm in diameter? Just about anything. A disk has to be thought of as a painter's palette; one who often mixes adjacent colors to create special shades. Many disks I have done use only one type of timbre, but some of them gather as many as possible. Same with intonations: some disks play one single drone developed in various textures, while others simply develop a scale, or eventually do not present a single stable tone at all.

With the years, I tend to make disks in which timbres and intonations are more and more strongly interwoven, with several listening levels (tones, undertones, overtones, inharmonics, noises) and gradual spectral developments, rather than purely linear melodic scales or transpositions of one single timbre.

A Selection of Disks (1984–1990)

The introduction explains why this selection stops in 1990. This is not a selection of the most spectacular disks I made, but only complementary examples chosen to explain how patterns have been conceived to achieve specific sounding results.

Figure 35.8 (a) shows the simplest waveform available – equal alternation of black and white at frequencies chosen to reproduce the traditional chromatic scale of a "27-keys Erman" Barbarie organ (mechanical organ very popular in Europe, using perforated cardboards). Very often, in my photosonic Barbarie organ (described later), I have replaed this disk with a similar disk tuned to a Javanese slendro scale, while playing the cardboard of a Mozart piece – to give it a change.

Figure 35.8 (b) shows the same simple waveform, producing some kind of a flute tone, applied to a "glissando", for the same instrument. This is realized by increasing the frequency of every new ring by one unit (two units in the second octave) on very thin rings.

Each radius has been calculated to follow a logarithmic progression. That means that whatever the transposition of a chord, it will be represented on the cardboard by equal distances in width between holes.

Figure 35.8 (c) This is another glissando performed in the "grillon" type of waveform (see Figure 35.6). This one was not meant for melodies, nor tuning facilities, but just the interest of the timbre. It's often been used in the "Balai-magique" automated instrument (Figure 35.11).

Figure 35.8 (d) One of the early cardboard disks. As the shape of the openings suggests, this is a "comb filter" – the more you get away from the center, the higher are the overtones you hear, based on the fundamental tone of this disk, of frequency 6.

Figure 35.8 (e) A "double" disk, meant to be played with two lights. In its center are four tones (at frequencies of 9 - 10 - 12 - 15) and their developed spectrum (using the technique of the comb filter as in Figure 35.8 (d)). Toward the outer edge is a

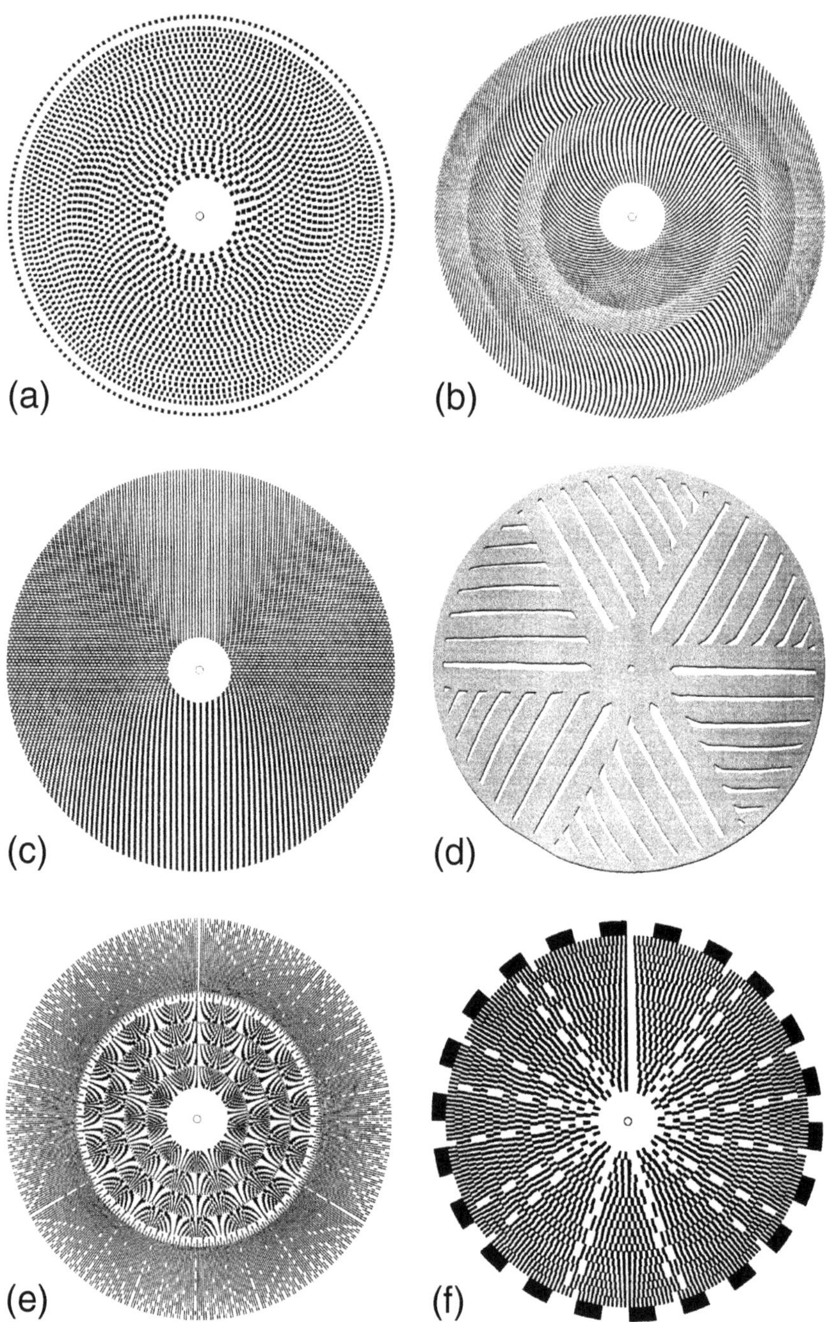

Figure 35.8 Photosonic disk patterns.

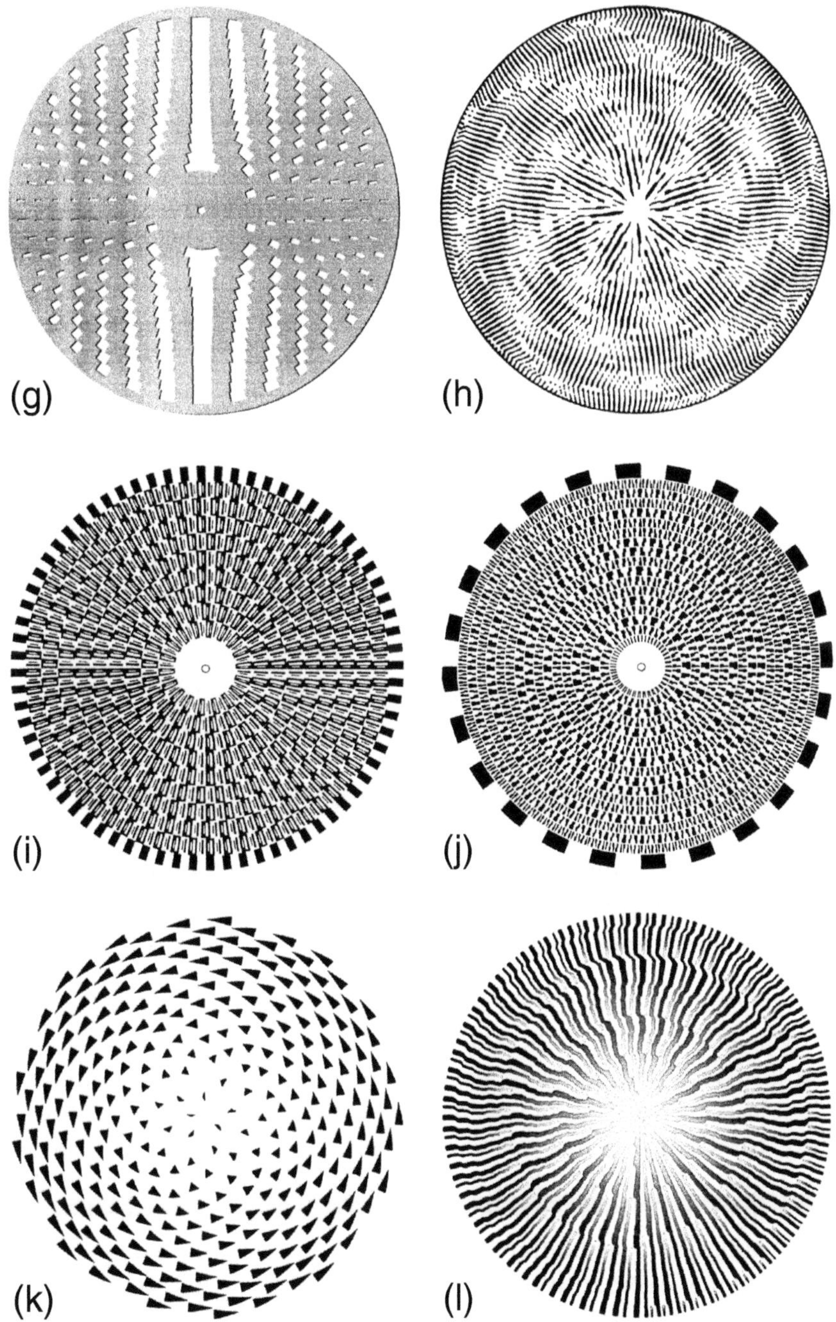

Figure 35.8 Photosonic disk patterns.

pattern using the "omission" technique from a fixed frequency of 540 in all its available dividers , generating a sub-harmonic series.

Figure 35.8 (f) The omission technique again, on the tones of a "mohajira" scale. The "omitted" frequencies, 9 and 11, are well visible herein the shape of a star. I used this disk in "Sumer" where my needs were of those two bass tones only.

Figure 35.8 (g) With cardboard disks, having to cut openings manually, I was also using high rotation speeds in order to reduce the number of waveform repetitions within each ring. Here is one of those disks, based on a spectral development (the first 16 harmonic overtones intermodulating two by two) of a single fundamental tone of frequency 2, which means that it could be audible as a bass sound only if the motor's speed was higher than 12 revolutions per second approximately. At that speed the disk would create a wind which I allowed to blow randomly on the flame of a candle, used as light source.

Figure 35.8 (h) This disk, used in "Fleurs de lumiere", shows another example of intermodulation, between Fibonacci and other golden series of numbers. This special tuning in interaction with precise phase relationships between consecutive rings creates interesting alignments, and ultimately, the scale becomes a timbre texture.

Figure 35.8 (i) A fractal waveform of a type I call "Clar" because of its clarinet sound, in its first developments only (three and five lines per period). The scale is a slendro issued from the Fibonacci numbers 3 - 8 - 21 - 55 - 144, very close to one of Lou Harrison's favorite slendro scales.

Figure 35.8 (j) Diatonic scale geometrically "linking" the fractal waveform of the disk in Figure 35.8 (i) with another one based on the number Phi (1.618). Both are chosen to have their major overtones (3 and 5) coherent with the scale.

Figure 35.8 (k) A sequencer disk, based on a modelization of the heart of the sunflower. The triangular shape of the openings are meant to produce a percussive envelope on each note. This poly rhythmic sequence, based on Fibonacci numbers, can be heard in the beginning of track 2 ("Tournesols") of my CD. If used as a tonal disk, this disk also generates the fractal waveform Phi, used intensively in the same piece.

Figure 35.8 (l) The same fractal Phi waveform again, but modulated in a glissando transposition of a 1.618 ratio over one full revolution. With each spiral revolution, the wave gradually changes into its precedent fractal unfoldment, and the result is a continuously increasing sound (or decreasing, depending on the direction of rotation) that has yet no audible beginning. If the disk turns rapidly enough, we hear a kind of a "pink noise" bubbling. I have used this disk successfully with a light itself rapidly turning at various speeds, causing long cycles of pitches determined by speed differences between the disk and the light motors. It is a nice, "inharmonic" alternative to my usual FM disks, whose frequency descends one octave per revolution, while splitting gradually towards a 2nd harmonic in the "grillon" way (Figure 35.6).

Some of the Instruments

How do you play a disk, and what are the controls you may use to transform the sounds you play? The photosonic process provides many types of gestural controls, which I will review here for each instrument.

With my first prototype, called the *lumiphone* (Figures 35.1 and 35.2), I was already experimenting with most of the main features of the photosonic controls today, such as mobile lights, optic filtering, and a sequencer. The light sources of the lumiphone were on the right side of the main disk, small lamps manually held, powered by 6v DC, or even candles. The absence of mechanical inertia in these mobile lights is something unbelievable, if you never experienced it. By gliding from left to right you run through the scale, making chords if you use several lights, while moving lights vertically you add vibrato or phasing effects, and by changing the distance to the disk you can control intensity but also resonance. With a candle, the sound is more diffuse and has a pleasant flickering musical quality.

On the left side an array of five small lamps that could be progressively masked played with the *optic sequencer*, a slow-turning opaque disk with holes designed to allow light from the main disk through according to specific rhythmic patterns. A ventilator used to spin a lamp connected to a battery box completed the light sources options.

When the right hand is not used to hold a light, it usually holds an *optic filter* of some sort, the simpler one being a transparent film with more or less contracted stripes (Figure 35.12); some others use optical convergent elements. This tool selects harmonics from the disk's projected shadow on its surface, that are in concordance with its own patterns. It is used gesturally like a bow, but one that would control the timbre, attack, decay, and vibrato of the sounds.

The main disks could be single or double, concentric or not, which would allow different intermodulations. In front of the photoelectric cell a special disk was used, providing an interesting stereo phase shift between two photovoltaic silicon cells, directly plugged in the channels L & R of an ordinary stereo amplifier. Later on, I found that microphone impedance transformers could help prevent speakers from burning out, an accident which happened to me many times at the beginning, the direct current sometimes delivered by the cells being not well accepted by normal speakers.

Double interception was performed another way in the instrument called *Sirius* (Figure 35.9), playing with all the possible ring intersections between two disks permanently turning together, thereby exploring all the difference tones available between their combined frequencies.

Sirius is played either manually or with the help of a small keyboard. An automated system also allowed it to play loops of fast twinkling sequences that were determined by the placement of the removable lights inside a perforated plate of Plexiglas. This was very Christmassy-looking.

Figure 35.9 Sirius (1985).

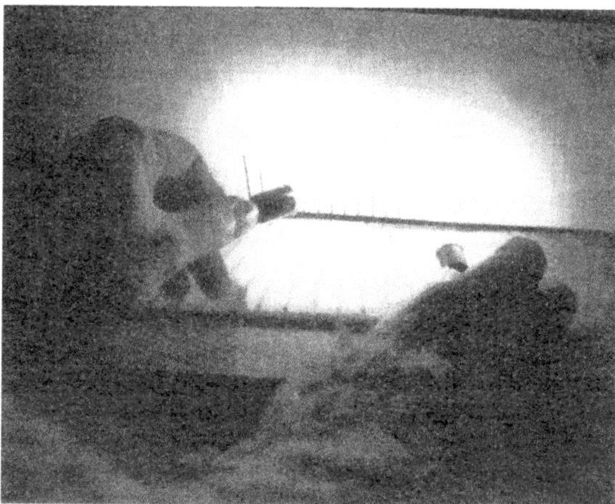

Figure 35.10 Tablette à percussions (1985).

In the *tablette à percussions* (Figure 35.10), a simple but pleasant toy-like instrument, an opaque screen hides the lights until they touch a soft piece of felt, colored with lines for each corresponding tone. This lets you play percussive or organ sounds, while anticipating pitches and controlling intensity.

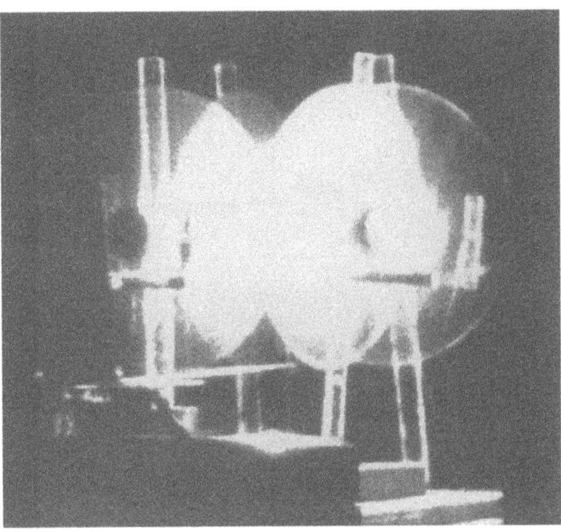

Figure 35.11 Balai-Magique (1985).

Balai-magique (Figure 35.11) is the first automated photosonic prototype I made: it uses one single light motorized between four disks. Its evolution is programmed with the help of two levers pulling it in two directions at a right angle. Because of the chosen speed of both levers, the trajectory of the light follows specific Lissajous figures that produce an interesting interpretation between the four discs.

Photon is still today the instrument I use the most in my concerts. It is basically a compact lumiphone with an integrated electric power source inside the same box that holds together the motor and the cell (Figure 35.12), and the option of assemblage between different modules on the same socket.

Photon's sequencer, one of these modules, is of a new kind, using large horizontal cells capturing one full radius of the whole disk's shadow. While in the lumiphone the distance between the lights and the sequencer disk did not vary, photon allows it to vary, bringing lots of additional interpretations. In Figure 35.13, photon is seen used with such a sequencer (left side of the main disk), and a rotating light sweeper.

Photon also allows the use of tonal disks on both sides (left and right) of the optical sequencer, in order to pulse associated rhythmic patterns coming from the same sequencer disk simultaneously into different scales and timbres.

Sequencer disks are still cut manually today from cardboard sheets, by first drawing a certain number of radii corresponding to the desired cycle, then by cutting holes at different places to create rhythmic patterns. By cutting "triangular" holes (Figure 35.13), we create a decay in the sound's envelope, such as plucked strings or

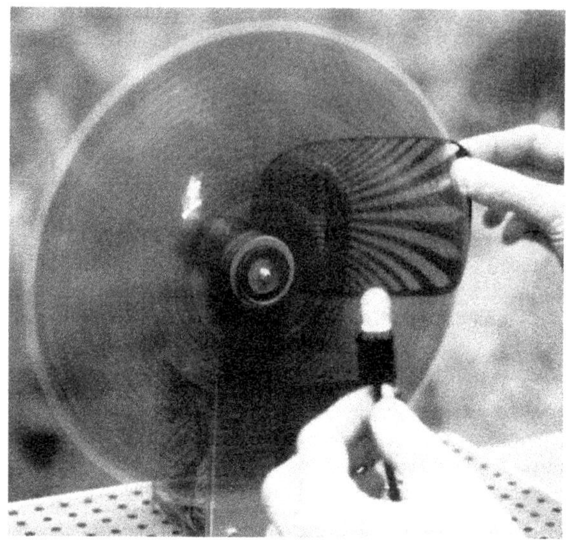

Figure 35.12 Photon (1986).

Figure 35.13 Photon's optical sequencer and rotating sweeper.

percussions. Many other kinds of patterns, rhythmic or not, can be applied. Certain tonal disks, or their superimposition, also provide various polyrhythms.

Some of the selected tonal disks (Figures 35.8 (f), 35.8 (i), and 35.8 (j)) show an outer ring with a repeated square signal. This was used to tune several disks playing together, with the help of a synchronization speed circuitry that was designed by Daniel Arfib. The frequency of that outer ring was detected by a small infrared opto-electronic fork, which would make the circuitry send more or less current to the motor depending on whether it would be lower or higher than a fixed frequency chosen among a set of dividing frequencies provided by a quartz oscillator.

Another feature of photon is the possibility of laying a piece of paper under the light's stand, on which you can draw a map, taking the form of converging lines, to serves as a visual guide representing different light positions for the disk you are playing and other information. With the *tablette à percussions,* this is the only system allowing you to play the tones you decide on before the sound starts.

The *photosonic Barbarie organ* (Figure 35.14) uses a technique similar to the sequencer disk in order to read the 13.0 mm-wide perforated cardboards used with traditional Barbarie organs.

The tonal disk (such as the disks in Figures 35.8 (a) or (b)) is removable from inside the box (Figure 35.15 and can be replaced to change scales or timbres. Among other features, this photosonic version of the Barbarie organ can freeze on one chord, or read in reverse, as the crank is used here only to move the cardboard, and not to operate the bellows as with the traditional Barbarie organ. Figure 35.16 shows different styles of cardboards played by this instrument. The one above is

Figure 35.14 Photosonic Barbarie organ (1987).

Figure 35.15 Photosonic Barbarie organ (inside).

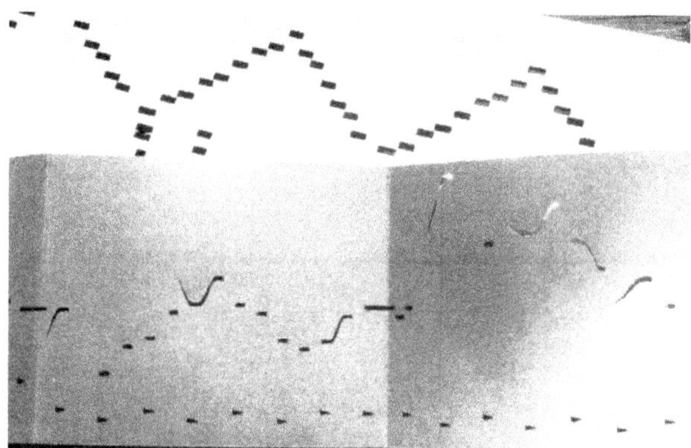

Figure 35.16 Perforated cardboards for photosonic Barbarie organ.

in the traditional "27-keys Ennan" format (one of the standard formats provided with thousands of songs, classical and jazz pieces in Europe). Below is one of my own carboards integrating intensity control, dynamic envelopes and glissandi, not permitted in the normal Barbarie organs.

Aton (Figure 35.17) is, to this day, the only photosonic instrument using the sun as a light source.

Figure 35.17 Aton (1990).

To be played outdoors, its disk is kept inside a transparent box that protects it from the wind. As you know the sun can't be moved, it's up to the player to move the solar cell instead. He stands therefore behind the disk, and his eyes are exposed to the pulsated beams, which can be totally hypnotic.

Playing Aton allows very different techniques, such as changing the solar cell orientation, or using the fingers as a filter and to play rhythmic patterns. Because the sun's rays are almost parallel, the size of the disk's shadow is not amplified as in other instruments, and large disks (30 cm here) are more comfortable. I also dream of making giant disks to be used in sunlight installations.

In the early morning or the end of the afternoon (when the sun is low in the sky), Aton can be played directly in front of the sun, but using a mirror allows it to be played at anytime of the day.

The *D.R.I.C.* (double rotofiltre à interprétation coronale) was designed for interactive installations, with a new type of optic filtering produced by two *rotofilter disks* concentric to the main disk (Figure 35.18), radically transforming its initial sound.

These rotofilters are high-frequency modulation disks slowly turning at slightly different speeds, in order to create changing moiré patterns that filter but also transpose the sounds with continuous modulations. Figure 35.19 simulates the intermodulation of the rotofilters near to their phase, as it can be seen at a certain moment of the full cycle.

Figure 35.18 Double rotofiltre à interprétation coronale (1997).

Figure 35.19 Simulation of the intermodulation of rotofilters.

In addition, the light and the photocell are moving according to specific trajectories, developing the spectral dimension of the timbres by a permanent sweep of the resonance.

With the help of a remote control, the public turns the light on with a time-switch, plays with the ratofilters and changes the read radius of the disks, making the instrument interpret other sounds.

The music of the D.R.I.C. depends to some extent on the tonal disk played, but it can be generally described as aleatoric gliding voices with surprising overtone melodies, with bass murmurs and whispers. It can never play the same thing twice, even without the public's help.

Next Projects

One of our main actual projects is to work on sound resynthesis through photosonic disks. We have just succeeded in the transcription of several vowels, and this will enable some of our next disks to sing in a manner close a human voice, the object I had in mind when this adventure started. We are also working on graphically blending sound with more and more precision, to approach cross-synthesis effects. Our aim is now to be able to produce ever more creative disks, in a wider range of timbres and with the highest printing quality. With the help of progress in the field of computer graphics, and the research we've been doing during these last 15 years, we feel that optical sound generation comes today to a new actuality, with lots of potentialities still unexplored. Though we have created nearly 600 different disks, we realize that we are only at the very beginning of our experimentation with light as a sounding material.

But this allows us to consider the time to be ripe for the next step, which should be the commercialization of the photosonic disk – probably by ourselves, since no serious proposition has come to us up to now – and that could be the best situation anyway.

If you wish to play the photosonic disk yourself, and if you are patient enough, wait until the first instrument, or at least till the first disks are commercialized. If you can't wait, you may try your own experiments with a simple photocell, a motor, and a light. Keep in mind that disks require a high degree of precision in order to produce good sounds, and be cautious about possible damage to your speakers in your first trials.

Notes

1 Sri M.R. Gautam, formerly Director of the Vocal Department of Benares Hindu University, now retired in Calcutta.

2 Related by Tom Nunn in "An encounter with Jacques Dudon" in *Experimental Musical Instruments* Vol.3 #5 (February, 1988). See also "Jacques Dudon: Instruments of Water and Light". by Bart Hopkin, in *Gravikords, Whirlies & Pyrophones*, Ellipsis Arts, New York, 1996.

3 The term "photosonic disk" has therefore two meanings: the full instrument, and one part of that instrument. In French we also call the machine alone (without the disk) a "lecteur" which means "reader" or "drive."

4 Digital standard designed by Adobe, used in high-quality image printing.

5 To know more about sirens, read Bart Hopkin's article in *Experimental Musical Instruments* Vol. 12 #4 & Vol. 13 #1 (June & Sept. 1997) [also included in this collection].

Related articles

One other article pertaining to the instruments of Jacques Dudon appeared in *Experimental Musical Instruments*: "Jacque Dudon's Music of Water and Light" by Tom Nunn in EMI Vol. 3 #5. This article can be accessed through the Internet Archive at https://archive.org/details/emi_archive/.

36

Journey Through Sound and Flame

A Ceramic Musical Instrument Maker

Brian Ransom

This article first appeared in *Experimental Musical Instruments* Volume 14 #4, June 1999.

The following is a brief collection of memories, thoughts, and anecdotes of a musical instrument maker. My first instrument making experience was in 1973. I was regularly playing the flugelhorn in an avant-garde jazz ensemble (whose name I now have forgotten); the only other player whom I distinctly remember was a brilliant young composer and trumpet player named Stephen Haynes (I later named my first son as his name-sake). We were all students at the time, at the Rhode Island School of Design, and making early classes after late night performances was always a problem. As a young composer myself, I was perpetually engaged in searching for musical sounds, tonalities, and scales which weren't available on conventional instruments. In retrospect, I think the first flutes I made were partly designed to appease my ceramics class requirements and partly to satisfy my composing desires.

The first instruments I made were a couple of ceramic harmonic flutes. (At the time, I didn't realize that their scales were based on harmonic overtones – I felt as if I was the lone discoverer of these tonalities!) Several distinct impressions remain with me in utter clarity from the experiences of making those first ceramic flutes. First was the amazing tone color of the ceramic. I knew that different materials affected the resonance and production of soundwaves; however, I was astonished by the unusually hollow and haunting sound these ceramic musical instruments produced. My second realization spurred by creating those first clay flutes was that I had stumbled upon a great abyss, a frontier of undiscovered tonalities! Through infinite choices of pitches and sounds, I found my way into the mysterious world of microtonality. The most troubling aspect of ceramic musical instrument making I found was coming to terms with the problem of shrinkage. The tuning phase of most clay instruments (with the exception of idiophones and membranophones) is in the leather-hard working state, which is before the firing process. In the firing process that follows, different amounts of shrinkage occur in small and large

Figure 36.1 Ceramic Dragon Congas.

instruments. In small instruments, the pitch rises a small amount; in larger ones, the pitch rises increasingly. I have spent many subsequent years developing a cannon of shrinkage for my specific clay body.

In the last two years of the 1970s, I was fortunate enough to receive a Fulbright/Hays congressional Fellowship for the study of pre-Columbian musical instruments in Peru. My approach to how and why I make instruments was significantly changed in many ways during the two years I spent in South America. Perhaps the most important revelation I came to was an empathy for the ancient Peruvian idea of animism which adheres to the belief that life force, in the form of spirits, can be found in all things, animate and inanimate. My studies included hundreds of hours in museums and collections documenting ancient instruments as well as ethnographic field work in rural communities throughout Peru.

In Portland, Oregon, during the early 1980s, after I had returned from living in Peru, I formed my first complete ceramic ensemble. The musicians in the first ensemble were unforgettable. After being enticed by the unique sounds of my ceramic drums, I was joined by master Ghanaian drummer Obo Addy. Other musicians included bassist Patrick O'Hearn, saxophonist Rich Halley, Bruce Sweetman, Brian

Figure 36.2 Ceramic Shakers.

Davis, and Bruce Smith. Our first performances were art events which melded dancers, original music, elaborate stage lighting, and some theatrics.

As the years progressed, so did my academic accolades, but mostly, so did my ceramic musical instruments. Most of the instruments I have made, have gone through countless generations of improvement. Voicings in the ceramic ensemble through the years include single, double, and multiple whistles and flutes, a variety of invented horns (in particular, the ceramic flugelhorn), reeds (my favorite, the sax-o-snake), strings (including low, bass-like instruments as well as higher pitched harps which utilize silk koto strings), microtonal bells, percussion ranging from bullroarers, rattles and rainsticks, to directional congas, djimbes, and invented percussive forms.

Some of the ceramic instruments I make are interactive, and some are based on the natural resonance of chambers. The first of these interactive series were "Activated Ceramic Resonators" which were pieces intended to be touched and played by viewers. Some were activated by electronics, some by blown air using internal fans, and others by the action of moving water. The most recent series of this type that I made were ceramic resonator vessels which I installed into a specific space (so far,

Figure 36.3 Triformation VIII. Whistling water vessel.

Figure 36.4 Triformation IX. Whistling water vessel.

two different art galleries). These resonators activate the standing wave or resonant pitch in the chamber and are powered by tone generators. The result to the viewer is a space which is geographically perceived through sensations of sound. When you walk through the installation, you sense the physical waves of sound as an environment – the viewer is actually composing by choosing a walking path throughout the sounding resonators.

Throughout the late 1980s and early 90s, I lived in Los Angeles where the ceramic ensemble flourished. I was joined by many eminent musicians performing and recording tapes, CDs, musical soundtracks, and dance scores; members included Chris Darrow, Rico Garcia, Ben Harper, Ernesto Salcedo, Hearn Stearns, Norma Tanega, and others.

I have a long history of exhibiting ceramic instruments and sounding sculptures. While living in Los Angeles, I became affiliated with the Couturier Gallery where I have had many exhibitions, including a series of sculptures which I started making in the early 1990s, *The Deities of Sound.* They first appeared in a series of dreams as images of beings whose existence was otherworldly, whose voices were full of knowledge, whose origins were beyond the earthly plane. To date, these works are some of my most tenuous. In their making, I have had to suspend judgement. They

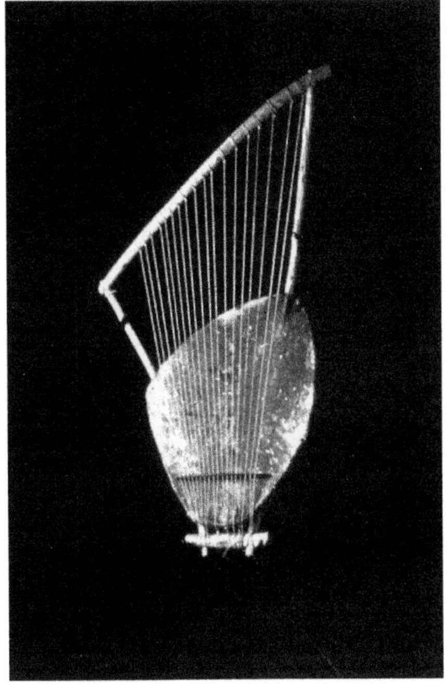

Figure 36.5 Ceramic Harp (Kora).

Figure 36.6 Double nest of Pot Flutes.

Figure 36.7 Automatic Ceramic Resonator III.

Figure 36.8 Automated Whistling Pads (electric wind instrument).

Figure 36.9 Deities of Sound I, "Singing Deity"

Figure 36.10 Deities of Sound XVI, "Defining".

Figure 36.11 Left: Deities of Sound XIII, "Ponderance".

Figure 36.12 Deities of Sound XIV, "Traveler".

are made at the insistence of my subconscious and are meant to be a testament of what I have learned in their making.

I am currently living in St. Petersburg, Florida, where I accepted a teaching position at Eckerd College. My latest musical project is the worldbeat ensemble "Common Ground," a collaboration with my colleague Joan Epstein and many talented students. I anticipate creating many future musical inventions and projects.

Related articles

Following here are other articles pertaining to ceramic instruments that appeared in *Experimental Musical Instruments* These articles can be accessed through the Internet Archive at https://archive.org/details/emi_archive/.

"Sharon Rowell[1]s Clay Ocarinas" by Sharon Rowell in Vol. 1 #2

"The Ceramic Whistles, Flutes, Ocarinas and Mirlitons of Susan Rawcliffe" by Bart Hopkin in Vol. 1 #6

"Sounds In Clay" by Ward Hartenstein in Vol. 4 #6

"Udu Drum: Voice of the Ancestors": Frank Giorgini in Vol. 5 #5

"Earthsounds" by Ragnar Naess in Vol. 7 #2

"Complex Acoustics in Pre-Columbian Flute Systems" by Susan Rawcliffe in Vol. 8 #2

"The Flutes and Sound Sculptures of Susan Rawcliffe" by Susan Rawcliffe in Vol. 11 #2, also included in the current collection

Index

Page numbers in **bold** indicate tables, page numbers in *italic* indicate figures and page numbers followed by n indicate notes.